A Fresh Start for Collegiate Mathematics

Rethinking the Courses below Calculus

The National Science Foundation provided support for the conference "Rethinking the Preparation for Calculus" and partial support for preparation of this volume under Grant No. NSF/DUE 0136162. Any opinions, findings, and conclusions or recommendations expressed in this volume are those of the authors and do not necessarily reflect the views of the NSF.

A Fresh Start for Collegiate Mathematics

Rethinking the Courses below Calculus

Edited by

Nancy Baxter Hastings
Dickinson College

Associate Editors

Florence S. Gordon
New York Institute of Technology

Sheldon P. Gordon
Farmingdale State University of New York

Jack Narayan
State University of New York at Oswego

Published and Distributed by
The Mathematical Association of America

The MAA Notes Series, started in 1982, addresses a broad range of topics and themes of interest to all who are involved with undergraduate mathematics. The volumes in this series are readable, informative, and useful, and help the mathematical community keep up with developments of importance to mathematics.

MAA Notes

14. Mathematical Writing, by *Donald E. Knuth, Tracy Larrabee, and Paul M. Roberts.*

16. Using Writing to Teach Mathematics, *Andrew Sterrett*, Editor.

17. Priming the Calculus Pump: Innovations and Resources, Committee on Calculus Reform and the First Two Years, a subcommittee of the Committee on the Undergraduate Program in Mathematics, *Thomas W. Tucker*, Editor.

18. Models for Undergraduate Research in Mathematics, *Lester Senechal*, Editor.

19. Visualization in Teaching and Learning Mathematics, Committee on Computers in Mathematics Education, *Steve Cunningham and Walter S. Zimmermann*, Editors.

20. The Laboratory Approach to Teaching Calculus, *L. Carl Leinbach et al.*, Editors.

21. Perspectives on Contemporary Statistics, *David C. Hoaglin and David S. Moore*, Editors.

22. Heeding the Call for Change: Suggestions for Curricular Action, *Lynn A. Steen*, Editor.

24. Symbolic Computation in Undergraduate Mathematics Education, *Zaven A. Karian*, Editor.

25. The Concept of Function: Aspects of Epistemology and Pedagogy, *Guershon Harel and Ed Dubinsky*, Editors.

26. Statistics for the Twenty-First Century, *Florence and Sheldon Gordon*, Editors.

27. Resources for Calculus Collection, Volume 1: Learning by Discovery: A Lab Manual for Calculus, *Anita E. Solow*, Editor.

28. Resources for Calculus Collection, Volume 2: Calculus Problems for a New Century, *Robert Fraga*, Editor.

29. Resources for Calculus Collection, Volume 3: Applications of Calculus, *Philip Straffin*, Editor.

30. Resources for Calculus Collection, Volume 4: Problems for Student Investigation, *Michael B. Jackson and John R. Ramsay*, Editors.

31. Resources for Calculus Collection, Volume 5: Readings for Calculus, *Underwood Dudley*, Editor.

32. Essays in Humanistic Mathematics, *Alvin White*, Editor.

33. Research Issues in Undergraduate Mathematics Learning: Preliminary Analyses and Results, *James J. Kaput and Ed Dubinsky*, Editors.

34. In Eves' Circles, *Joby Milo Anthony*, Editor.

35. You're the Professor, What Next? Ideas and Resources for Preparing College Teachers, The Committee on Preparation for College Teaching, *Bettye Anne Case*, Editor.

36. Preparing for a New Calculus: Conference Proceedings, *Anita E. Solow*, Editor.

37. A Practical Guide to Cooperative Learning in Collegiate Mathematics, *Nancy L. Hagelgans, Barbara E. Reynolds, SDS, Keith Schwingendorf, Draga Vidakovic, Ed Dubinsky, Mazen Shahin, G. Joseph Wimbish, Jr.*

38. Models That Work: Case Studies in Effective Undergraduate Mathematics Programs, *Alan C. Tucker*, Editor.

39. Calculus: The Dynamics of Change, CUPM Subcommittee on Calculus Reform and the First Two Years, *A. Wayne Roberts*, Editor.

40. Vita Mathematica: Historical Research and Integration with Teaching, *Ronald Calinger*, Editor.

41. Geometry Turned On: Dynamic Software in Learning, Teaching, and Research, *James R. King and Doris Schattschneider*, Editors.

42. Resources for Teaching Linear Algebra, *David Carlson, Charles R. Johnson, David C. Lay, A. Duane Porter, Ann E. Watkins, William Watkins*, Editors.

43. Student Assessment in Calculus: A Report of the NSF Working Group on Assessment in Calculus, *Alan Schoenfeld*, Editor.

MAA Service Center
P.O. Box 91112
Washington, DC 20090-1112
1-800-331-1MAA   FAX: 1-301-206-9789

In memory of

Janet Andersen

*Your passion for mathematics and for
helping students learn will always be with us.*

Preface

This volume is an outcome of the NSF-funded conference, "Rethinking the Preparation for Calculus," which took place in Washington, DC, in October 2001. Approximately 50 mathematicians were invited to attend the conference, which was organized by Jack Narayan (SUNY Oswego), with support from members of the steering committee: Steven Dunbar (University of Nebraska-Lincoln), Sheldon Gordon (Farmingdale State University of New York), Christopher Hirsch (Western Michigan University), Jo Ann Lutz (North Carolina School of Science and Mathematics) and myself, Nancy Baxter Hastings (Dickinson College).

As the name of the conference implies, the purpose was to rethink the preparation for calculus. All of us were aware, as Jack Narayan writes in his paper describing the conference, "in general, only a small percentage of students who take precalculus courses ever go on to take calculus and many of them who do are not particularly well-prepared for calculus and never complete the course." As organizers of the conference, we hoped to establish some principles for changing the precalculus offerings, provide guidance to the mathematics community, and focus attention on the problems and needs in the area of precalculus. In preparation for the conference, we invited participants to submit discussion papers organized around the following themes: Precalculus Reform, Student Learning and Research, Changes in College Algebra, Transition from High School to College, Needs of Other Disciplines, Technology, Implementation, and Influencing the Mathematics Community. Our plan was to collect these papers in a proposed volume for the *MAA Notes Series*. These papers provided a framework for the discussions that took place at the conference and form the core of this volume.

Following the conference, it became apparent that our intention to focus on precalculus courses that are not terminal—that is, to focus on courses that serve as a prerequisite for calculus—was too narrow. We do need to rethink how we prepare students for calculus, but we also need to rethink the mathematical experiences of students in courses *below* calculus. For the vast majority of college and university-level students, the courses below calculus are the last mathematics courses that they take. These are the courses that students need for use in other disciplines. These are the courses that supposedly prepare students to be informed citizens. As a result, we expanded our vision to include courses in quantitative literacy and college algebra, and we encouraged colleagues to contribute papers in these areas. And we changed the name of the proposed volume. Actually we changed it several times from *Rethinking the Preparation for Calculus*, to *Rethinking the Road to Calculus*, and finally to *A Fresh Start for Collegiate Mathematics: Rethinking the Courses below Calculus*.

Although the papers in the volume are organized around the general themes that formed the basis for discussion at the conference, we added a new section: "Ideas and Projects that Work." A number of people have rethought their precalculus and college algebra courses and have developed materials reflecting those new visions. We invited some of them to write short papers describing their visions and how they developed them into text materials for inclusion in this section. This new section was developed in response to participants who attended the conference. Participants, who for the most part were members of the choir, wanted specific suggestions about how to refocus traditional, computational-based courses. They wanted to know what was being done, and they wanted to see examples. (Sounds like our students, doesn't it?!)

Although I have served as the primary editor, this was certainly not a one-person project. Whenever I was perplexed by a particular situation or needed some help guiding a prospective author, Shelly Gordon was always there to assist me. He was the idea man and the problem solver. He solicited additional authors to help provide a balance to the volume and suggested the final catchy title. Flo Gordon solicited the short papers for the section "Ideas and Projects that Work." She worked closely with the authors, helping them write informative descriptions of their projects, which express their enthusiasm for what they are doing (without sounding too self-promotional) and include supportive examples. Jack Narayan kept us organized and focused. Andrea Marsh (SUNY Oswego, class '04) made a first pass at converting the papers to LaTex.

I would like to thank the members of the MAA Notes Committee, especially members of the review team, Jack Bookman, Paul Fishback, Barbara Reynolds and Sharon Ross, for their helpful suggestions, for their support and for their guidance. Because of their efforts, the papers are better written, more informative, and organized in a more coherent fashion. As I mentioned earlier this was a dynamic project—the table of contents kept changing. Papers were added. Papers were eliminated. Papers were moved around. The review team received the papers in batches and never saw the complete picture until the very end. Yet, they kept encouraging us.

Finally, I want to thank the most important people of all—the authors of the 49 papers that appear in this volume. I really enjoyed working with you and getting to know you. I am proud of what you have done and appreciate all your hard work. Together, we can provide improved learning experiences for our students. Together, we can help things change. And we will.

Nancy Baxter Hastings, Editor
Dickinson College
baxter@dickinson.edu

Contents

Introduction

The calculus renewal effort continues to have a significant impact on undergraduate mathematics education. Today, most new editions of calculus texts, even so-called traditional ones, incorporate significant themes and problems developed as part of the calculus reform movement. On the other hand, the mathematics community has paid little attention to the courses that precede calculus, most notably precalculus and college algebra. Moreover, only a small percentage of students who take precalculus ever go on to take calculus, and many who do are not well prepared and never complete the next course. In response to these concerns, the conference, *Rethinking the Preparation For Calculus*, was held in October, 2001. Since the papers presented at the conference form the basis for this volume, the volume begins with an overview of the conference co-authored by Jack Narayan, the principle organizer of the conference, and with the text of the keynote address given by Lynn Steen. The twenty questions about precalculus that Lynn posed guided many of the discussions at the conference. As you read about the conference in the initial papers in this section and in the papers that follow, please keep in mind that the focus of the conference was on precalculus courses that are not terminal—that is, on courses that prepare students to take calculus. During the conference it became apparent that this focus was too narrow and consequently the focus of this volume was extended to include the courses below calculus. Also please keep in mind that four years have passed between the conference and publication of the volume.

> The purpose of the conference was to rethink the preparation for calculus, with the following considerations: (1) Students are having different mathematical experiences in high school. The routine use of graphing calculators is standard, there is a greater emphasis on group work and collaborative learning, and there is a growing emphasis on conceptual understanding and realistic problem solving, not just skill development. (2) Calculus in college is placing different expectations on students, particularly an emphasis on conceptual understanding and the use of technology. (3) New technologies provide a wider selection of tools for both the teaching and learning of mathematics. (4) College algebra courses are in the process of changing. [The intended outcomes of the conference included:] articulating principles for changing precalculus offerings and providing guidance to the mathematical community.
>
> *The Conference: Rethinking the Preparation for Calculus*
> Jack Narayan and Darren Narayan

> Approximately fifteen years ago, a workshop similar to this one took place at Tulane University where a merry band of reformers sought to make calculus lean and lively. I had the opportunity to address that workshop with a list of twenty questions for calculus reformers. Thus I thought it appropriate to take a similar approach to this current workshop, to help launch your work by asking twenty questions about precalculus.... Clearly precalculus (and its alter ego college algebra) is the single most common mathematics course in undergraduate education. Data aside, it also appears to be the rock on which college students' mathematics education most often founders. That dark secret is why we are all here.... To reflect on the issues, to "rethink precalculus," as this workshop intends, nothing can beat the journalist's simple questions, [what, who, why, when, where, and how?]
>
> *Twenty Questions about Precalculus*
> Lynn Arthur Steen

1

The Conference: Rethinking the Preparation for Calculus

Jack Narayan   and   **Darren Narayan**
State University of New York, Oswego   *Rochester Institute of Technology*

Editor's note: The papers presented at this conference form the basis for this volume. This article, which was written shortly after the conference was held in October 2001, gives the rationale for the conference. For a more up-to-date view on some of the issues discussed here, please see the article, "Where Do We Go from Here: Creating a National Initiative to Refocus the Courses below Calculus," by Sheldon Gordon, later in this volume.

Rationale for the conference

During the last decade, calculus renewal efforts occured at all levels of post-secondary institutions as outgrowths of the Tulane Conference in 1987 and the subsequent national conference on *Calculus for A New Century*, hosted by the National Academy of Sciences. An MAA special report, *Assessing Calculus Reform Efforts* [1], estimated that "at least 150,000 students or 32% of all calculus enrollments in the spring of 1994 were in reform courses." Since 1994, several reform calculus texts have been among the highest selling nationally, and the number of institutions utilizing one or more aspects of reform in their calculus courses continues to rise. The calculus renewal movement continues to have a significant impact on undergraduate mathematics education. Instructors have experimented with alternative teaching methods that included the use of technology, collaborative learning, and out-of-class projects. These methods were integrated into new curricula with an increased emphasis on conceptual understanding. Today, all new editions of calculus texts, even so-called traditional ones, incorporate significant themes and problems developed as part of the calculus renewal movement.

One major, although unanticipated, outcome of the calculus renewal effort is the development of comparable efforts to revise college algebra and developmental mathematics offerings. There have also been several efforts to rethink precalculus courses, most notably those by Baxter Hastings, Connelly et al, and Gordon, et al. However since the publication of the volume, *Preparing for a New Calculus* [2], the mathematics community has paid insufficient attention to courses that bridge precalculus courses with calculus courses. The need to address this issue is essential since, as Lynn Steen points out, "Clearly precalculus (and its alter ego college algebra) is the single most common mathematics course in undergraduate education." The enrollment data in the fall of 2000 supports Steen's statement. In particular, the precalculus enrollment then was twice the enrollment of calculus I at all types of institutions and four times the enrollment of calculus I at two-year schools [12]. It is now time to renew a national dialogue on these issues.

The *Rethinking the Preparation for Calculus* project focused on precalculus courses that are not terminal—that is, those that are requirements for some type of calculus. All of us are aware, in general, that only a small percentage of students who take precalculus courses ever go on to take calculus and that many of them who do are not particularly well-prepared for calculus and never complete the course. But this has never been carefully documented. In a comprehensive study done at the University of Lincoln at Nebraska, Steve Dunbar found just how small that percentage actually is [13]. As a result, large numbers of students lose the opportunity to pursue mathematics or mathematics based disciplines. Mathematics instructors can readily identify with the observation noted in [3]:

> Students who were substantially underprepared reported more conceptual problems and feelings of being overwhelmed in the early stages of their major... Not only did most of these students abandon their ambition to continue with a S.M.E. (Science, Mathematics, and Engineering) major, they also suffered emotional damage by attempting what proved an impossible task.

Several colleges and universities, for example, University of Michigan, University of Texas at El Paso, and SUNY Farmingdale have recognized the need for rethinking the precalculus curriculum and have implemented completely different approaches.

Furthermore, given the importance that two and four-year schools attach to the development of articulation agreements among two and four-year schools, it is essential that there should be some serious discussion on the topic in conjunction with any discussion of changing precalculus courses.

Overview of the conference

At the 2001 joint meetings in New Orleans, Jack Narayan discussed the idea of having an NSF-funded conference focusing on the precalculus curriculum and was encouraged to develop a proposal. Shortly thereafter a steering committee was formed consisting of Jack Narayan (chair), Steve Dunbar, Sheldon Gordon, Nancy Baxter Hastings, Christopher Hirsch, and Jo Ann Lutz.

The committee proposed to organize a special invited conference to bring together mathematicians with a deep interest in this topic. The purpose of the conference was to rethink the preparation for calculus, with the following considerations:

1. Students are having different mathematical experiences in high school. The routine use of graphing calculators is standard, there is a greater emphasis on group work and collaborative learning, and there is a growing emphasis on conceptual understanding and realistic problem solving, not just skill development.

2. Calculus in college is placing different expectations on students, particularly an emphasis on conceptual understanding and the use of technology.

3. New technologies provide a wider selection of tools for both the teaching and learning of mathematics.

4. College algebra courses are in the process of changing.

The expected outcomes of the conference included:

- articulating some principles for changing precalculus offerings

- providing guidance to the mathematics community

- developing a cohesive effort among those individuals who have done groundbreaking work in this area to make a larger impact on the mathematics community

- focusing attention on problems and needs in the precalculus area that will lead to new funding programs/opportunities from NSF and other funding agencies

- publication of the conference proceedings as a volume in the MAA Notes series

The proposal for the conference was funded by the National Science Foundation (DUE 0136162) and the Calculus Consortium for Higher Education (CCHE). In the fall of 2001, fifty-five mathematics educators participated in the invited conference in Arlington, VA to rethink the preparation for calculus. The major themes for the conference included:

- Transition from high school

- Changes in college algebra

- Precalculus reform projects

- Technology

- Implementation issues

- Research in student learning

- Influencing the mathematics community

Invited position papers for each theme were presented and discussed. Participants were then encouraged to help identify challenges and make recommendations. This MAA Notes volume represents the contributions of many of the participants.

The discussions were based on a series of basic principles about precalculus courses that are stated and addressed by Nancy Baxter Hastings and Sheldon Gordon (see [14], [15]):

- Precalculus courses serve two distinct student populations: the overwhelming majority for whom precalculus is a terminal course and the relatively small minority for whom it is a gateway to higher mathematics. The needs of both populations should be met.

- Precalculus courses need to prepare students for calculus both *conceptually* and *algebraically*. It is not enough just to emphasize the development of manipulative skills; students need help to learn how to understand and apply the basic calculus concepts. Very few students have the ability to develop those conceptual connections on their own.

- Calculus is no longer the first mathematics course that is considered a prerequisite for courses in other quantitative disciplines. Precalculus and college algebra are now prerequisites for (non-calculus-based) courses in many fields. The mathematical needs of those fields are often not satisfied by standard, algebra skills-oriented precalculus/college algebra courses.

- Students need to see an emphasis on mathematical modeling to learn how mathematics is connected to the real world. The basic mathematical concepts and methods should be developed in contexts to help the students transfer their learning outside the mathematics classroom.

- Precalculus courses should help students learn to use modern technology wisely and appropriately.

Moreover, current research into the learning process has much to tell us about how students acquire fundamental precalculus (mathematical) concepts. Only a small minority of students learn mathematics the way their professors did.

The original intent for the conference was to focus exclusively on precalculus courses that are intended as the immediate precursors to calculus. However, the discussions at the conference almost immediately demonstrated that it was impossible to separate such precalculus courses from all precursor courses, especially college algebra. As a consequence, most of the comments in this article, as well as the current volume in its entirety, reflect this broader vision.

The principal recommendation from the conference was to collect extensive data from many different types of institutions to identify which students take precalculus (and college algebra) courses and why they take these courses. In addition, it was recommended that data be collected concerning the success rates in these courses, which successor courses the students actually take, and how they do in successor courses. The conference participants felt that such data is critical for convincing the mathematics community at large that precalculus courses need to change, as well as to acquaint potential funding agencies of the magnitude and implications of the problem. Moreover, the participants felt that any efforts to rethink precalculus should involve high school mathematics teachers and faculty in client disciplines.

Some conclusions

The task facing the mathematics community is challenging. Although changes in the curriculum are essential, the need to develop a process to improve learning and teaching, particularly at these introductory levels, is greater. In many ways, the "teaching gap" is as readily evident in the higher education mathematics curriculum as it is in the school system [4]. The authors observe that:

> American teachers aren't incompetent, but the methods they use are severely limited, and American teaching has no system in place for getting better. It is teaching, not teachers that must be changed... Teachers present definitions of terms and demonstrate procedures for solving specific problems. Students are then asked to memorize the definitions and practice the procedures. In the United States, the motto is "learning terms and practice procedures."

Mathematics instruction in traditionally-taught courses appear to be locked in a cycle. Many incoming college students are underprepared for calculus. College professors, despite good intentions of stressing conceptual understanding and mathematical modeling, have often yielded to the teaching of procedures because of time and syllabi constraints. As a result of this culture, the next generation of teachers adopts similar teaching strategies. In turn, their students arrive at college expecting to be taught in the same way they learned in high school. This cycle must be broken simultaneously at all levels. NCTM's efforts to encourage adoption of its Standards have made major gains at breaking the cycle at the school level, but it must be addressed at the college level as well.

The conference on *Rethinking the Preparation for Calculus* attempted to lay the foundation for change by reaching out to all constituents. This includes groups that have been instrumental in bringing about comprehensive change in the secondary curriculum, change in the calculus curriculum, adoption of technology at all levels of mathematics instruction, and a more informed view of mathematics education among college and university administrators. This collaborative effort will have a good chance of developing a process upon which we can improve in time. The papers in this MAA Notes volume describe the problem, begin the process of data collection, give examples of successful practices using mathematical modeling, new technologies, alternative learning strategies, assessment of student learning, and outline possible next steps.

On the positive side, mathematics educators, mathematical organizations, and politicians are beginning to recognize the need to go beyond curriculum renewal and focus instead on developing plans to improve learning and teaching. Recent publications provide the theoretical basis to develop a more efficient and effective process for the improvement of learning (see [5], [6], [7], [8], [9], [10]).

The individuals who sparked the movement to revitalize the teaching of calculus stressed that calculus should be "a pump, not a filter." Taking precalculus should be a positive experience for all students, not just the handful who pursue mathematically intensive fields. This requires rethinking precalculus. After all, a pump is only as good as the motor that powers it.

References

1. Leitzel, James. R.C. and Alan C. Tucker, *Assessing Calculus Reform Efforts: A Report to the Community*, MAA Special Report, Mathematical Association of America, Washington, DC, 1994.

2. Solow, Anita E., ed., *Preparing for a New Calculus: Conference Proceedings*, MAA Notes #36, MAA, Mathematical Association of America, Washington, DC, 1994.

3. Seymour, Elaine, and Nancy M. Hewitt, *Talk about Leaving: Why undergraduates leave the disciplines*, Westview Press, Boulder, Colorado, 1997.

4. Stigler, James W., and James Hiebert, *The Teaching Gap, The Free Press*, New York, NY, 1999.

5. National Research Council, *Knowing What Students Know: The Science and Design of Educational Assessment*, National Academy Press, Washington, DC, 2001.

6. Rogers, Elizabeth C. et al., *Cooperative Learning in Undergraduate Mathematics: Issues that Matters & Strategies that Work*, MAA Notes #55, Mathematical Association of America, Washington, DC, 2001.

7. Delong, Matt, and Dale Winter, *Learning to Teach and Teaching to Learn Mathematics: Resources for Professional Development*, MAA Notes #57, Mathematical Association of America, Washington, DC, 2002.

8. Tucker, Alan C., ed., *Models that Work: Case Studies in Effective Undergraduate Mathematics Programs*, MAA Notes #38, Mathematical Association of America, Washington, DC, 1995.

9. Ewing, John, ed., *Towards Excellence: Leading a Doctoral Mathematics Department in the 21st Century*, AMS, Washington, DC, 1999.

10. National Research Council, *How People Learn: Brain, Mind, Experience, and School*, National Academy Press, Washington, DC. 2000.

11. Steen, Lynn, "Twenty Questions about Precalculus," in this volume.

12. McGowen, Mercedes, "Who are the Students Who Take Precalculus?," in this volume.

13. Dunbar, Steve, "Enrollment Flow to and From Courses below Calculus," in this volume.

14. Gordon, Sheldon P, "Preparing Students for Calculus in the Twenty-First Century," in this volume.

15. Baxter Hastings, Nancy, "Refocusing Precalculus: Challenges and Questions," in this volume.

2

Twenty Questions about Precalculus

Lynn Arthur Steen
St. Olaf College

Editor's note: This paper is the text of the keynote address given by Lynn Steen in October 2001 at the conference Rethinking the Preparation for Calculus. (For an overview of the conference, please see the preceding paper in this volume by Jack Narayan and Darren Narayan.)

Introduction

Approximately fifteen years ago a workshop similar to this one took place at Tulane University where a merry band of reformers sought to make calculus lean and lively. I had the opportunity to address that workshop with a list of twenty questions for calculus reformers. Thus I thought it appropriate to take a similar approach to this current workshop, to help launch your work by asking twenty questions about precalculus. (For comparison, I reproduce in Appendix A the questions that I put before the calculus reformers at Tulane. There you will find not 20 but 28 questions, the extra eight being added to the manuscript as a result of issues raised during the workshop. The full text with elaborations on each question can be found in [3].)

At the time of the Tulane workshop I was President of the Mathematical Association of America, and in that capacity had some degree of oversight responsibility for MAA's many committees. Even as the Tulane rebels were training their sights on calculus, I was well aware that then, as now, more college students study precalculus than calculus. On several occasions I asked the CUPM subcommittee on the First Two Years (later to be renamed CRAFTY—Calculus Reform and the First Two Years) whether in order to fulfill the mission implied by their title they might be interested in looking at the mathematics course that is the most common of all taken during students' first two years in college, namely precalculus. Their answer was consistently negative: precalculus, in their judgement, was an unfortunate leftover from high school mathematics. Despite enrollment evidence, they said, college mathematics begins with calculus.

With this fifteen-year-old experience as backdrop, I checked current data to see what enrollments look like now. Figure 1 offers a sobering portrait of undergraduate mathematics prepared by combining recent data from two sources—the (forthcoming) quinquennial CBMS 2000 survey [2] and the annual AMS survey [1]. (Enrollments included in this figure are predominantly in departments of mathematical and statistical science. They do not count the many statistics, computer science, and applied mathematics courses found outside departments of mathematics or statistics.) Clearly precalculus (and its alter ego college algebra) is the single most common mathematics course in undergraduate education. Data aside, it also appears to be the rock on which college students' mathematics education most often founders. That dark secret is why we are all here.

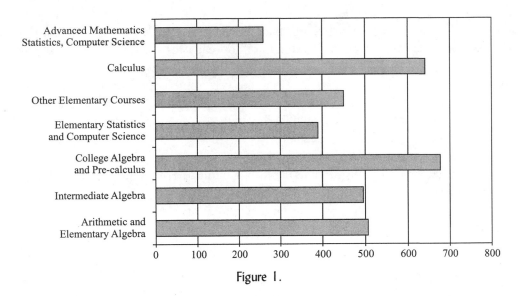

Figure 1.

One can approach the challenges of precalculus from several perspectives. For example, a managerial perspective would suggest a cycle of setting goals, developing strategies, implementing changes, assessing outcomes, reflecting on results, and making adaptations. A journalist's paradigm, in contrast, seeks insight by asking questions: what, who, why, when, where, and how? To actually make changes that improve student learning, the managerial paradigm is really the only effective option: set a goal, make some changes, look at the results, and then regroup. But to reflect on the issues, to "rethink precalculus" as this workshop intends, nothing can beat the journalist's simple questions.

What?

What exactly is precalculus? Is it the same as college algebra? (In this analysis, I ignore whatever differences there may be between them.) Does precalculus have an intellectual core like geometry or calculus? Does it have a center or a town square? Or is it more like a mathematical strip mall that just fills space between high school and college?

What is the real goal of precalculus? Is it really to prepare students for calculus, or does it have other purposes, either benign or sinister? Isn't it also, *de facto*, a ubiquitous prerequisite for a wide range of quantitatively-oriented college courses, a steady source of tuition revenue that reliably exceeds marginal costs, and an accepted means of screening students for access to the economic rewards of higher education?

What effect does calculus have on the nature of the precalculus course? What differences are there in preparation for reformed calculus, for traditional calculus, for mainstream calculus, or for non-mainstream (business) calculus? Can a single course provide suitable preparation for all flavors of calculus? Can precalculus possibly be made lean and lively?

Who?

Who takes precalculus? Is its clientele students who are reviewing (or relearning) what they once learned, students who did not learn what they once studied, students who never had the opportunity to learn precalculus topics, or students who declined the opportunity? In most courses, the answer is "all of the above." Can a single course really serve all these different students?

Who should take precalculus? Does precalculus serve well the quantitative needs of students preparing for fields that do not require calculus? Does it offer any lasting benefit for students who never take any further mathematical or quantitative course? For that matter, does precalculus really benefit the students it was created to serve—those who need calculus but are not ready for it?

Who should teach precalculus? University mathematicians? Teaching assistants or adjuncts? Experienced secondary school teachers (who perhaps teach the very same course during the day to high school students)? What about on-line tutorials? Is a Ph.D. in mathematics an appropriate credential for teaching precalculus? Might mathematicians' uncommon facility with elementary mathematics make them peculiarly inappropriate as empathetic teachers of precalculus?

Who benefits from precalculus? Who loses? Does precalculus have disparate impact on at-risk populations? For whom, if anyone, does precalculus serve as a pump? For whom is it a filter? Some believe its primary beneficiary is the budgets of mathematics departments for whom it serves as a cash cow. Maybe it is just a means of shifting tuition income from a required large enrollment course to low enrollment advanced electives—that is, from the mathematically weak to the mathematically strong.

Why?

Why is calculus so important for under-prepared undergraduates? Is preparing for calculus really a wise use of college students' time and energy? Might the siren call of calculus replace more important goals for students who enter college unprepared for calculus? Shouldn't more under-prepared undergraduates be steered in other quantitative directions?

Why do students take precalculus? Is it to prepare for calculus, to meet the prerequisite of a particular course or program of study, to fulfill a general education option, or to fulfill a graduation requirement? Are any of these reasons defensible, or are they simply traditional?

Why is precalculus so often part of general education? Does precalculus advance students' mathematical or quantitative literacy? Does anyone believe that precalculus is the right mathematics course to prepare students well for lives in the 21st century? Does it reveal important insights into the nature, power, and beauty of mathematics? Can precalculus possibly serve two masters—calculus and culture?

Why should students take precalculus? Does precalculus have value for the majority of students who take the course? Are its concepts and skills independently useful apart from their role in calculus? How many ever use the skills they learn in precalculus? Is precalculus a sensible choice for the last mathematics course a student ever takes?

Why do so many prospective elementary school teachers take precalculus? In the majority of departments, precalculus (or college algebra) is the second most common course taken by students preparing for K–3 certification [2]. Does this make any sense? Does precalculus really provide teachers with deep understanding of the mathematics they will be teaching?

When?

When should students take precalculus? Is there an optimal window for learning precalculus? Isn't precalculus taught and learned better in high school? Currently only about 25% of high school graduates take precalculus in high school, even though over 60% enroll in some form of postsecondary education. Shouldn't higher education tell students and schools that it is more important for more students to finish precalculus in high school than for more students to finish calculus?

Where?

Where do precalculus students come from? What have been their mathematical backgrounds? What are their major programs of study or career interests? How many are returning after having interrupted their study of mathematics? How do students' prior mathematical experiences influence their views of mathematics, their confidence in their own abilities, and their likelihood of success with precalculus?

Where do precalculus students go? How many precalculus students eventually take calculus? (*Answer*: Relatively few.) How many take other courses that utilize ideas from precalculus? (*Answer*: A few more.) How many complete a year of calculus with good grades and incentive to continue their study of mathematics? (*Answer*: Embarrassingly few.) For how many is precalculus the end of their study of and interest in mathematics? (*Answer*: Far too many.)

How?

How should the changing role of mathematics influence the nature of precalculus? In the last two decades mathematical practice has become increasingly algorithmic and digital. New applications range from genomics to cinema, from manufacturing to Wall Street. How, if at all, should the content of precalculus reflect this expanded interface of mathematics with the rest of the world?

How do articulation agreements constrain precalculus? Are inter-institutional agreements on syllabi and standards essential instruments of quality control? Or do tight curriculum specifications lead to curricular sclerosis? Are the transparency benefits of articulation agreements worth the cost of inflexibility and stifled innovation? On balance, do students gain or lose from these protocols?

How well aligned is precalculus with common placement tests? Do commercial or homegrown placement tests reflect the same level and type of performance expectations as a precalculus course? Do they accurately place students into or out of precalculus? Are they fair to students?

How should technology influence precalculus? Is technology a means or an end? Is its role to help students learn traditional mathematics, or is technology now so much part of the way mathematics is practiced that it has itself become an important goal of instruction? Is the use of numerical, graphing, and CAS systems a prerequisite to learning calculus?

How do you measure success? This may be the toughest question of all. Fewer than one in four students, perhaps as few as one in ten, achieve the *prima facie* goal of precalculus: to succeed in calculus. Without clarity about goals, it is impossible to gauge success. Without data on students' future academic careers, success is unknowable. And without external validation, precalculus may never improve.

Conclusion

These questions suggest an overwhelming agenda for a course of enormous importance, but a course that is all but invisible to the mathematical community. I wonder how much has really changed in the last fifteen years since CRAFTY's predecessor declined to take up the challenge? Neither enrollment patterns, course prerequisites, nor general education requirements have changed very much. Nor, I suspect, have mathematicians' attitudes about what constitutes appropriate college mathematics. Does the mathematical profession now consider precalculus a challenge worth working on, or do they still see it as a peripheral problem best ignored? Can any mathematician earn tenure by teaching or improving precalculus? (That's a rhetorical question.)

In addition to seeking answers to the twenty questions I have suggested, the merry band of reformers assembled for this conference will need to think hard about where precalculus fits into the agenda of mathematics, of science, and of our nation. Rethinking precalculus may lead to some surprising conclusions.

References

1. Loftsgaarden, Don O., James W. Maxwell, and Kinda Remick Priestley. "2000 Annual Survey of the Mathematical Sciences (Third Report)." *Notices of the American Mathematical Society*, 48:8 (September 2001) 819–828.

2. Lutzer, David, *et al. CBMS 2000: Statistical Abstract of Undergraduate Programs in the Mathematical Sciences in the United States.* Providence, RI: American Mathematical Society, 2002.

3. Steen, Lynn A. "Twenty Questions for Calculus Reformers." In *Toward a Lean and Lively Calculus: Report of the Tulane Calculus Conference.* Ronald G. Douglas, Editor. Washington, DC: Mathematical Association of America, 1986, pp. 157–165.

Appendix A

Twenty Questions for Calculus Reformers
Lynn Arthur Steen, January, 1986 (From [3])

1. Should fewer students study calculus?
2. Is calculus an appropriate filter for the professions?
3. Will computer science dethrone calculus?
4. Do students really learn the major ideas of calculus?
5. Has calculus become a cookbook course?
6. Does calculus focus excessively on closed-form formulas?
7. Should calculus students learn to use or to imitate computers?
8. What new topics are essential for calculus in a computer age?
9. Which topics in calculus are no longer essential?
10. Do engineers still need the traditional calculus?
11. Should calculus be a laboratory course?
12. Is there any reason to teach high school calculus?
13. Why do U.S. students perform so poorly on international tests?
14. Is there any value to precalculus remedial programs?
15. Why do calculus books weigh so much?
16. Can one design a good calculus course from a survey?
17. Is calculus a good course to train the mind?
18. Can calculus courses convey cultural literacy?
19. Does calculus contribute to scientific literacy?
20. What will calculus be like in the year 2000?

Added after workshop discussion:

21. Do students ever read their calculus books?
22. Should precalculus be a prerequisite for calculus?
23. Is teaching calculus most like teaching a foreign language?
24. Should the student-faculty ratio for calculus be limited?
25. Do student evaluations favor calculation-based courses?
26. Are there enough qualified calculus teachers?
27. Who will be the calculus teachers in the year 2000?
28. Should calculus be taught only by experienced teachers?

Background

It is one thing to *claim* we have a problem and another thing to *show* we have a problem. Based on their own experiences and observations, participants at the conference, *Rethinking the Preparation for Calculus*, felt that the number of students who continue on to calculus from precalculus was small and their success rate was low. However, they realized that in order to convince colleagues, administrators, state legislators, funding agencies, and book publishers that changes need to be made, they needed data that clearly indicate the extent of the problem. They also realized that much could be learned from the calculus renewal efforts. The papers in this section by Mercedes McGowen and Steven Dunbar provide some preliminary data in response to some important questions, while the papers by Deborah Hughes Hallett and Susan Ganter summarize the lessons learned from the calculus initiative and discuss future directions.

Mercedes McGowen analyzes the enrollment in mathematics courses at two- and four-year colleges and at universities over the past twenty years, and she describes the demographic profile of students at William Rainey Harper College, "a two-year college whose mathematics course enrollment closely parallels enrollment at two-year colleges nationally." She examines the questions: "Who are the undergraduate students who enroll in precalculus courses? What courses do students take after completing a precalculus course?"

> In fall 2000, more than three million students attending two- and four-year colleges and universities were enrolled in mathematics courses taught in departments of mathematics and mathematical sciences. Twenty-two percent of these three million students (653,000) enrolled in precalculus courses—courses many had completed previously in high school. As undergraduates, students often find themselves repeating their high school courses, receiving no general education credit in college for these courses. Explanations given include: (a) students didn't retain what they learned in high school; (b) the algebraic competencies and understandings were insufficient foundation for success subsequently; (c) the college placement exam inappropriately placed students. Whatever the reason(s), it appears that precalculus courses are yet another effective filter—not a pump—for many students.
>
> *Who are the Students Who Take Precalculus?*
> Mercedes McGowen

Steven Dunbar describes the results of a comprehensive study he undertook over 20 consecutive semesters at the University of Nebraska-Lincoln, where he sought partial answers to the questions: "What mathematics courses (calculus and otherwise) will students take after completing courses supposedly intending to prepare them for calculus? What mathematics courses have students studied before calculus preparation courses and how recent is their knowledge?"

> The conclusion to be drawn from this study is that at least at the University of Nebraska-Lincoln, a sizable majority of the students in the precalculus course prepare to start calculus. The students from the precalculus course constitute a sizable fraction of the students taking calculus, although not a majority. A fairly insignificant fraction of the students taking precalculus at the university ever take three semesters of calculus. Another conclusion is that only a minority of students taking college algebra go on to start the three semester sequence of calculus, although a considerable fraction do take the one semester business calculus course.
>
> *Enrollment Flow to and from Courses below Calculus*
> Steven R. Dunbar

13

Deborah Hughes Hallett raises the question: "What has been the impact of [the calculus reform movement]?" She describes the rationale for the movement and some of the subsequent changes in the way calculus is taught. She discusses the impact of emerging technologies, the emphasis on conceptual understanding, the utilization of new pedagogies, and the increased cooperation with client disciplines.

In order to evaluate the impact of the calculus reform movement, we first need to recall its goals. Although different people may phrase it differently, everyone involved agreed that they were trying to improve the teaching of calculus. Some would say they wanted more student involvement; others would say they wanted to take advantage of technology; others would say they wanted to emphasize problem solving and modeling. Most would agree that they wanted to improve conceptual understanding.... What has been the impact of this effort?

What Have We Learned from Calculus Reform?
Deborah Hughes Hallett

Susan Ganter asks: "Where have we been and where are we going with calculus and, more importantly, the entire scope of introductory college mathematics?" Susan bases her observations in part on two studies she conducted for the National Science Foundation in the late 1990s: one to determine the national impact of the calculus reform movement since 1988, and a second to examine the subsequent NSF Institutional Reform (IR) Program.

For better or worse, the ideas of calculus reform are deeply embedded in the conversations of the mathematics community. Technology, cooperative learning, student projects, applications—the elements of this movement—have become a part of the vocabulary in mathematics departments across the country. Efforts to change the nature of the calculus course at the undergraduate and secondary levels have sparked discussion and controversy in ways as diverse as the actual changes. Such interactions range from "coffee pot conversations" to university curriculum committee agendas to special sessions on undergraduate education at regional and national conferences. ... But what is the significance of these activities?

Calculus and Introductory College Mathematics
Susan L. Ganter

3

Who are the Students Who Take Precalculus?

Mercedes A. McGowen
William Rainey Harper College

Introduction

This paper examines the questions: Who are the undergraduate students who enroll in precalculus courses? What courses do students take after completing a precalculus course? These questions are addressed by an analysis of enrollment in mathematics courses at two- and four-year colleges and at universities from 1980 to 2000, followed by a demographic profile of students at a Midwestern two-year college whose mathematics course enrollment closely parallels enrollment at two-year colleges nationally [1].

What is "precalculus"?

As we begin the task of rethinking the precalculus curriculum, we first need to examine our assumptions about precalculus. Do we, as well as members of our own departments, colleagues in other institutions, and those teaching in the high schools mean the same thing when we characterize a course or courses as "precalculus"? When colleagues are asked what "precalculus" is, they give the following responses:

- the courses before calculus (college algebra, trigonometry, and precalculus/elementary functions, and modeling)
- a specific course that combines college algebra (including the topics of induction, theory of equations, sequences, and series) and trigonometry
- 2nd year high school algebra and a brief introduction to trigonometry
- a high school algebra and trigonometry course for students in the accelerated track.

Answers to the question: What is "college algebra"? also vary. Some respond that it is a third algebra course taken in high school—a one-semester course taken after algebra I and algebra II have been completed. Others believe it is the second year of high school algebra renamed because it is taken in college by under-prepared undergraduates. As Humpty Dumpty pointed out to Alice in *Through the Looking Glass*: "You see, it's like a portmanteau—there are two meanings packed up into one word." In this instance, there are more than two meanings. Clarification of the terms "precalculus" and "college algebra" is needed if we are to answer the questions: What is precalculus? Which students take precalculus? What courses do they take after completing a precalculus course? In this paper, the term "precalculus" refers to those courses intended to prepare students to take calculus: college algebra, trigonometry, the combined college algebra/trigonometry course or a precalculus/elementary functions course.

What mathematics courses do undergraduate students enroll in?

Sixty-eight percent of incoming freshman at four-year colleges and universities took four years of mathematics in high school in 1997. However, many who complete three or four years of "rigorous" high school mathematics are unsuccessful in subsequent college-level mathematics courses [2, 3]. In fall 2000, more than three million students attending two- and four-year colleges and universities were enrolled in mathematics courses taught in departments of mathematics and mathematical sciences. Twenty-two percent of these three million students (653,000) enrolled in precalculus courses—courses many had completed previously in high school. As undergraduates, students often find themselves repeating their high school courses, receiving no general education credit in college for these courses. Explanations given include: (a) students didn't retain what they learned in high school; (b) the algebraic competencies and understandings were insufficient foundation for success subsequently; (c) the college placement exam inappropriately placed students.

There are no national studies that provide answers to the questions:

- Who are the students who take precalculus as undergraduates?

- Why do they take precalculus?

- How appropriately is the present precalculus curriculum aligned with the needs of students who enroll in these courses?

- What do other programs want students to know upon completion of precalculus courses?

- How many of the students who enroll in precalculus courses intend to enroll in calculus?

There is a growing consensus that the precalculus courses currently offered are not meeting the needs of students who intend to major in mathematics or take math-intensive programs [4,5,6,7]. Many seriously question whether the precalculus courses currently offered are the appropriate prerequisite for students in non-mathematics intensive programs. They are interested in knowing how many students take a precalculus class solely because the course is a required prerequisite for their particular program of study which does not require any additional mathematics-intensive coursework. How many of the students who take precalculus courses are successful in subsequent mathematics courses—assuming they take another mathematics course—remains an open question (see Figure 1).

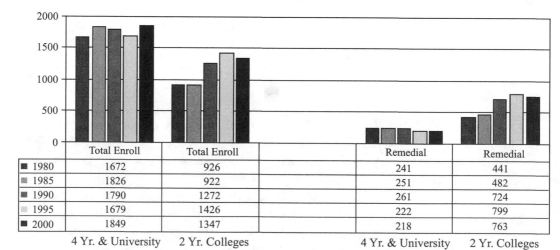

	Total Enroll	Total Enroll		Remedial	Remedial
■ 1980	1672	926		241	441
▦ 1985	1826	922		251	482
■ 1990	1790	1272		261	724
1995	1679	1426		222	799
■ 2000	1849	1347		218	763

4 Yr. & University   2 Yr. Colleges   4 Yr. & University   2 Yr. Colleges

Data Source: CBMS Survey Fall 2000: Table A-1 and Table TYR.3.

Figure 1. Total mathematics and statistics enrollment and remedial enrollment (in thousands)

Mathematics course enrollment: 1980–2000

Mathematics course enrollments taught in departments of mathematics at colleges and universities have increased 23% since 1980. Most of that growth occurred at two-year colleges where mathematics enrollment increased 45% during the past twenty years—four times the 11% increase at four-year colleges and universities. In fall 2000, 41% of all undergraduates enrolled in mathematics courses attended two-year colleges.

The growth in enrollment at two-year colleges has included increasing numbers of students who begin college taking remedial courses—that is, arithmetic, introductory algebra, intermediate algebra, and geometry. Enrollment in these courses has skyrocketed since 1980—increasing 73%—while overall mathematics course enrollments at two-year colleges increased 45% between 1980 and 2000. In fall 2000, remedial course enrollment constituted 12% of the total mathematics course enrollment at four-year colleges and universities and 57% of total mathematics course enrollment at two-year colleges.

Non-remedial mathematics courses also experienced increased enrollment over the past twenty years at four-year colleges and universities, as well as at two-year colleges. It is of interest to note in which courses increases in enrollment occurred. Our particular interest is in the precalculus course enrollment. Since preparation for calculus is the goal of precalculus and significant efforts have gone into reforming both the curriculum and instructional approaches utilized in mainstream calculus courses, let us examine the national calculus enrollment data and the effect of calculus curriculum initiatives on mathematics enrollment during the past decade before turning our attention to the precalculus enrollment data.

The Tulane Conference in 1986 reaffirmed the importance of calculus in the curriculum [8]. The primary outcome of the conference was the development of alternative curricular materials and instructional strategies, with the goal of transforming calculus from a filter into a pump. Building on that initiative, various changes have been introduced into the mainstream calculus curriculum during the past ten years. At four-year colleges and universities:

- Use of graphing calculators in mainstream calculus I classes has grown from 3% in 1990 to 51% in 2000.

- Computer assignments were required in 9% of these classes in 1990. By 2000, 31% of the calculus I classes included computer assignments.

- Group projects and writing assignments were regular features in 5% of these courses in 1990. Ten years later, 27% of calculus I classes include writing assignments and 19% include group projects.

Two-year colleges have seen similar reform initiatives incorporated into the calculus I curriculum.

- Graphing calculators are now used in 78% of the classes and 35% include computer assignments.

- Twenty-seven percent of classes currently assign group projects and 31% include a writing component [9].

What effect have the changes introduced into the calculus curriculum had on enrollment? Despite ongoing efforts to transform calculus from a filter into a pump with implementation of various reform initiatives during the past fourteen years—*the actual number of students enrolled in mainstream calculus courses declined more than 12%* since 1985, the year before the Tulane Conference. Comparing the percent of students enrolled in calculus of the total number enrolled in mathematics courses in 1980 with the percent enrolled in calculus in 2000 reveals an even more discouraging story—the percent of students enrolled in mainstream calculus courses also declined. In 1980, 405,000 students were enrolled in calculus courses at four-year colleges and universities—24% of the total mathematics enrollment. As of fall 2000, calculus enrollment had dropped to 352,000 students—19% of the overall fall 2000 mathematics course enrollment. Mainstream calculus enrollment at two-year colleges also declined—from 8% of the total mathematics enrollment in 1980 to 6% in 2000 (see Figure 2).

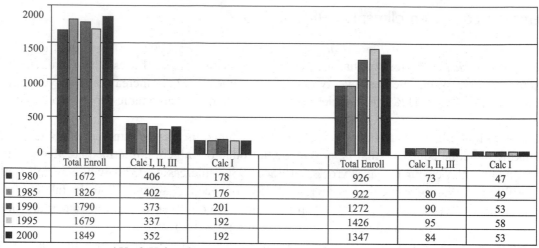

	Total Enroll	Calc I, II, III	Calc I		Total Enroll	Calc I, II, III	Calc I
■ 1980	1672	406	178		926	73	47
▨ 1985	1826	402	176		922	80	49
■ 1990	1790	373	201		1272	90	53
▨ 1995	1679	337	192		1426	95	58
■ 2000	1849	352	192		1347	84	53

4 Yr. & University   2 Yr. Colleges

Data Source: CBMS Statistical Abstract Table A-1 and Table TYR.3.

Figure 2. Calculus enrollment compared to total mathematics enrollment (in thousands)

Calculus I enrollments have remained relatively constant since 1980 at 11% of the total number enrolled in mathematics courses each year. Two-year college calculus I enrollment has also remained fairly constant, ranging from 4% to 5% of the total mathematics enrollment over the past twenty years.

Non-mainstream calculus enrollment has shrunk since 1990, decreasing 29% from a high of 163,000 in 1990 to 115,000 in 2000. Two-year college non-mainstream calculus courses experienced an even greater loss of enrollment. In 1990, two-year enrollment peaked at 34,000. Only 17,000 students were enrolled in non-mainstream calculus courses—a 50% decrease (see Figure 3).

Given that the total number of undergraduates taking mathematics courses has increased 23%, while mainstream and non-mainstream calculus enrollments declined—and the number of students in calculus I remained relatively constant over the past twenty years—*what mathematics courses are students taking?* The answer is: *non-calculus based mathematics courses.* As of fall 2000, the CBMS Survey data indicate

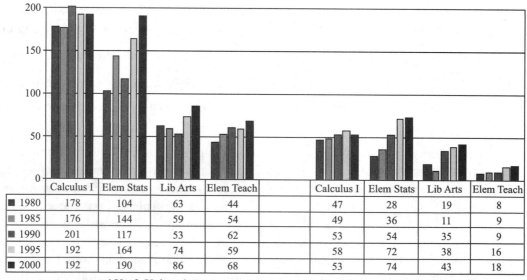

	Calculus I	Elem Stats	Lib Arts	Elem Teach		Calculus I	Elem Stats	Lib Arts	Elem Teach
■ 1980	178	104	63	44		47	28	19	8
▨ 1985	176	144	59	54		49	36	11	9
■ 1990	201	117	53	62		53	54	35	9
▨ 1995	192	164	74	59		58	72	38	16
■ 2000	192	190	86	68		53	74	43	18

4 Yr. & University   2 Yr. Colleges

Data Source: CBMS Statistical Abstract Table A-1 and Table TYR.3.

Figure 3. Enrollment in calculus vs. non-calculus based courses (in thousands)

that enrollment in elementary statistics courses is approaching the calculus I enrollment at four-year colleges and universities.

The enrollments in the non-calculus-based courses have continued to grow over the past twenty years:

- Elementary statistics courses enrollment increased 83%.

- Content course(s) enrollment for elementary teachers increased 55%.

- Liberal arts mathematics course enrollment increased 37%.

The growth in two-year college enrollment of the non-calculus based courses is even greater than that in the four-year colleges and universities. Non-calculus-based elementary statistics course enrollment grew by 164% and now exceeds enrollment in calculus I. The liberal arts and the math content courses for elementary teachers enrollments have more than doubled during the past twenty years.

Precalculus enrollment: 1980–2000

Since 1980, enrollment in the various precalculus courses increased 17% at four-year colleges and universities. During the past decade, enrollment increased in all precalculus courses except trigonometry, which declined 11%. Since 1990, enrollment in the precalculus/elementary functions course increased 35%, more than eight times the growth in enrollment in college algebra. Despite a much smaller increase of 4% growth during the past ten years, college algebra enrollment is still double the precalculus/elementary functions course enrollment. Combined college algebra/trigonometry course enrollment experienced a 6% increase during this time.

Two-year college precalculus enrollment reveals a similar picture over the past ten years. College algebra remains the course with the greatest enrollment, experiencing a 13% increase—more than three times the increase in the course at four-year colleges and universities. The precalculus/elementary functions course, like its counterpart at four-year colleges and universities, has had the greatest growth over the past decade, with enrollment increasing 37%. Both trigonometry and the combined college algebra/trigonometry enrollment at two-year colleges declined—23% in trigonometry and 11% in the combined course.

The increasing enrollment in the precalculus/elementary functions course could be interpreted as an attempt to strengthen the preparation of students intending to take calculus by directing students who plan to pursue a mathematics-intensive program into a more rigorous course—recognizing that the college algebra course includes large numbers of students who are not intending to take calculus subsequently.

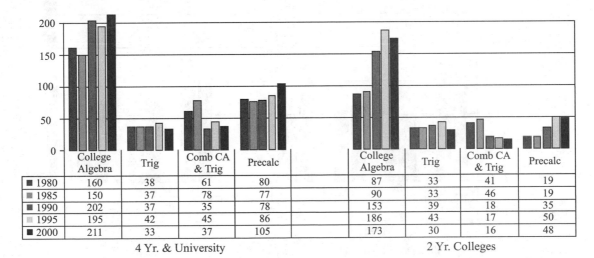

	College Algebra	Trig	Comb CA & Trig	Precalc		College Algebra	Trig	Comb CA & Trig	Precalc
■ 1980	160	38	61	80		87	33	41	19
■ 1985	150	37	78	77		90	33	46	19
■ 1990	202	37	35	78		153	39	18	35
■ 1995	195	42	45	86		186	43	17	50
■ 2000	211	33	37	105		173	30	16	48

4 Yr. & University   2 Yr. Colleges

Data Source: CBMS Statistical Abstract Table A-1 and Table TYR.3.

Figure 4. Enrollment in precalculus courses (in thousands)

A comparison of enrollment for precalculus courses and calculus I over the past twenty years reveals that the increasing precalculus enrollment has had little, if any, impact on the calculus I enrollment. Between 1980 and 1990, total calculus enrollment exceeded the precalculus enrollment at four-year colleges and universities. Since 1980, calculus enrollment has declined, precalculus enrollment has increased, and commencing in 1995, the precalculus enrollment has exceeded the 4-year college and university calculus enrollment.

A comparison of precalculus and calculus enrollment at two-year colleges documents an even sharper contrast of growth in precalculus enrollment and the declining growth in calculus enrollment since 1995.

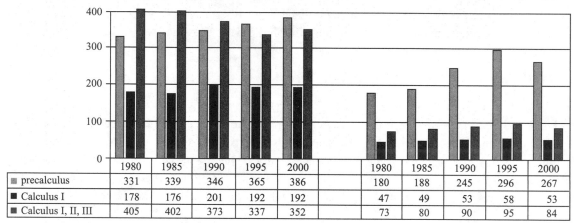

Data Source: CBMS Survey Fall 2000 Table A-1 and Table TYR.3.

Figure 5. A comparison of precalculus and calculus enrollment (in thousands)

If we disaggregate the data by type of institution, we note that mainstream calculus enrollment exceeded precalculus enrollment in 1995 and 2000 *only at PhD granting institutions*. At MA and BA institutions and at two-year colleges, precalculus enrollment exceeded mainstream calculus enrollment, with the gap between precalculus and calculus I enrollment widening over the past five years.

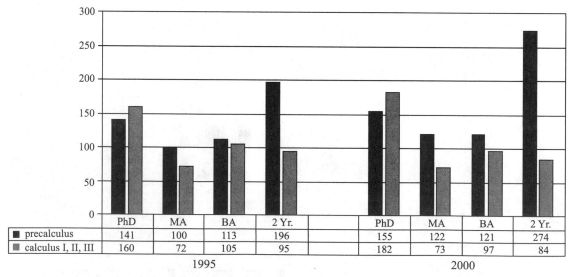

Data Source: CBMS Survey Fall 2000 Table A-1.

Figure 6. Precalculus and calculus enrollment by type of institution (in thousands)

A comparison of calculus I enrollment with precalculus enrollment in 1995 and 2000 shows that precalculus enrollment exceeds calculus I enrollment at all types of institutions.

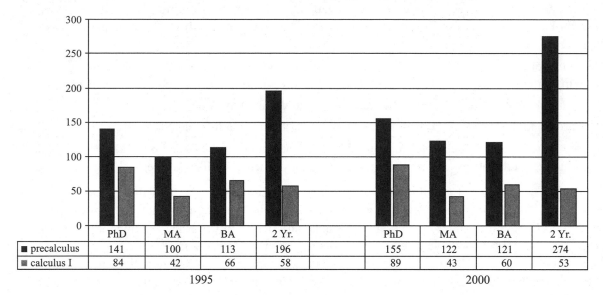

	PhD	MA	BA	2 Yr.		PhD	MA	BA	2 Yr.
■ precalculus	141	100	113	196		155	122	121	274
■ calculus I	84	42	66	58		89	43	60	53
			1995					2000	

Data Source: CBMS Survey Fall 2000 Table A-1.

Figure 7. Precalculus and calculus I enrollment by type of institution (in thousands)

Meeting the needs of our students: A two-year college profile

An examination of national enrollment data suggests that the present precalculus curriculum does not meet the needs of many of the students enrolled in precalculus courses—those who enroll in precalculus courses intending to take calculus, as well as those who take a precalculus class only because the course is a required prerequisite for their program of study. It appears that large numbers of students do not have learning experiences in precalculus courses that generate enthusiasm and lead to subsequent enrollment in calculus I and a program of studies in mathematics or other math-intensive fields.

Aggregate data summarize various quantifiable factors and indicate trends. They do not tell the whole story. Individual institutions also have stories to tell—stories that confirm the national profile and stories that contradict or modify the national profile. We all know of institutions that have introduced new curriculum materials and incorporated changes in their instructional strategies. We exchange stories of the successes and failures of these efforts to transform the curriculum. Each story provides information that can be utilized to make informed decisions and contributes to our understanding of who takes precalculus. We now consider the story of one mid-western community college and the efforts of the mathematics faculty at that institution to address some of the questions raised in this paper.

During spring 1998, the Department of Mathematical Sciences at William Rainey Harper College began a large-scale project to improve the effectiveness of the mathematics curriculum. The goal was to assess, revise, and restructure the curriculum to meet the needs of students for the 21st century and the needs of the various disciplines served by the department. One of the first tasks was to collect demographic information, enrollment data, and longitudinal data on student success in subsequent courses to determine how well each course is presently meeting its goals and objectives. Harper College Foundation grants provided funds for reassigned time for full-time faculty and stipends for adjunct-faculty members to work on the project.

A demographic survey was administered twice in all mathematics courses—during spring 1999 ($n = 2286$) and fall 1999 ($n = 1609$) semesters. The survey responses indicate that:

- Nearly 90% of students enrolled in mathematics courses plan to transfer to four-year colleges or universities.
- The vast majority of students enrolled in mathematics courses are white, non-Hispanic (76%), with Asian/Pacific islanders as the second largest ethnic group (13%), followed by Hispanics (8%), typical of the overall college enrollment patterns.
- Seventy-one percent of students enrolled in mathematics courses at Harper were full-time students (12 credit hours or more) in the fall semester (63% were full-time during spring semester).
- The numbers of male and female students were approximately equal.
- Approximately 15% of Harper's students were enrolled in math-intensive courses.
- More than 60% of all students were enrolled in mathematics courses that do not satisfy state undergraduate general education requirements for graduation (remedial courses, college algebra, trigonometry).

Harper College course enrollment approximates the national enrollment pattern of two-year colleges. Harper has a smaller percent of total mathematics enrollments in remedial mathematics and a larger percent of students in calculus I than do two-year colleges nationally. Enrollment in elementary statistics course increased 28% since 1998—well below the increase nationally (see Figure 8).

The precalculus enrollment constitutes 18–20% of Harper's total mathematics enrollment, with the percent of students who take mathematics courses at Harper enrolled in calculus I each of the past four years ranging from 5% to 7%. The mainstream calculus enrollment at Harper was 9% of total enrollment

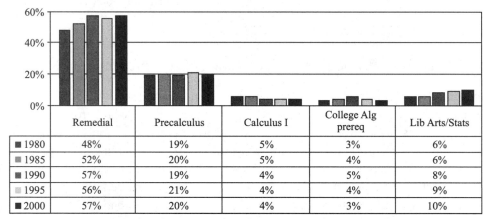

	Remedial	Precalculus	Calculus I	College Alg prereq	Lib Arts/Stats
■ 1980	48%	19%	5%	3%	6%
■ 1985	52%	20%	5%	4%	6%
■ 1990	57%	19%	4%	5%	8%
■ 1995	56%	21%	4%	4%	9%
■ 2000	57%	20%	4%	3%	10%

Two-Year Colleges: 1980–2000

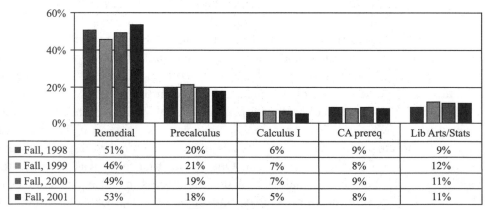

	Remedial	Precalculus	Calculus I	CA prereq	Lib Arts/Stats
■ Fall, 1998	51%	20%	6%	9%	9%
■ Fall, 1999	46%	21%	7%	8%	12%
■ Fall, 2000	49%	19%	7%	9%	11%
■ Fall, 2001	53%	18%	5%	8%	11%

Harper College: 1998–2001

Data Source: CBMS Survey Fall 2000 Table TYR.3; Harper Institutional Research (2001).

Figure 8. Two-year and Harper College enrollment as a percent of total math enrollment

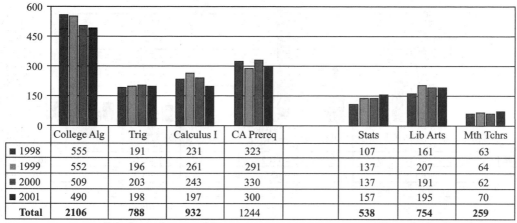

	College Alg	Trig	Calculus I	CA Prereq		Stats	Lib Arts	Mth Tchrs
■ 1998	555	191	231	323		107	161	63
■ 1999	552	196	261	291		137	207	64
■ 2000	509	203	243	330		137	191	62
■ 2001	490	198	197	300		157	195	70
Total	**2106**	**788**	**932**	1244		**538**	**754**	**259**

Calculus and prerequisites   Non-calculus based

Data Source: Harper Institutional Research (2001).

Figure 9. Precalculus, calculus and non-calculus based enrollment: 1998–2001 (actual)

(1998–2001)—one and one-half times the national average for two-year colleges. Students at Harper, like their counterparts at other two- and four-year colleges and universities, enroll in college algebra in far greater numbers than in calculus I. The number of students enrolled in college algebra over the past four years was more than twice the number of students who took calculus I (see Figure 9).

Faculty most often cite students' lack of knowledge of trigonometry and inability to interpret and make sense of their responses as greater problems than the lack of algebraic skills, which are often quite proficient. The data reveal that many calculus I students are not coming from precalculus course(s). Where do the calculus I students come from? The department analyzed data on enrollment and success based on method of placement, hoping to answer this question.

At Harper, three different methods of placement are used: a student's ACT or SAT score; COMPASS (a computer-generated placement test developed by ACT); or successful completion of the prerequisite course at Harper (a grade of C or better). The COMPASS exam is the means by which most students are placed into developmental courses. Successful completion of intermediate algebra is the prerequisite

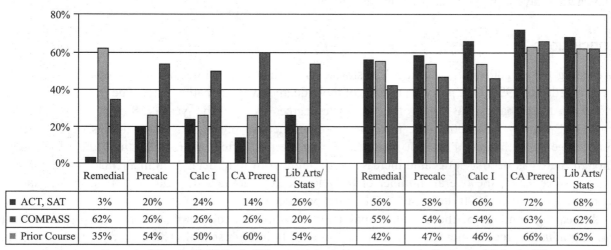

	Remedial	Precalc	Calc I	CA Prereq	Lib Arts/Stats		Remedial	Precalc	Calc I	CA Prereq	Lib Arts/Stats
■ ACT, SAT	3%	20%	24%	14%	26%		56%	58%	66%	72%	68%
■ COMPASS	62%	26%	26%	26%	20%		55%	54%	54%	63%	62%
■ Prior Course	35%	54%	50%	60%	54%		42%	47%	46%	66%	62%

Percent of Course by Method of Placement   Success in Course by Method of Placement

Data Source: Harper Institutional Research (2001).

Figure 10. Harper College: Percent of course by placement and success by placement

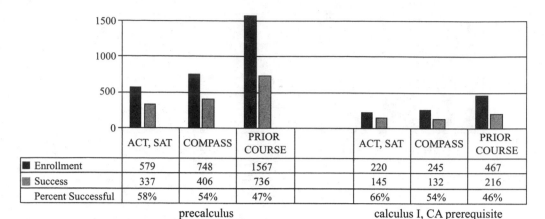

	ACT, SAT	COMPASS	PRIOR COURSE		ACT, SAT	COMPASS	PRIOR COURSE
■ Enrollment	579	748	1567		220	245	467
■ Success	337	406	736		145	132	216
Percent Successful	58%	54%	47%		66%	54%	46%
		precalculus				calculus I, CA prerequisite	

Data Source: Harper Institutional Research (2001).

Figure 11. Comparison of enrollment and success via method of placement 1998–2000

for students who enroll in the precalculus courses. College algebra and trigonometry are the prerequisites for calculus I. The number of students who placed into calculus I via ACT/SAT score or COMPASS exceeds the number who enter via Harper's precalculus courses. Since 1998, 54% of all students enrolled in precalculus courses also took the prerequisite course, intermediate algebra, at Harper.

Precalculus courses continue to be major stumbling blocks for many students at Harper. Nearly twenty-nine hundred students enrolled in precalculus courses since 1998—yet only one-third of the 1,479 students who successfully completed the precalculus course(s) enrolled in calculus I. The number of students who placed into calculus I via ACT/SAT score and COMPASS equals the number who enter via Harper's precalculus courses (see Figure 10).

A student's ACT (or SAT) score has been the most accurate predictor of success in precalculus, calculus I, and courses that have college algebra as the prerequisite, i.e., finite mathematics and business calculus, as well as the mathematics courses generally taken by liberal arts majors—elementary statistics (non-calculus based), quantitative literacy, and the mathematics content courses for elementary teachers. With all high school students in Illinois now required to take the ACT test, it is expected that placement by ACT scores will become the primary method of placement in the precalculus course and in subsequent courses that have a precalculus prerequisite (see Figure 11).

Conclusions

According to the 2000 U.S. Statistical Abstracts, more than one million bachelor's degrees (1,164,792) were awarded in 1996. One percent (13,143) of those degrees were in mathematics. The number of associate degrees in mathematics was 0.1% of all associate degrees awarded that year (758 of 555,216). Based on the CBMS Survey Fall 2000 data, mathematics enrollment in precalculus and calculus courses leading to a degree in mathematics has not improved during the past fifteen years since the Tulane Conference. Though there has been a noticeable increase in the number of joint degrees in mathematics and computer science since 1995, the overall number of bachelor degrees granted through mathematics and statistics departments in 1999–2000 decreased 7.5% in the past decade.

Increasing numbers of these students pay college tuition for courses that do not satisfy general education graduation requirements—courses taken previously in high school. These courses move along at a pace many students find impossible to maintain. Already over-taxed algebraic skills, combined with time constraints due to unrealistic commitments of full-time enrollment (12 semester hours) and 20 or more hours of outside employment per week on the part of many of these students doom them to yet another unsuccessful mathematical experience.

An analysis of enrollment trends of the past twenty years provide us with a "reality check" on various assumptions and reveals the following:

- Calculus enrollment is declining, both in actual numbers and as a percentage of the total undergraduate mathematics enrollment at the same time that mathematics enrollment overall is increasing.
- Courses that have experienced the greatest growth during the past decade are the non-calculus based courses such as elementary statistics, mathematics for liberal arts (quantitative literacy), and mathematics for elementary teachers.
- Even though enrollment in the precalculus/elementary functions course is growing at a much faster rate, the vast majority of students take college algebra.
- The increasing precalculus enrollment has had little impact on calculus enrollment.
- Graphing calculators are now used in more than 50% of sections teaching calculus I but in less than one-third of the developmental algebra sections.

Many factors need to be taken into consideration when analyzing the data presented in this paper. In order to interpret the data meaningfully, one should also examine student behaviors in the context of the classroom environment—taking into consideration the sequence and methods of instruction; the topics on which emphasis is placed; students' beliefs and attitudes; and the means by which conceptual knowledge and skill competencies are assessed.

Prior factors such as students' backgrounds, attitudes, cognitive preferences, and their mathematical competencies also need to be considered. Many students' prior experiences with mathematics have led them to believe that mathematics is a collection of meaningless rules and procedures to be memorized [10, 11]. Often, the focus has been on instruction that contributes to instrumental understanding—*"rules without reason"* [12]. Interviews with students reveal their inadequate understanding of mathematics. Recently, when asked to explain the difference between solving an equation and evaluating a function or expression, a graduate student who works as an on-line tutor replied:

> If a book asks you to evaluate $x^2 - 2x + 1$, what they are asking for is a simplified version of this polynomial, which would be $(x - 1)^2$. Solving an equation or expression is actually plugging in a particular value to come up with a solution.

> For example:

> $f(x) = x^2 - 2x + 1$. Solve for $f(4)$.
> $f(4) = 4^2 - 2(4) + 1 = 16 - 8 + 1 = 9$.

> Is this helping you feel a little bit better about the difference between the two?

The *intended curriculum*—the course content as outlined in the syllabus or based on the textbook—is not necessarily the *implemented curriculum*—what is actually taught. The number of sections in the text that students actually study, the sequence in which topics are studied, and the time spent investigating various topics significantly impact the formation of students' knowledge. Assessment choices—what is assessed, the methods and artifacts of assessment—place an emphasis on certain aspects of the curriculum at the expense of other parts of the curriculum.

How instructors incorporate technology and how students use the technology are factors that impact student learning and their understanding of concepts. Incorporation of technology into a course changes the nature of the learning process—the sequence of instruction as well as the skills students need to learn. Students who are already having difficulties coping with learning new mathematical concepts and procedures tend to view the graphing calculator as a tool they reject—it necessitates the learning of more procedures, along with the mathematics they are already struggling to learn. These students elect not to add to their cognitive burden—continuing to depend on rote-learned algorithms using pencil and paper as their primary tools. Contrast the learning experiences of two students in the same class:

Another process that was very helpful in understanding algebra (specifically factoring) was using a graph to find the x-intercepts to find the zeros of an equation. This is a procedure I had never seen before, but I was able to connect it to my prior knowledge. I found the graphing calculator very useful to graph equations to find the number of solutions (finding zeros), and also to find equations when they are unknown (using the graphing calculator as a data process machine). *Student A.*

I find the graphing calculator to be very confusing. I feel as if everything is thrown at me at once. I have never used the graphing calculator before this class, and now I find it difficult to adapt to using it. The only thing I can do without too much difficulty is put a table into the calculator. After that I don't know what to do. *Student B.*

We also need to take into account the impact on college calculus enrollment of the increasing number of students who successfully complete Advanced Placement calculus courses in high school. In 1999, 127,744 students took the Calculus AB exam, with 64% of those who attempted the AB exam receiving a grade of 3 or better. Forty percent received a grade of 4 or 5. The following year, 137,000 students took the AB exam, with 63% receiving a grade of 3 or better. In 2001, the number of students taking the Calculus AB exam increased to 146,771 students and 63% received a score of 3 or higher. The number of students who took the AP BC exam also continues to increase. The number of students taking the BC exam increased from 30,724 students in 1999 to 34,142 students in 2000, with more than 75% of these students receiving a score of 3 or better. In 2001, 38,134 students took the Calculus BC exam, with 67% of those students receiving a score of 3 or better [13].

If more and more students are taking Advanced Placement calculus in high school, it is reasonable to assume that they are not taking precalculus or calculus I in college. What mathematics courses do these students take as undergraduates? It is probable that many of the students enrolling in precalculus courses as undergraduates are students for whom learning mathematics has not been easy. If this is true, what are the implications for precalculus courses as preparation for advanced study of mathematics?

It is evident that much additional information is needed if we are to effect meaningful changes in the precalculus curriculum. The precalculus data, like the calculus enrollment data, generate many questions for which we do not have answers. We need answers to the questions:

- Who takes precalculus courses and for what reasons?

- Are students who take the precalculus course(s) successful in subsequent courses?

- Why do more students continue to enroll in college algebra than in all of other precalculus courses combined?

- How many students take precalculus to satisfy required prerequisites for a program of study other than mathematics?

- What do other programs of study expect students who complete precalculus to know?

- Why has the precalculus/elementary functions course—a course designed for students intending to take calculus—experienced such large increases in enrollment while calculus I enrollment declined at four-year colleges and universities and remained relatively constant at two-year colleges since 1990?

Perhaps, more importantly, as we seek answers to these questions, we need to determine why we are attracting fewer and fewer students into our mathematics-intensive programs.

References

1. The author wishes to thank Dave Lutzer and Stephen Rodi for providing pre-publication data cited in this paper and published in Lutzer, D., Maxwell, J. and Rodi, S. *Statistical Abstract of Undergraduate Programs in the Mathematical Sciences in the United States: Fall 2000 CBMS Survey.* American Mathematical Society. 2002.

2. National Center for Education Statistics (NCES). *Findings From Education and the Economy: An Indicators Report*, Washington, DC: U.S. Government Printing Office, http://nces.ed.gov/pubs97/97939.html. 1997.

3. U.S. Department of Labor, Bureau of Labor Statistics. *Occupational Outlook Handbook*, Washington, DC: U.S. Government Printing Office, http://stats.bls.gov:80/oco2003.htm. 1997.

4. Dubinsky, E., Schoenfeld, A., and Kaput, J. *Research In Collegiate Mathematics Education. I.* Providence, RI: American Mathematical Society. 1994.

5. Thompson, P., Students, Functions, and the Undergraduate Curriculum. In Dubinsky, E., Schoenfeld A. and Kaput, J. J., (Eds.). *Research in Collegiate Mathematics Education, I. CBMS Issues in Mathematics Education.* Vol. 4. 21–44. 1994.

6. Tall, D., *Understanding the Processes of Advanced Mathematical Thinking.* An Invited ICMI presentation at the International Congress of Mathematicians. Zurich. Available online at: http://www.davidtall.com. 1994.

7. Harel, G and Dubinsky, E., *The Concept of Function Aspects of Epistemology and Pedagogy.* Washington, DC: Mathematical Association of America. 1992.

8. Douglas, R. G., *Toward a Lean and Lively Calculus.* Mathematical Association of America, MAA Notes No. 6. 1986.

9. Lutzer, D., Maxwell, J. and Rodi, S., *CBMS Survey Fall 2000 Statistical Abstract of Undergraduate Programs in the Mathematical Sciences in the United States.* American Mathematical Society, Table SFY.20, p. 40. 2002.

10. Davis, G. and McGowen, M., "Function Machines & Flexible Algebraic Thought." *Proceedings of the 26th International Group for the Psychology of Mathematics Education.* University of East Anglia, Norwich, U.K., Vol. 2, 273–280. 2002.

11. Tall, D., Gray, E., Ali, M. Crowley, L., DeMarois, P., McGowen, M., Pitta, D., Pinto, M., Thomas, M., and Yusof, Y, "Symbols and the Bifurcation between Procedural and Conceptual Thinking." *Canadian Journal of Mathematics, Science and Technology Education*, Vol 1, No. 1, pp. 81–104. 2000.

12. Skemp, R., *The Psychology of Learning Mathematics: Expanded American Edition.* Hillsdale, NJ: Lawrence Erlbaum & Associates, Publishers. 1987.

13. College Entrance Examination Board. http://www.collegeboard.org/ap/calculus/html/grade01.html, html/grade00.html, html/grade99.html. 2001.

4

Enrollment Flow to and from Courses Below Calculus

Steven R. Dunbar
University of Nebraska-Lincoln

The intent of the analysis

In rethinking the courses below calculus, two questions naturally arise:
- What mathematics courses (calculus and otherwise) will students take after completing courses supposedly intending to prepare them for calculus?
- What mathematics courses have students studied before calculus and calculus preparation courses, and how recent is their knowledge?

This report attempts to partially answer these two questions based on a study that tracked the actual enrollment of students in precalculus, calculus, and non-calculus based courses over 20 successive semesters, from fall 1992 to spring 2002. This is a study of the enrollment flow of students on a microscopic level, in that it tracks the enrollment of individual students at the University of Nebraska-Lincoln, a large midwestern university. This study allows us to examine actual student behavior along with changes and trends in that behavior over time. This is in contrast to national studies tracking the yearly number of students enrolled in generically labeled groups of courses (e.g., precalculus which encompasses college algebra, college algebra and trigonometry, precalculus, etc.) which gives a macroscopic view of trends in enrollment. This is also an examination of the probable success of calculus preparation curricula which plan student movement from course to successive course in sequential semesters in that the study examines actual student behavior, which may vary from the ideal.

The data source

Each semester the registrar generates a "correction roster" for the Department of Mathematics and Statistics. The correction roster lists the students enrolled in every course offered by the department as of the third week of the semester. This roster represents the "true" beginning enrollments of courses, after the drop-and-add period.

The correction roster has a line for each student enrolled in every mathematics or statistics course. The line contains the following information: the course number, the section number, the credit hours, the pass-no pass option, the student's name, the unique student identification number, the student's college of enrollment, and the class-year (freshman, sophomore, junior, or senior by credit hours accumulated) of the student.

The correction roster for a semester is large, about 4,000 to 6,500 lines. Altogether, there are well over 100,000 total lines of data to be sorted and analyzed, one for each of the students who took a mathematics or statistics course in the decade being studied.

The data source is authoritative since it comes from the registrar to generate information used early in the semester for billing and later in the semester for assigning grades. Less officially, the information on the correction roster always closely matches instructor class lists and in fact is called the correction roster because it is used to correct informally maintained class lists. Thus, the information from the correction roster can be trusted as an accurate representation of class enrollments at the time of generation. However, the data analysis from the correction rosters has some limitations.

The analysis only counts students enrolled in the third week of the semester; it does not count students who finish the course successfully with a passing grade. In fact, the data does not even count students who finish the course because the correction roster analysis counts as enrolled students who later drop a course. There are also a few students who change (with permission) from, for instance calculus to precalculus, or from precalculus to algebra after the third week, although this number is probably insignificant on the scale of all students considered.

The analysis does not count students who took the analyzed courses in summer sessions, typically between 30 and 100 students each year. The analysis for the three years from fall 1992 to spring 1995 also does not count students enrolled in the night courses. Because of a change in the Student Information System generating the correction roster, the analysis since fall 1995 does count students enrolled in a night section of the analyzed courses. The number of students taking one of these courses in a night section is 30 or less in each semester. The analysis also does not count students who take or re-take a course outside of the university, for example at a local community college, and then return to the university to take a subsequent course. It is unknown what the number of such students is, but I expect it is low.

Occasionally, a few students will register simultaneously for a course and its prerequisite, such as trigonometry and calculus during the same semester, even though this should be impossible under department placement policies. This will appear in the analysis as some small number of students taking a course and its prerequisites in the same semester.

The totals may involve some double-counting. For example, in the output analysis, the tables track the number of students going from precalculus in a particular semester to calculus in some subsequent semester. A possible double counting would occur, for example, when a student succeeds in precalculus, takes calculus in the subsequent semester and fails, then retakes calculus again in the subsequent semester. The output accounting counts such a student twice.

Because of these data limitations, I estimate the analysis gives totals that probably have "error bars" of ± 15 students at worst. Comparing the data from my "correction roster" analysis to another analysis performed independently for another purpose by Professor Leo Chouinard using the math placement database supports this estimate. For example, the correction roster analysis says that 129 students went from fall 1993 precalculus-calculus to spring 1994 calculus. The Math Placement database says that 141 students went from fall 1993 precalculus to spring 1994 calculus. The "error bars" of the counting will affect the percentages accordingly. For example, for the college algebra totals, the counting errors affect percentages by $\pm 4\%$ at worst.

I round percentages to the nearest integer percent. Probably the best use of the data is to indicate which quartile or decile the percentages are in and use the quartiles to suggest appropriate policies.

Description of the courses

In the following analysis, I refer to courses by their title or educational intention, such as college algebra, or precalculus. In the tables, I refer to courses by the University of Nebraska-Lincoln course number with an abbreviated title. However, a course labeled as "precalculus" at one institution may be called "algebra and trigonometry" elsewhere, and may be called by yet another name at another institution.

I describe each of the courses in Appendix A by course number, then course title, and then quote directly from the course catalog description of the course. The course catalog descriptions have remained unchanged over the course of the study. Naturally, some of the course content has changed over the decade studied, and so whenever the information is available I have listed the textbook used as an additional indicator of the course content and level. Finally, the pedagogical format of each course is indicated. While this may vary depending on individual instructor style, it gives a sense of the course. Together, the characteristics described here indicate the degree to which the courses surveyed in this enrollment flow analysis are similar to or different from analogous courses taught elsewhere.

See Figure 1 in Appendix B for a diagram of the prerequisite dependencies and typical flow between courses.

The data analysis

Output analysis: Looking downstream

One could consider the group of students taking, say college algebra in fall 1996, and ask "How many of these students will eventually (within eight or twelve semesters) take mainstream calculus I?" That is, one can summarize this output analysis as being analogous to the conditional probability

$$\text{Pr(to mainstream calculus I | in college algebra fall 1996).}$$

Tables 1-6 in Appendix C summarize the answer to this specific question and other similar questions.

The base semesters selected for display are fall 1996, allowing eleven semesters for students to eventually take another math class, and fall 1998, allowing seven semesters for students to eventually take another math class. I also examine the output from spring 1996 allowing eleven semesters for students to take another math class. These are representative semesters with a long enough time-line to see whole college careers. They provide the most up-to-date glimpse of the mathematical careers of the cohort of students who entered in fall 1998 and could have graduated in the nominal four years of college, or as is sometimes more typical now, the cohort of students who entered in fall of 1996 and could have graduated within six years. The tables for students in a base math class in spring 1996 give a glimpse of the differences between fall and spring math careers.

In each table, the entry in the first row and first column is the number of students enrolled in the base or originating course as of the correction roster date, about the third week of classes, after the drop-and-add period. The columns list the courses that students can ultimately take, including retaking the current base course for unsuccessful students. The next rows, by successive semesters, list the number of students from the base course enrolled in the destination course in that semester.

Tables 1 and 2 summarize the output from precalculus in fall 1996 and fall 1998 respectively. Table 3 summarizes the output for precalculus in spring 1996 to illustrate the differences between a fall cohort and a spring cohort. Table 4 and Table 5 summarize the output for college algebra in fall 1996 and fall 1998. For comparison, Table 6 summarizes the output for college algebra in spring 1997.

Examining the data from *all* the semesters from fall 1992 to spring 2002, one can conclude the precalculus course, which is intended as preparation for calculus, is generally succeeding in its mission. First, considering the students who took the precalculus course in a fall semester, a large majority go on to take calculus, but generally only one semester of calculus. Specifically,

- An average of 58% (with a high of 68% in fall 1996 to a low of 50% in fall 2000) of fall precalculus students eventually take mainstream calculus I. The recent low of students eventually taking calculus is probably due to the fact that they have not yet had enough time to take calculus and show up in the accounting.

- An average of 50% (with a high of 55% in fall 1996 to a low of 44% in fall 2000) of fall precalculus students immediately take mainstream calculus I in the succeeding semester.

- A minority of about 17% of fall precalculus students ever take the third semester of calculus.

An average of 8% (from a high of 16% in fall 1992 to a low of 5% in fall 1998) of fall precalculus students eventually retake precalculus. This compares favorably with a typical retake rate of 20% in most other lower-division courses.

For comparison, the precalculus course in a spring semester is not as successful in its intended mission of preparing students for calculus because more retake precalculus, a smaller percentage go on to take calculus eventually, a smaller percentage go on to calculus immediately, and a smaller percentage still take three semesters of calculus. Considered in the same order as for the fall semesters above, the comparative statistics are:

- An average of 50% (with a high of 79% in spring 1997 to a low of 33% in spring 1998) of spring precalculus students eventually take mainstream calculus I.
- An average of about 30% (with a high of 34% in spring 1997 to a low of 22% in spring 1998) of spring precalculus students immediately take mainstream calculus I in the succeeding semester.
- A minority of about 12% of spring precalculus students ever take the third semester of calculus.

An average of 13% (with a high of 19% in spring 1999 to a low of 3% in spring 2000) of spring precalculus students eventually retake precalculus.

The population of students taking precalculus has a different composition than the students taking college algebra. Specifically for the fall semesters examined in this study:

1. An average of 5% of students in precalculus are enrolled in the College of Business Administration, compared to an average of 25% of the students in college algebra.
2. An average of 15% of students in precalculus are enrolled in the College of Engineering, compared to an average of less than 2% of the students in college algebra.
3. An average of 44% of students in precalculus are enrolled in the College of Arts and Sciences, compared to an average of 20% of the students in college algebra.

Observe that half or more of the students in precalculus do get to a calculus course. However, students taking the college algebra course are fragmented into many different succeeding courses. First, I will look at the statistics for students who took college algebra during the fall semester. Students in college algebra retake the course at a greater rate than do students in precalculus. Specifically, an average of 20% (with a high of 22% in fall 1998 to a low of 18% in fall 1999) of fall college algebra students eventually retake college algebra. An even lower rate of college algebra students immediately take trigonometry, the next pre-requisite course before calculus. In fact, an average of 18% (with a high of 23% in fall 1999 to a low of 13% in fall 2000) of fall college algebra students immediately take trigonometry in the succeeding semester. Only a minority of college algebra students ever take any form of calculus, specifically:

- An average of 21% (with a high of 24% in fall 2001 to a low of 18% in fall 1993) of fall college algebra students immediately take the business calculus course in the succeeding semester.
- An average of 11% (with a high of 16% in fall 1995 to a low of 7% in fall 2000) of fall college algebra students eventually take the main sequence calculus I course. The recent low of students eventually taking calculus is certainly due to the fact that they have not yet had enough time to take calculus and show up in the accounting.
- Only an insignificant minority of about 1% of fall college algebra students ever take the third semester of calculus.

Now let's look at the students who took college algebra during spring semesters. In the measure of retaking the course, these students are slightly more successful, since an average of 17% (with a high of 20% in spring 1996 to a low of 10% in spring 1998) of spring college algebra students eventually retake Math 101. However, in the measure of taking successive courses, these students are less successful. For instance, an average of about 11% (with a high of 15% in spring 1995 to a low of 7% in spring 2001) of spring college algebra students immediately take trigonometry in the succeeding semester. Lower rates of students in college algebra in the spring semesters take calculus in any form than their peers in the fall semesters. In fact:

- An average of 15% (with a high of 18% in spring 1997 to a low of 12% in spring 1994) of spring college algebra students immediately take business calculus in the succeeding semester.

- An average of 7% (with a high of 12% in spring 1995 to a low of 5% in spring 1999) of spring college algebra students eventually take mainstream calculus I.

- Only an insignificant minority of about 1% of spring college algebra students ever take the third semester of calculus.

Another surprising conclusion of the study is that it is possible to find a few students who enrolled in college algebra or precalculus in 1992 or 1993 and who are still taking some lower-division mathematics classes as much as eight years later!

Input analysis: Looking upstream

Suppose one is teaching calculus. Then a reasonable question to ask about the class is "How many students previously took a specific precalculus course and when?" One can briefly summarize this as the conditional probability

$$\Pr(\text{ from precalculus } | \text{ in mainstream calculus I })$$

and so on. Tables 7 through 10 provide four examples of that data. The first table does the input analysis for several courses from the fall semester of the 2001–2002 academic year with respect to our precalculus course in previous semesters. The second table does the input analysis for spring semester of the 2001-2002 academic year. The two tables yield different input analyses since the students taking a course in the spring semester are likelier to have had a prerequisite course at UN-L than students taking the same course in the fall semester, who would likelier have had the prerequisite course from high school or a feeder institution.

The top of a column is a particular course number. The next line gives the total enrollment from the correction roster of that course in the base semester. This provides the base enrollment, or if you will, the denominator of the conditional probability. The next lines, in reverse chronological order of preceding semesters, give the number of students coming from the prerequisite course (precalculus or college algebra) in that semester. Next is the total number of students who have taken the prerequisite course in all the preceding semesters. This is the numerator of the conditional probability. The conditional probability, expressed as a percentage, is at the bottom of the column.

I have tables analogous to Tables 7 through 10 for each course in each semester. After summarizing all the tables, some conclusions about input from precalculus to subsequent calculus courses appear. First, considering fall semesters, one conclusion is that relatively few of the students in precalculus or any level of calculus have previously been in a precalculus course. This is not surprising because in lower-division mathematics courses in a fall semester many students are entering freshmen enrolling in their first math course. Even for mainstream calculus II in the fall where students could have taken precalculus in the previous fall, followed by calculus I in the previous spring, and finally calculus II in the fall, the percentage only averages 12%, and it is always less than 22%. In this sense, the fall precalculus course is not a precursor to calculus. Specifically:

- In a fall semester, an average of 6% (ranging from a high of 10% in fall 1995 to a low of 4% in fall 2001) of the students in precalculus previously took precalculus.

- In a fall semester, an average of 7% (ranging from a high of 11% in fall 1995 to a low of 4% in fall 1999) of the students in calculus I previously took precalculus.

- In a fall semester, an average of 12% (ranging from a high of 22% in fall 1995 to a low of 1% in fall 2001) of the students in calculus II previously took precalculus. These students likely took precalculus in the fall semester of the year before, then took calculus I in the spring semester of the year before, and ultimately enrolled in calculus II in the year examined.

- In a fall semester, an average of 6% (ranging from a high of 10% in fall 1995 to a low of 4% in fall 2001) of the students in calculus III previously took precalculus.

The picture is somewhat different in spring semesters. A higher percentage of students from a prior precalculus course are either retaking precalculus or taking calculus I. Nevertheless, the constituency of a calculus II or calculus III course is not much different than in a fall semester. This indicates that students in precalculus are taking one semester of calculus, but not much more. The detailed picture is that

- In a spring semester, an average of 32% (ranging from a high of 45% in spring 2001 to a low of 22% in spring 1999) of the students in precalculus previously took precalculus.

- In a spring semester, an average of 37% (ranging from a high of 42% in spring 1995 to a low of 31% in spring 2000) of the students in calculus I previously took precalculus.

- In a spring semester, an average of 7% (ranging from a high of 10% in spring 1997 to a low of 5% in spring 2000) of the students in calculus II previously took precalculus.

- In a spring semester, an average of 11% (ranging from a high of 18% in spring 1996 to a low of 8% in spring 2001) of the students in calculus III previously took precalculus.

Some different conclusions about input from college algebra and intermediate algebra to subsequent courses appear from all of the input data in fall semesters. A large percentage of students in trigonometry and business calculus classes have previously been in the college algebra and intermediate algebra courses. I note that the percentages are large by comparison to some of the other input figures but still less than half of the students in business calculus have previously been in college algebra. However, relatively few of the students in any level of calculus have previously been in a college algebra course or intermediate algebra course. The precise figures for various courses in a fall semester are:

- An average of 13% (ranging from a high of 16% in fall 1998 to a low of 10% in fall 2000) of the students in college algebra previously took college algebra or intermediate algebra.

- An average of 59% (ranging from a high of 69% in fall 1996 to a low of 51% in fall 1999) of the students in trigonometry previously took college algebra and intermediate algebra.

- An average of 43% (ranging from a high of 50% in fall 1995 to a low of 38% in fall 2000) of the students in business calculus previously took college algebra and intermediate algebra.

- An average of 13% (ranging from a high of 18% in fall 1996 to a low of 9% in fall 2000) of the students in calculus I previously took college algebra and intermediate algebra.

- An average of 5% of the students in calculus II and calculus III previously took college algebra and intermediate algebra.

The differences between spring and fall constituencies of some courses stand out when the data for input into various courses from the algebra courses is summarized. A much higher percentage of students are repeating college algebra having previously been in college algebra. The vast majority of students in trigonometry and business calculus have taken an algebra course. Still a minority of students in any level of mainstream calculus have taken college algebra. The precise numbers in a spring semester are:

- An average of 29% (ranging from a high of 34% in spring 1997 to a low of 24% in spring 1995) of the students in college algebra previously took college algebra.

- An average of 87% (ranging from a high of 94% in spring 1997 to a low of 78% in spring 2000) of the students in trigonometry previously took college algebra and intermediate algebra.

- An average of 77% (ranging from a high of 84% in spring 1996 to a low of 73% in spring 2000) of the students in business calculus previously took college algebra and intermediate algebra.

- An average of 19% (ranging from a high of 24% in spring 1998 to a low of 17% in spring 1995) of the students in calculus I previously took college algebra and intermediate algebra.

- An average of 5% of the students in calculus II and calculus III previously took college algebra and intermediate algebra.

Conclusions

The conclusion to be drawn from this study is that at least at the University of Nebraska-Lincoln, a sizable majority of students in the precalculus course prepare to start calculus. The students from the precalculus course constitute a sizable fraction of the students taking calculus, although not a majority. A fairly insignificant fraction of the students taking precalculus at the university ever take three semesters of calculus.

Another conclusion is that only a minority of students taking college algebra go on to start the three semester sequence of calculus, although a considerable fraction do take the one semester business calculus course. In the spring semester a vast majority of the students in trigonometry and the business calculus course have taken the college algebra course at the university. Only an insignificant number of students taking college algebra ever go on to complete three semesters of calculus. For comparison, an average of 22% of fall mainstream calculus I students go on to take three semesters of calculus.

Finally, it appears that many students do not take these calculus preparatory math classes and then calculus in successive semesters. For example, an average of 29% of college algebra students eventually take business calculus, but about a quarter to a third of those students do so at least one semester after they have finished college algebra. Of the approximately 11% of college algebra students eventually taking mainstream calculus, a little less than half take calculus in a semester later than they could have. While an average of 58% of precalculus students eventually take mainstream calculus, about one-eighth do so a semester or more after finishing precalculus. In fact, some students remain taking lower-division math classes as much as 10 years after taking college algebra or precalculus.

Appendix A
Description of the courses

Math100A Intermediate Algebra "A review of the topics in a second-year high-school algebra class taught at the college level. Topics include: real numbers, 1st and 2nd degree equations and inequalities, linear systems, polynomials and rational functions, exponents and radicals. Hours earned in Math 100A do not count toward degree requirements."

Graduate TAs teach the course, typically in their second (or later) year. The class is three credit-hours, with three contact hours, generally taught three days per week. Class sizes are at most 35 students. The presentation style is generally lectures, with three or four hour exams, quizzes or graded homework, and a common final exam. The textbook used recently is *Intermediate Algebra*, 6th Edition, by McKeague.

Math101 College Algebra "Real and complex numbers, exponents, factoring, linear and quadratic equations, absolute value, inequalities, functions, graphing, polynomial and rational functions, exponential and logarithmic functions, analytic geometry, systems of equations."

Graduate TAs teach the course, typically in their second (or later) year. The class is three credit-hours, with three contact hours, generally taught three days per week. Class sizes are at most 35 students. The presentation style is generally lectures, with three or four hour exams, quizzes or graded homework, and a common final exam. The textbook used recently is *College Algebra, Graphing and Data Analysis*, by Sullivan and Sullivan.

Math102 Trigonometry "Trigonometric functions, identities, trigonometric equations, solution of triangles, inverse trigonometric functions, graphs, logarithms, and exponential functions."

Graduate TAs teach the course, typically in their second (or later) year. The class is two credit-hours, with two contact hours, generally taught two days per week. Class sizes are at most 35 students. The presentation style is generally lectures, with three or four hour exams, quizzes or graded homework, and a common final exam. Since 1996, group projects consisting of an extended homework exercise are part of the evaluation. The textbook used recently is *Trigonometry and its Applications*, by Goldstein.

Math103 College Algebra and Trigonometry "First and second degree equations and inequalities, absolute value, functions, polynomial and rational functions, trigonometric functions and identities, laws of sines and cosines, applications, polar coordinates, graphing, conic sections."

Graduate TAs teach the course, typically in their third year (or later). The class is five credit-hours, with five contact hours, generally taught five days per week. Class sizes are at most 35 students. The presentation style is lectures with a mix of hands-on or in-class activities including worksheets, with three or four hour exams, quizzes or graded homework, and a common final exam. Group projects consisting of an extended homework exercise are also part of the course evaluation. The textbook used recently is *Functions Modeling Change: A Preparation for Calculus*, by Connally, Hughes-Hallett, Gleason, et al.

Math104 Calculus for Managerial and Social Sciences "Rudiments of differential and integral calculus with applications to problems from business, economics, and social sciences."

Faculty teach the course in large lectures with about 100-120 students. The class is three credit-hours, with three contact hours, taught two or three days per week. The presentation style is generally lectures, with three or four hour exams, quizzes or graded homework, and a common final exam. The textbook used recently is *Brief Calculus and its Applications*, 2nd Edition, by Bernice.

Math106 Analytic Geometry and Calculus I "Functions of one variable, limits, differentiation, exponential, trigonometric and inverse trigonometric functions, maximum-minimum, and basic integration theory with some applications."

The course has lecture sessions three days per week with a faculty lecturer in large lecture halls with about 100-125 students. The presentation style is generally lecture with a mixture of hands-on or calculator exercises. Small recitation sections of 25 students meet with a first-year teaching assistant twice per week. The presentation is problem-solving, group work, some hands-on learning, with some explanation. Instructors base grades on three hour exams, quizzes or graded homework, and a common final exam. As well, group projects requiring writing and analysis are part of the evaluation. The textbook used over the period of the study is *Calculus, Single and Multivariable*, 2nd Edition, by Hughes-Hallett, Gleason, McCallum, et al.

Math107 Analytic Geometry and Calculus II "Integration theory, techniques of integration, applications of definite integrals, basics of ordinary differential equations."

The course has lecture sessions three days per week with a faculty lecturer in large lecture halls with about 100-125 students. The presentation style is generally lecture with a mixture of hands-on or calculator exercises. Small recitation sections of 25 students meet with a first-year teaching assistant twice per week. The presentation is problem-solving, group work, some hands-on learning, with some explanation. Instructors base grades on three hour exams, quizzes or graded homework, and a common final exam. As well, group projects requiring writing and analysis also are part of the evaluation. The textbook used over the period of the study is *Calculus, Single and Multivariable*, 2nd Edition, by Hughes-Hallett, Gleason, McCallum, et al.

Math200 Mathematics for Elementary School Teachers "Fundamental mathematical concepts basic to the understanding of arithmetic."

Faculty and some advanced graduate TAs teach the course. The class size is typically 30 or less. The presentation style is a mixture of lecture and hands-on learning activities.

Math201 Geometry for Elementary School Teachers "Fundamental mathematical concepts basic to the understanding of geometry."

Faculty and some advanced graduate TAs teach the course. The class size is typically 30 or less. The presentation style is a mixture of lecture and hands-on learning activities.

Math203 Contemporary Mathematics "Applications of quantitative reasoning and methods to problems and decision making in management, statistics, and social choice. Topics include networks, critical paths, linear programming, sampling, central tendency, inference, voting theory, power index, game theory and fair division."

Faculty and some advanced graduate TAs teach the course. The class size is typically 30 or less. The presentation style is a mixture of lecture and hands-on learning activities, along with some videotape presentations.

Math 208 Analytic Geometry and Calculus III "Vectors and surfaces, parametric equations and motion, functions of several variables, partial differentiation, maximum-minimum, Lagrange multipliers, multiple integration, vector fields, path integrals, Green's theorem and applications."

Faculty teach the course. The class is four credit-hours, with four contact hours, generally taught four days per week. Class sizes are at most 40 students. The presentation style is lectures with a mix of hands-on or in-class activities, with three or four hour exams, quizzes or graded homework, and a common final exam. Typically, group projects consisting of an extended homework exercise also are part of the course evaluation. The textbook used over the period of the study is *Calculus, Single and Multivariable,* 2nd Edition, by Hughes-Hallett, Gleason, McCallum, et al.

Stat 180 Elements of Statistics "Finite probability, random variables, probability distributions, statistical inference, estimation, and testing of hypotheses."

Faculty and some advanced graduate TAs teach the course. The class size is typically 40 or less. The presentation style is a mixture of lecture and hands-on learning activities, with three or four hour exams, quizzes or graded homework, and a common final exam. Typically, group projects consisting of an extended homework exercise also are part of the course evaluation. The textbook used recently is *The Basic Principles of Statistics,* 2nd Edition, by Moore.

Appendix B
Diagram of course dependencies

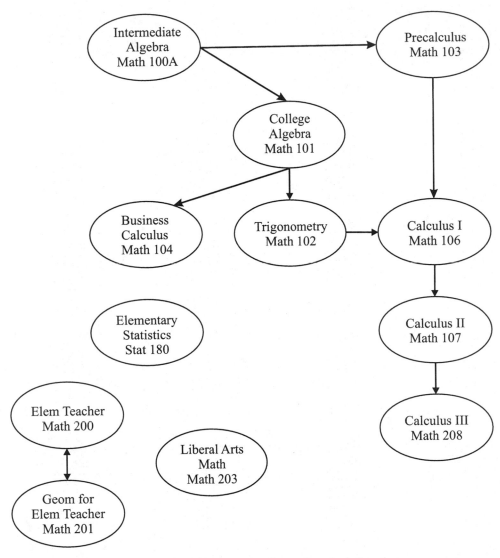

Figure I. Diagram of prerequisite dependencies and typical flow between courses.

Appendix C
Tables

Prec 103	Prec 103	BCal 104	CalI 106	CalII 107	CalIII 208	MET 200	GET 201	LAM 203	Stat 180
F96	233	0	0	0	0	0	0	0	0
S97	16	10	129	1	0	0	0	2	2
F97	7	1	18	33	4	0	0	2	9
S98	1	1	13	15	14	2	0	0	18
F98	3	1	6	4	7	0	0	1	3
S99	1	2	7	4	7	0	0	1	4
F99	1	0	2	3	4	0	1	1	4
S00	0	0	2	2	1	0	0	0	4
F00	1	0	2	1	2	0	0	1	0
S01	0	0	1	2	1	0	0	0	2
F01	0	0	1	0	0	0	0	0	1
S02	0	0	0	0	1	0	0	0	1

Table 1. The number of students in subsequent courses who took precalculus in fall 1996.

Prec 103	Prec 103	BCal 104	CalI 106	CalII 107	CalIII 208	MET 200	GET 201	LAM 203	Stat 180
F98	198	0	0	0	0	0	0	0	0
S99	4	6	107	0	0	0	0	0	12
F99	2	4	15	25	2	0	0	0	13
S00	1	4	11	7	17	1	0	1	6
F00	2	2	2	2	4	0	0	3	3
S01	2	1	1	2	3	1	1	2	3
F01	0	0	2	1	0	0	1	0	3
S02	0	0	1	1	1	0	0	0	2

Table 2. The number of students in subsequent courses who took precalculus in fall 1998.

	Prec 103	BCal 104	CalI 106	CalII 107	CalIII 208	MET 200	GET 201	LAM 203	Stat 180
Prec 103									
S96	77	0	0	0	0	0	0	0	1
F96	6	3	22	2	0	0	0	1	3
S97	0	1	7	8	1	0	0	2	2
F97	2	1	2	2	2	0	0	1	1
S98	1	2	2	0	3	2	0	0	2
F98	0	1	1	0	0	1	1	1	1
S99	0	1	0	1	0	1	0	0	1
F99	0	0	0	0	1	0	2	0	1
S00	0	0	0	0	0	0	0	0	0
F00	0	0	0	0	0	0	0	0	0
S01	1	0	0	0	0	0	0	0	0
F01	0	0	1	0	0	0	0	0	0
S02	0	0	0	0	0	0	0	0	0

Table 3. The number of students in subsequent courses who took precalculus in spring 1996.

CA 101	CA 101	Trig 102	Prec 103	Bcal 104	CalI 106	CalII 107	CalIII 208	MET 200	GET 201	LAM 203	Stat 180
F96	793	0	0	0	0	0	0	0	0	0	2
S97	106	139	0	161	13	0	0	0	0	7	20
F97	25	32	0	47	43	10	0	1	0	11	37
S98	18	21	0	21	20	14	2	0	0	7	25
F98	8	9	0	14	14	7	4	1	0	4	8
S99	5	4	0	14	5	5	6	1	0	9	13
F99	2	3	0	6	1	5	1	2	1	2	3
S00	3	4	0	2	1	0	1	0	0	2	6
F00	2	1	0	1	3	1	0	1	0	1	1
S01	1	2	0	0	0	0	0	0	1	2	3
F01	0	0	0	2	2	0	1	0	1	0	1
S02	0	2	0	1	2	1	0	0	0	1	1

Table 4. The number of students in subsequent courses who took college algebra in fall 1996.

CA 101	CA 101	Trig 102	Prec 103	Bcal 104	CalI 106	CalII 107	CalIII 208	MET 200	GET 201	LAM 203	Stat 180
F98	784	0	0	0	0	0	0	0	0	0	3
S99	126	122	0	180	17	0	0	0	0	7	27
F99	20	31	0	68	39	8	2	1	0	19	32
S00	13	15	0	29	18	8	2	1	1	18	28
F00	3	9	0	14	9	5	4	3	0	11	8
S01	4	5	0	10	8	2	3	5	1	9	13
F01	2	4	0	4	7	2	1	4	1	4	5
S02	4	3	0	2	6	2	2	0	0	2	6

Table 5. The number of students in subsequent courses who took college algebra in fall 1998.

CA 101	CA 101	Trig 102	Prec 103	Bcal 104	CalI 106	CalII 107	CalIII 208	MET 200	GET 201	LAM 203	Stat 180
S96	665	1	0	0	0	0	0	0	0	1	1
F96	63	70	1	82	12	1	0	0	0	15	14
S97	33	33	0	50	13	3	0	1	0	11	21
F97	15	13	0	18	9	2	1	6	0	11	18
S98	7	10	0	14	9	2	0	2	0	8	7
F98	6	3	0	8	7	2	0	1	1	6	6
S99	4	3	0	11	3	3	0	0	1	5	4
F99	1	3	1	5	2	2	0	0	0	3	6
S00	2	4	0	4	2	1	0	0	0	2	1
F00	0	1	0	2	0	1	1	0	0	0	2
S01	1	1	0	1	0	0	2	0	0	0	2
F01	0	2	0	1	2	1	0	0	0	1	2
S02	0	0	0	1	2	0	1	0	0	0	1

Table 6. The number of students in subsequent courses who took college algebra in spring 1996.

	Prec 103	BCal 104	CalI 106	CalII 107	CalIII 208	MET 200	GET 201	LAM 203	Stat 180
F01	215	519	730	364	280	71	52	353	506
S01	2	2	13	0	0	0	0	0	2
F00	4	6	25	15	0	0	0	1	16
S00	0	0	0	1	1	0	0	0	0
F99	1	0	7	2	4	0	0	0	4
S99	0	0	0	0	0	0	0	0	0
F98	0	0	2	1	0	0	1	0	3
S98	1	0	0	0	0	0	0	0	0
F97	0	0	0	0	2	0	1	0	1
S97	0	0	0	0	0	0	0	0	1
F96	0	0	1	0	0	0	0	0	1
S96	0	0	1	0	0	0	0	0	0
F95	0	0	0	1	0	0	0	0	0
S95	0	0	0	0	0	0	0	0	0
F94	0	0	2	0	0	0	0	0	0
S94	0	0	0	0	1	0	0	0	0
F93	0	0	0	0	0	0	0	0	0
S93	0	0	0	0	1	0	0	0	0
F92	0	0	0	0	1	0	0	0	0
Total	8	6	51	20	10	0	2	1	28
%	4%	1%	7%	1%	4%	0%	4%	0%	5%

Table 7. The input analysis from precalculus to courses in fall 2001.

	Prec 103	BCal 104	CalI 106	CalII 107	CalIII 208	MET 200	GET 201	LAM 203	Stat 180
S02	53	440	349	312	252	68	42	322	518
F01	20	10	95	0	0	0	0	2	8
S01	2	0	7	0	0	0	0	0	2
F00	0	3	8	6	12	1	1	0	5
S00	0	2	1	1	0	0	0	0	1
F99	1	2	3	2	3	0	0	2	4
S99	0	0	0	0	0	0	0	0	0
F98	0	0	1	1	1	0	0	0	2
S98	0	0	1	0	0	0	0	0	0
F97	0	0	4	0	0	0	1	0	1
S97	0	0	0	0	0	0	0	0	0
F96	0	0	0	0	1	0	0	0	1
S96	0	0	0	0	0	0	0	0	0
F95	0	0	0	0	2	1	1	0	0
S95	0	0	0	0	1	0	0	0	0
F94	0	0	0	0	1	0	0	0	0
S94	0	0	0	0	0	0	0	0	0
F93	0	0	0	0	0	0	0	0	0
S93	0	0	0	0	0	0	0	0	0
F92	0	0	0	0	0	0	0	0	0
Total	23	17	140	10	21	2	3	4	24
%	43%	4%	40%	3%	8%	3%	7%	12%	5%

Table 8. The input analysis from precalculus to courses in spring 2002.

	CA 101	Trig 102	Prec 103	BCal 104	CalI 106	CalII 107	CalIII 208	MET 200	GET 201	LAM 203	Stat 180
F01	799	174	215	519	730	364	280	71	52	353	506
S01	58	39	1	82	8	1	2	2	1	20	25
F00	25	26	0	66	31	5	0	0	0	24	32
S00	8	8	0	17	7	4	1	2	2	6	10
F99	3	7	0	23	9	4	1	4	0	8	18
S99	2	5	0	3	3	2	2	1	1	4	5
F98	2	4	0	4	7	2	1	4	1	4	5
S98	3	3	0	1	2	1	0	1	2	2	4
F97	0	1	0	0	3	0	0	1	2	2	2
S97	1	1	0	1	2	0	0	0	1	0	2
F96	0	0	0	2	2	0	1	0	1	0	1
S96	0	2	0	1	2	1	0	0	0	1	2
F95	0	1	0	0	0	0	1	0	0	1	0
S95	0	1	0	0	1	0	0	0	0	0	0
F94	1	1	0	0	0	0	0	0	0	0	1
S94	1	0	0	0	0	0	0	0	0	0	0
F93	0	1	0	0	0	0	0	0	0	0	1
S93	0	2	0	1	0	0	0	0	0	0	0
F92	0	1	0	0	0	0	0	0	0	0	0
Total	104	103	1	201	77	20	9	15	11	72	108
%	13%	59%	0%	39%	11%	5%	2%	21%	21%	20%	21%

Table 9. The input analysis from college algebra to courses in fall 2001.

	CA 101	Trig 102	Prec 103	BCal 104	CalI 106	CalII 107	CalIII 208	MET 200	GET 201	LAM 203	Stat 180
S02	518	184	53	440	349	312	252	68	42	322	518
F01	96	102	0	191	15	0	0	0	0	20	27
S01	18	13	0	49	11	3	0	2	0	25	22
F00	6	16	0	42	18	8	3	1	0	18	30
S00	7	6	0	18	2	2	1	2	2	3	6
F99	4	6	0	12	9	2	1	0	2	9	8
S99	4	4	0	4	2	2	0	1	0	9	6
F98	4	3	0	2	6	2	2	0	0	2	6
S98	2	2	0	1	1	2	0	0	1	2	1
F97	0	2	0	0	1	0	0	0	1	5	3
S97	0	2	0	1	3	0	0	0	0	0	2
F96	0	2	0	1	2	1	0	0	0	1	1
S96	0	0	0	1	2	0	1	0	0	0	1
F95	0	1	0	0	1	1	0	0	0	1	0
S95	1	0	0	1	2	0	0	0	0	0	0
F94	1	0	0	0	1	0	0	0	0	0	0
S94	0	0	0	0	0	0	0	0	0	1	0
F93	0	0	0	0	0	0	0	0	0	2	1
S93	0	1	0	1	0	0	0	0	0	0	0
F92	0	0	0	0	0	0	0	0	0	0	0
Total	143	160	0	324	76	23	8	6	6	89	114
%	28%	87%	0%	74%	22%	7%	3%	8%	14%	28%	22%

Table 10. The input analysis from college algebra to courses in spring 2002.

5

What Have We Learned from Calculus Reform?
The Road to Conceptual Understanding

Deborah Hughes Hallett
University of Arizona

Goals

In order to evaluate the impact of calculus reform, we first need to recall its goals. Although different people may phrase it differently, everyone involved would agree that they were trying to improve the teaching of calculus. Some would say they wanted more student involvement; others would say they wanted to take advantage of technology; others would say they wanted to emphasize problem solving and modeling. Most would agree that they wanted to improve conceptual understanding. What has been the impact of this effort?

Background to calculus reform: Rationale for change

The teaching of calculus came under scrutiny in the 1980s for several reasons. One was concern over the students' apparent lack of understanding of the subject, especially when asked to use it in an unfamiliar situation. Faculty outside mathematics frequently complained that students could not apply the concepts they had been taught. In some instances, ideas were being used in other fields in ways that were sufficiently different from the way they are used in mathematics that it was not surprising that students did not make the connection. For example, the minimization of average cost was done symbolically in mathematics, if at all, whereas it is usually done graphically in economics.

However, students also had difficulty recognizing mathematical ideas that were presented the same way as in mathematics. A small difference in notation or the absence of familiar clues—such as "largest" or "smallest" in an optimization problem—easily threw students off. This striking difficulty in transferring knowledge between fields suggested that students' understanding was not sufficiently robust.

In addition, many students came to college believing that mathematics centers on manipulative techniques, rather than interpretation and understanding.[1] These students spent little energy thinking about where ideas came from or how they were used. Besides being a disappointment to faculty, these students never saw the power of mathematics to unite disparate fields.

Materials used in the middle 1980s suggested that mathematicians were doing little to challenge students' views that equated mathematics to applying formulas. Exam questions were often of the form

---

[1]See, for example, Deborah Hughes Hallett "Are We Encouraging Our Students to Think Mathematically?" in *How to Teach Mathematics* by Steven G. Kranz, 2nd edition, American Mathematical Society, 1999.

"Use method X to do Y"; problems in the text were usually to be done by the formula most recently presented. Consequently students got little experience on choosing a method. Even the choice of variables seldom needed thought. It was not uncommon for an entire set of exercises in a text to be written as functions in terms of x, with at most a couple that involved t or θ. Since an unfamiliar variable is a real stumbling block to students and x is virtually never used outside mathematics, this lack of variety significantly limited students' ability to apply their mathematics.

Changes in the teaching of calculus

When efforts to improve the teaching of calculus started in the late 1980s, there was great variation amongst the projects that were undertaken. Some were technology-driven, redesigning curriculum and pedagogy using the power of the new computer tools—for example, *Calculus&Mathematica*, from the University of Illinois, and Project CALC, from Duke. Others used no technology, but incorporated extended applications, group work, and collaborative learning—for example, the New Mexico State Project. Over the next decade, there was remarkable convergence among the projects. Effective components in one project were adapted and incorporated by other projects—and eventually by mainstream authors. The most fundamental change made by the new calculus texts in the 1990s was the introduction of many more nonstandard problems. By the mid 1980s, before calculus reform got underway, such problems had been relegated to the end of the exercise sets in the texts and were few enough in number that many students—and many instructors—ignored them. Newer calculus texts have a much wider variety of problems, and fewer "template" problems that can be solved by mimicking a worked example in the text.

Technology is incorporated in many current calculus courses. The Advanced Placement (AP) exam taken by many high school students now requires a graphing calculator, as does the International Baccalaureate (IB). Most high school calculus courses, and hence most high school precalculus courses, are taught with a graphing calculator. College courses are more varied. Some have not moved beyond the memorization and multiple choice of the 1980s; others are now Web or computer based.

Although using technology in calculus is the most visible, and perhaps the most controversial, change in the teaching of calculus, it is not the one that will have the most impact on student learning. Expecting conceptual understanding on homework and exams is more important. Although faculty often spend more time designing lectures than homework, most students learn more from homework than from lectures. Changes in homework and exams have a larger effect on student learning than changing lecture content. Requiring thinking is central to establishing the idea that mathematics is more than applying formulas. To the surprise of faculty, students often described the new courses as being "more theoretical" than the old. Although they are not using the word "theoretical" in the usual mathematical sense, from a student perspective they are right. The new courses require more reasoning, justification, and explanation. Just getting an answer is no longer enough.

Along with the increase in nonstandard problems and the use of technology, many new calculus courses emphasize open-ended problems that require extensive writing, often in cooperative groups. In the early 1990s, several books of calculus labs were published; the IB now requires all students to submit a portfolio as part of their final assessment.

Thus there is now both more variety in calculus courses and more emphasis on conceptual understanding.

Cooperation with client disciplines

An unexpected side benefit of calculus reform has been increased cooperation between mathematics and other fields, such as engineering, biology, physics, and economics. In their quest to learn how students used calculus, mathematicians talked to their colleagues in other departments, looked at their texts and exams, and listened carefully to their needs. The result was some new texts that genuinely reflected the needs of the sciences—for example, those from the University of Iowa and Iowa State—and much improved calculus texts for the social sciences. Students take a subject more seriously—and learn it better—if faculty from more than one field "conspire" (in the students' words) to teach it to them.

However, more important than the improved texts was the good will generated by these conversations. Efforts to make the teaching of calculus more responsive to other fields led directly to the MAA's Curriculum Foundations Project. In this project, faculty from outside mathematics were invited to have direct input into the current MAA curriculum review. Perhaps the most sobering aspect of the Curriculum Foundation Project was how surprised and grateful the faculty from other fields were to be contacted. This is an effort whose time was long overdue.

Impact of calculus reform

The most obvious measure of the impact of calculus reform is that the features initiated in the 1990s are now commonplace in "traditional" courses. Some of the new features have been transformed—in some cases their originators would say beyond recognition—but many have been adopted "as is" by standard textbooks. There is a wider variety of problems than before, and the "Rule of Three"—Every topic should be presented geometrically, numerically, and algebraically—which originated in calculus reform, has found its way into a large number of calculus texts. The wider variety of problems and the use of multiple representations (graphical, numerical, analytical, and verbal) make it harder for students simply to memorize template problems—though still not impossible—and hence encourage conceptual understanding.

Most texts now allow the use of technology, although often as an add-on, to accommodate a variety of faculty preferences. Open-ended problems and extended applications are found in many books, although often as an add-on at the end of the chapter. Students are now expected to write more often in calculus courses than they were in the 1980s, although seldom in university lecture courses with large enrollments. However instructors who get their students to write report that writing both deepens conceptual understanding and provides a window into their students' thinking processes.

Summary

The impact of calculus reform has been substantial. In spite of objections—sometimes vociferous—to any one particular aspect of the new courses (technology, conceptual, rather than formal, understanding), many aspects of calculus reform are now so embedded in the mainstream that they are considered mainstream.

The changes that have had the most impact are not those that were originally considered to be the most profound. For example, the topics covered have not been greatly impacted, in spite of the call for a"lean and lively" curriculum. However, the pedagogy and types of problems solved have been impacted. During the 1990s essentially every math department made some changes to their calculus courses. Some of these changes have persisted, some have not, but all have made the teaching of calculus a subject of discussion in many math departments where this was not the case previously.

In the long run, the largest impact of calculus reform is likely to be the creation of a community of mathematicians who innovate and reflect on their teaching—and who do so in collaboration with faculty in other disciplines and across institutional boundaries.

Acknowledgement: Parts of this paper are taken from "Calculus at the Start of the New Millenium," by Deborah Hughes Hallett, the *Proceedings of the International Conference on Technology in Mathematics Education*, Beirut, Lebanon, July 2000.

6

Calculus and Introductory College Mathematics: Current Trends and Future Directions

Susan L. Ganter
Clemson University

Introduction

For better or worse, the ideas of calculus reform are deeply embedded in the conversations of the mathematics community. Technology, cooperative learning, student projects, applications—the elements of this movement—have become a part of the vocabulary in mathematics departments across the country. Efforts to change the nature of the calculus course at the undergraduate and secondary levels have sparked discussion and controversy in ways as diverse as the actual changes. Such interactions range from "coffee pot conversations" to university curriculum committee agendas to special sessions on undergraduate education at regional and national conferences. But what is the significance of these activities? Where have we been and where are we going with calculus and, more importantly, the entire scope of introductory college mathematics?

The observations to follow are based in part on two studies conducted for the National Science Foundation (NSF) in the late 1990s: one to determine the national impact of the calculus reform movement since 1988, and a second that examined the subsequent NSF Institutional Reform (IR) Program. The study of calculus reform resulted in countless communications with the mathematics community and others about the status of calculus, while the IR study examined the ability of institutions to integrate science and mathematics across disciplinary boundaries and create an environment for learning that encourages science literacy for all students at the undergraduate level. The outcomes from these studies (reported in [1] and [2]) suggest the importance of considering the following series of questions:

1. What exactly have we learned about the teaching of introductory college mathematics?
2. What questions still need to be asked and answered, through carefully conducted research, to guide future changes?
3. What do we now know about the interaction of the discipline of mathematics with other courses in the sciences and beyond?
4. How can current knowledge about what students need to learn in introductory college mathematics—and how they learn it—be applied in positive ways to the organization and delivery of these courses?

What have we learned about the teaching and learning of introductory college mathematics?

The report from the study of calculus reform [1] includes information from more than 300 studies and writings about calculus reform over a ten-year period. Information for the report was collected to investigate what was learned about the effect of calculus reform on (1) student achievement and attitudes, (2) faculty and the mathematics community, and (3) the general educational environment.

Both mathematicians and colleagues in other disciplines have stated the need for revision and renewal in calculus that focuses on raising students' conceptual understanding, problem-solving skills, analytical and transference skills, while implementing new methods to reduce tedious calculations. Coordination with other disciplines in science and engineering and interactions between different sectors of the mathematical sciences community in the development of model curricula and prototypical instructional materials was encouraged through funding initiatives by NSF and others. The movement has been described as an effort to develop a vision for calculus that is challenging and stimulating, with the primary goal being to improve the quality of calculus courses and the level of learning by students in these courses at all types of institutions [1].

Calculus reform efforts have yielded mixed results in the areas of student achievement and student attitudes. The varying outcomes at different institutions are certainly not surprising, given the wide range of definitions of reform that have been implemented by each of the projects. Nonetheless, a number of elements are in fact common to many of the projects, including the use of computer technology and applications in the teaching of calculus. Many participants in the reform efforts also believe in the importance of emphasizing a student-centered environment, including discovery learning and cooperative activities that support a more conceptually-based course [1].

The existence of these common elements throughout the majority of the projects implies that *the relative success or failure of reform efforts is not necessarily dependent upon what is implemented, but rather how, by whom, and in what setting.* The consistent reactions of students from a wide variety of institutions point to several key components in the success of a reform environment. For example, instructors must communicate to students (and other faculty) the purpose of the changes being made in the calculus course. This is perhaps not as easy as it seems, as the reasons for the change must be seen as relevant and important to future success. It is unfortunate that many students believe mathematics is a static list of rules and algorithms to be memorized, a barrier to be overcome before they can do so-called real problems in other disciplines. Perhaps the most important role of reform efforts then is to challenge these beliefs and help students to see the many uses of calculus, both within the discipline of mathematics and in a wide range of problems from other disciplines. However, the means by which this can be communicated is not at all clear and needs to be addressed by the mathematics community as a critical part of the reform efforts.

The level of personal attention available to students also greatly affects their attitude and level of commitment in a calculus course. This is not a characteristic unique to reform courses, but one that has been highlighted as a result of the foreign environment that a reform course introduces. Specifically, the elements that define such a course are often ones that students have never experienced in mathematics (or perhaps any other discipline) and therefore additional support is required as the students adjust their learning styles. This is likely to be the reason that many of the projects report a dramatic shift in student (and faculty) attitudes after they have experienced a couple of calculus courses using reform ideas.

One area of particular concern is the adamant opposition to reform of many students who have excelled in the traditional environment. It is important that the reform efforts not cause these students to lose their interest in mathematics, just as it is important that others be encouraged through the wider variety of opportunities for success that reform courses offer. It is not necessarily the case that these two goals must be in opposition. As with anyone who is opposed to change (including faculty), it is likely that these students are simply reluctant to move from a learning environment in which they are successful to one which is unfamiliar. Although this can be overcome with open communication and creative opportunities that continue to challenge these students, it is certainly not something that can be deemed as unimportant because "these students will succeed no matter what we do with them." A primary goal of the reform efforts is the creation of a course that makes calculus real and interesting for *all* students, including the ones for whom the course has always been, as well as those for whom it has not.

What questions still remain?

The first several years of the calculus reform movement were characterized by a whirlwind of ideas about the organization of the course and the associated curriculum. Soon after came a dissemination phase, where other institutions eager to participate and improve their own courses "jumped on the bandwagon." Throughout, skeptics of the changes—and even many of the "reformers" themselves—began more and more to question the justification for these changes and the implications for faculty development, recruitment of mathematics majors, and the place for mathematics within the undergraduate curriculum. Even though many mathematicians conceded that change was necessary, they questioned the fact that the current changes were being made without first carefully examining—as a community, rather than individually by project— the overriding objectives and goals for students enrolled in introductory college mathematics courses. Many claimed that the development of such goals is critical to the discipline of mathematics because these introductory courses will in fact be the last formal mathematical training for a large majority of students.

Given these valid concerns, it is important that the mathematics community develop, research, and form workable solutions for many difficult questions. One effort to address this issue [3] poses and discusses a small subset of the questions prompted by the reform movement, including the following:

- What are the contributions that mathematics can—and should—make in the world of science, engineering, and technology? How can we better nurture our relationships with colleagues in partner disciplines?

- What mathematical skills and knowledge should an undergraduate student have after completing an introductory college mathematics course?

- What has been learned about the cognitive processes involved in learning mathematics through the extensive research conducted in this area during the past decade? How can we use this research to better inform the development of introductory college mathematics courses?

- How does mathematics fit into the broader context of student learning; e.g., what are the life skills that our students will need to succeed in the workplace?

- What is the appropriate role of technology in the teaching and learning of mathematics?

- How does an introductory college mathematics course contribute to the overall mathematics education of our students? How can the changes made in these courses be used to improve other mathematics courses?

- What is the appropriate role of colleges and universities in supporting curricular change? How can administrators provide an environment that is conducive to change and enables faculty to develop the necessary skills for supporting such change?

- What are the appropriate mechanisms by which the mathematics community can evaluate progress and thereby better inform continuing change? What changes ultimately help students to better understand mathematics and to have an appreciation for the importance of mathematics in our society?

Some of these questions will be discussed in the following sections.

What do we know about the interaction of introductory college mathematics with partner disciplines?

Mathematics can and should play an important role in the education of undergraduate students. Unfortunately, students often do not see the connections between mathematics and their chosen disciplines; instead, they leave mathematics courses with a set of skills that they are unable to apply in non-routine settings and whose importance to their future careers is not appreciated. Indeed, the mathematics many students learn often is not the most relevant to their chosen fields. For these reasons, faculty outside mathematics often perceive the mathematics community as uninterested in the needs of non-mathematics majors, especially those in entry level courses.

The Mathematical Association of America (MAA) has gathered input from partner disciplines through a series of eleven workshops held across the country from November 1999 to February 2001, followed by a final summary conference in November 2001 (See theme 5 in this volume, [4] and [5]). Each workshop produced a report summarizing its recommendations and conclusions. The reports were written by representatives of the partner disciplines and directed to the mathematics community. The following are some major recommendations from the partner disciplines — in the sciences and beyond — about the learning that their students need to achieve in introductory college mathematics courses.

1. Emphasize conceptual understanding Mathematics courses in the first two years should focus on understanding broad concepts and ideas. This emphasis should be used to assist students in the development of precise, logical thinking, requiring them to reason deductively from a set of assumptions to a valid conclusion. Proofs should be presented only when they enhance understanding — and even then at an elementary level. When presented, the fundamental connection between proof and understanding must be highlighted.

2. Emphasize problem-solving skills and mathematical modeling Introductory college mathematics courses should develop the fundamental computational skills the partner disciplines require, but also emphasize integrative skills; i.e., the ability to apply a variety of approaches to single problems, to apply familiar techniques in novel settings, and to devise multi-stage approaches in complex situations. Students should be expected to create, solve, and interpret mathematical models, including models from the partner disciplines. Interpretations should include opportunities for students to describe their results in several ways: analytically, graphically, numerically, and verbally.

3. Emphasize communication skills All courses (not just mathematics courses) should incorporate the development of reading, writing, speaking, and listening skills. Such skills include the ability to explain mathematical concepts and logical arguments in words. To develop these skills, they must be reinforced continuously throughout the curriculum.

4. Strive for depth over breadth Mathematics departments should continue to offer calculus and linear algebra in the first two years, but should make the curriculum more appropriate for the needs of the partner disciplines. For example, emphasize two- and three-dimensional topics, and pay attention to units, scaling, and dimensional analysis. Mathematicians should work with local colleagues from partner disciplines to decide what topics can be omitted. The remaining topics can then be taught in more depth.

How can current knowledge be used to improve introductory collegiate mathematics courses?

The current information-based economy demands from citizens different skills than those of just 20 years ago, ones that involve "the sophisticated use of elementary mathematics more often than elementary applications of advanced mathematics" [6]. Individuals are required "to reason, to make sense of real-world situations, and to make judgements grounded in data" [7]. Personal issues such as health, safety, taxes, budgets, credit, and financial planning (just to name a few) force citizens to face decisions that utilize quantitative information. It is difficult to function successfully without being able to interpret these numbers in order to assess risks, to manage finances, and to make critical decisions [8]. Such skills, broadly termed *quantitative literacy* (QL), are beginning to be recognized as critical to good citizenry in a democracy of the 21st century—skills equal in importance to reading and writing.

However, QL is rarely an explicit goal of education—at any level, in any discipline. This is due in part to the fact that a widespread need for QL has only recently been recognized, implying that the definition of what it means to be quantitatively literate is still unclear. But the problem also is grounded in the firmly established hierarchy within the mathematics curriculum, which implicitly states that if you are not in the prescribed mathematics sequence leading to calculus, you are not succeeding in the mathematics game. And traditionally for students who have not fared well in this sequence in high school, this has meant enrolling (or, more often than not, re-enrolling) in a college algebra course as their introduction to college-level mathematics. How can we improve this situation?.

1. Promote the importance of quantitative literacy to educational and public policy leaders. The civic responsibilities that imply a need for quantitative literacy need to be a high priority for education. To encourage such a transformation, quantitative literacy must be visible to the public. Widespread implementation will also imply the need to align quantitative literacy efforts with those of other educational organizations and initiatives. Therefore, discussions among faculty from multiple disciplines need to be initiated to determine the role of quantitative literacy in their discipline, in their courses, and in the goals and requirements of their institutions [8].

2. Determine the appropriate expectations for quantitative literacy, especially in relation to different levels of education. Mathematicians need to work with other educators to establish benchmarks for quantitative literacy appropriate for graduates at the high school, two-year college, and four-year college levels. Once established, these benchmarks will need to be supported with effective assessment of quantitative literacy, including a web-based collection of class-tests, QL resources, tasks, and assessment items. Existing large-scale tests also will need to be examined to determine the degree to which they do or do not support QL goals [8].

3. Replace traditional college algebra courses with courses stressing problem-solving, mathematical modeling, descriptive statistics, and applications in the appropriate technical areas. Students enrolled in college algebra usually leave these courses quantitatively illiterate and unable to apply any of the techniques they have learned to problems of real significance. Therefore, mathematics courses that teach them to think mathematically in context are critical to their continuing development as citizens [8]. College mathematics curricula should provide alternative routes to advanced mathematics, while ensuring that each student's level of QL continues to develop throughout college. Such curricula will de-emphasize intricate algebraic manipulation, provide non-calculus-based descriptive statistics and data analysis in the first two years, and include non-calculus-based discrete mathematics and mathematical reasoning as alternative introductory courses [4].

4. Use a variety of teaching methods, including team teaching and interdisciplinary courses. Since different students have different learning styles, mathematics instructors should encourage the use of active learning, including in-class problem-solving opportunities, class and group discussions, collaborative group work, and out-of-class projects [4]. Team teaching and interdisciplinary courses also can be used to model scholarly debate and cooperation; develop critical, process-oriented thinking; and provide different perspectives to problem-solving—even within a subject. Students respond positively to team teaching—the transitions between faculty add variety and emphasize relevance by connecting different disciplines [2]. Team teaching also improves faculty to faculty mentoring and provides an informal peer review. Appropriate technologies can be used to enhance all alternative teaching methods [4].

5. Emphasize the use of appropriate assessment. Mathematics departments need to establish mechanisms for the development, review, and dissemination of effective instructional materials and techniques, including collaborative efforts between mathematicians and partner disciplines that result in innovative instructional materials. Institutional assessment of programmatic changes also should be encouraged [4].

Implications of the proposed changes

Changes such as those discussed here necessitate an overhaul of current institutional structures — the conversion of a well-established educational system. Three critical components of affecting such institutional change are:

- *Faculty rewards* —Faculty are getting mixed signals from their administrators and are reluctant to participate in curriculum development. Scholarship in teaching needs to be built into the promotion and tenure guidelines.

- *Change as a priority* —Senior administrators need to provide the impetus and set the tone for change.

- *Instructional resources* —Limited resources are always a problem, especially for professional development for faculty. But solutions exist, and departmental budgets need to be tied to renewing the undergraduate curriculum [2].

1. Faculty rewards   Institutions must be honest about their expectations of faculty members and ensure that rewards match the stated requirements. There is concern that research-based tenure systems prevent faculty who are particularly good at teaching and mentoring from being retained and promoted. The challenge, however, is how to measure effectiveness when instructors promote learning through a diversity of teaching styles.

The tenure system discourages non-tenured faculty from engaging in activities that would detract from their own acquisition of tenure, including the development of non-traditional courses such as those previously discussed. In addition, tenured faculty are often resistant to change for fear they will lose the support of colleagues and administrators. Therefore, institutions need to provide incentives and rewards for all faculty in the form of positive recognition, tenure, and promotion. Departments and colleges need to have explicit conversations in order to agree on goals and establish priorities for the evaluation of teaching. Finding a balance for innovation in education is difficult. However, faculty must not feel their careers are at risk when experimenting with reform activities. Administrators must acknowledge and accept the potential for failure in innovation.

2. Change as a priority   Senior administrators must be involved early and often, with their enthusiasm kindled by concrete examples of successes from faculty that provide credibility for the changes. In addition, grass roots enthusiasm is essential—especially participation of young faculty who can continue the project in the future. Staying power is critical—don't give up. And, one very important lesson: there are many research issues associated with such work; faculty need to be involved in the development of a solid research base. Some implementation strategies include:

- consider and build from the core strength of your institution;

- as appropriate, collaborate with other institutions that have different core strengths;

- celebrate and build on past accomplishments;

- recognize that change involves stress, but this type of stress can motivate faculty and administrators to work together in securing resources;

- get changes that promote QL on college agendas so that they receive attention;

- create a council of administrators, faculty, students, and business people to discuss curricular needs;

- re-examine the model for the relative importance placed on teaching, research, and service;

- build inter-institutional collaborative links for pedagogical as well as traditional research;

- foster the well-organized use of undergraduates to assist faculty; and

- encourage faculty to motivate other faculty, either through one-on-one mentoring, team teaching, or by disseminating successful models [2].

3. Instructional resources Faculty need significant support in order to implement QL (and other curricular changes) in their courses. Professional development opportunities that support these new roles for faculty, housed within an infrastructure that can provide resources in quantitative literacy, need to be abundant. Such resources might include case studies of innovative QL programs and best practices [8]. Faculty development also requires resources for:

- release time from teaching duties,
- learning institutes and professional development workshops,
- cross-disciplinary experimentation, and
- development of new assessment instruments and mechanisms for evaluating effective teaching [2].

Ultimately, the creation of a structural home for quantitative literacy will be critical for widespread implementation [8]. Departmental cultures need to change to support elements of QL, such as interdisciplinary course development. Therefore, departments should work to function as cooperative units, rather than collections of individuals, and support programs in which faculty that have experience with interdisciplinary course development and other elements of QL mentor their colleagues [2].

Supporting QL in introductory college mathematics courses through networking

Experiences from curricular reforms in calculus and other disciplines reveal that a supportive network can be used to promote buy-in by all affected individuals and groups [9]. Therefore, an important development is the design and formation of the National Numeracy Network (NNN) to assist locales in which efforts are underway to translate QL from aspiration into educational practice, to disseminate promising practices, and to exchange information among existing and potential network sites. (Information about NNN, can be found at www.math.dartmouth.edu/~nnn.)

Planned activities of the network include professional development experiences and opportunities to learn about QL for educators and others and a Web site through which resources, information, and exchange of ideas regarding QL will be made accessible to the broadest possible audience. Critical to NNN are local projects and meetings that bring together schools, colleges, civic groups, the media, business, and industry. Additional outreach efforts include research, reports, and publications that increase understanding of QL and its significance in education, work, and private and civic life.

One part of NNN is a QL Resource Library that provides the opportunity for faculty to share ideas by developing an extensive collection of QL materials across a variety of disciplines. The resource library includes program descriptions, course syllabi, examinations, activities, laboratories, projects, readings, publications, and examples of student work. Such a collection of materials is very important to the work of NNN, serving as a resource for outreach efforts and as a means of teaching interested individuals and organizations about QL. In addition, the QL Resource Library soon will include a database of individuals, projects, and institutions involved in the development of QL curricula.

By focusing on different aspects of policy, practice, professional development, dissemination, and assessment, the National Numeracy Network will provide a catalyst for quantitative literacy, especially in grades 10 to 14. QL programs participating in the network already are working with organizations that can directly influence a wider audience to create public pressure for QL. NNN institutions and organizations are developing QL course materials and programs to share through professional development opportunities, the QL Resource Library, and the QL Web site. [9]

Conclusions

It is certain that the next decade will bring more change in introductory college mathematics. These changes will be tackled by a mathematics community that has experience with the issues, questions, and ideas presented here. We have come a long way in the development of an educational environment that is conducive to positive change. The true success of these efforts will be measured by the fervor with which the mathematics community and their colleagues in partner disciplines build upon these efforts and define the appropriate place for mathematics in the undergraduate curriculum.

References

1. S. L. Ganter, *Changing Calculus: A Report on Evaluation Efforts and National Impact from 1988 to 1998*, MAA Notes #56, Mathematical Association of America, Washington, DC, 2001.

2. S. L. Ganter and J. S. Kinder (Eds.), "Targeting Institutional Change: Quality Undergraduate Science Education for All Students," *Targeting Curricular Change: Reform in Undergraduate Education in Science, Math, Engineering, and Technology*, American Association for Higher Education, Washington, DC, pages 1–27, 2000.

3. S. L. Ganter (Ed.), *Calculus Renewal: Issues for Undergraduate Mathematics Education in the Next Decade*, Kluwer Academic/Plenum Publishers, New York, 2000.

4. S. L. Ganter and W. Barker, "A Collective Vision," in S. L. Ganter and W. Barker (Eds.), *A Collective Vision: Voices of the Partner Disciplines*, MAA Reports Series, Mathematical Association of America, Washington, DC, 2004.

5. S. L. Ganter and W. Barker (Eds.), *A Collective Vision: Voices of the Partner Disciplines*, MAA Reports Series, Mathematical Association of America, Washington, DC, 2004.

6. G. Nelson, "Quantitative Literacy: A Science Literacy Perspective," in B.L. Madison and L.A. Steen (Eds.), *Quantitative Literacy: Why Numeracy Matters for Schools and Colleges, National Council on Education and the Disciplines*, Princeton, NJ, 2003.

7. L. P. Rosen (with L. Weil and C. von Zastrow), "Quantitative Literacy in the Workplace: Making It a Reality," in B. L. Madison and L. A. Steen (Eds.), *Quantitative Literacy: Why Numeracy Matters for Schools and Colleges*, National Council on Education and the Disciplines, Princeton, NJ, 2003.

8. L. A. Steen, *Achieving Quantitative Literacy: An Urgent Challenge for Higher Education*, MAA Notes #62, Mathematical Association of America, Washington, DC, 2004.

9. S. L. Ganter, "Creating Networks as a Vehicle for Change," in B. L. Madison and L. A. Steen (Eds.), *Quantitative Literacy: Why Numeracy Matters for Schools and Colleges*, National Council on Education and the Disciplines, Princeton, NJ, 2003.

Theme 1. New Visions for Introductory Collegiate Mathematics

Courses below calculus face a number of challenges. They need to serve two populations: students who plan to continue their study of mathematics and students who do not. They need to meet the mathematical needs of today's population of students (which is different from when many of us were in school). They need to prepare students who do continue their study of mathematics to take a calculus course where they are expected to think and understand and not just do computations. The six papers in this section describe some of the major changes that are taking place in the courses below calculus and the challenges that need to be met. Nancy Baxter Hastings, Sheldon Gordon and Bernard Madison discuss changes and challenges at the precalculus level; Don Small and Scott Herriott discuss the state of college algebra; and Janet Andersen discusses a new approach to quantitative literacy.

Nancy Baxter Hastings seeks to initiate a dialogue about some of the challenges confronting the reform of precalculus. Nancy begins by claiming: "Introductory collegiate mathematics is in the midst of a revolution." She articulates some of the forces for change in introductory mathematics courses at the collegiate level and how these concerns are being addressed. She goes on to describe the distinguishing features of reform-based instructional materials and the pedagogical changes that the new materials have fostered in the teaching and learning environment. Sheldon Gordon describes changes in the population of students taking college-level mathematics, the impact of the growth of technology on teaching and learning, changes in the mathematical needs of students and changes in pedagogy. Sheldon stresses the need for conceptual understanding, arguing that today's students need to be "prepared for calculus intellectually, not just algebraically," and they need to be able to grapple with realistic problems. Bernard Madison observes that "preparation for calculus is the primary aim of grades 9–12 mathematics." Like Sheldon, he argues that this emphasis does not serve the current population of students well.

Don Small "presents a case for transforming traditional college algebra from a failed program attempting to prepare students for calculus to one that enables students to address the needs of society, the workplace, and the quantitative aspects of disciplines." He describes some "characteristics of improved college algebra programs, as well as the symbiotic relationship between a transformed college algebra and quantitative literacy." Scott Harriott describes several distinct models for a college algebra course, based on the content, organization, and pedagogy of representative textbooks, and he contrasts these curricula with the traditional college algebra course. He identifies some related issues of national and local educational policy.

Like the other authors in this section, Janet Andersen notes that a course designed to prepare students for calculus is not necessarily well designed for students who need general education credit or need to prepare to take courses outside the mathematics department. Janet describes the course she helped develop at Hope College, *Understanding Our Quantitative World*, which was designed around the question: "What skills and concepts are useful for informed citizens?"

When the calculus reform movement got under way in the late 1980s and early 1990s, the mathematics teaching profession was energized and enthusiastic. ... Changes in the course that feeds into calculus, precalculus, have followed much more slowly. New reform-based precalculus books are being published (for example, see the texts in the section "Ideas and Projects that Work"), and there are pockets of change, but there does not appear to be the wave of excitement that accompanied the changes in calculus. The good news is that a national movement to refocus precalculus and other courses below calculus is now getting under way. [This paper] articulates some challenges confronting the reform of precalculus and poses some associated questions. Some questions have easy answers. Some are hard. Many are addressed in the papers in this volume.

Refocusing Precalculus: Challenges and Questions
Nancy Baxter Hastings

Major changes have taken place in the mathematical education of students over the last decade. These changes have come about for a variety of reasons, including the changing demographics of the students taking college-level mathematics, the growth of technology and what it can provide for the teaching and learning of mathematics, and the changing mathematical needs among the people who use mathematics. All of these factors have major implications for what we teach, and how we teach it, both at the precalculus level and in all other mathematics offerings.

Preparing Students for Calculus in the Twenty-First Century
Sheldon P. Gordon

Preparation for calculus has for several decades been a primary aim of grades 9–12 mathematics. Often this preparation extends to grades 13 or 14. Consequently, calculus has taken on extreme importance and extreme responsibility in US mathematics education. . . . This importance and responsibility are inflated and serve neither calculus nor general education well. Neither students nor the general public see the value that calculus delivers and very few people ever have any use for calculus techniques. The focus on preparation for calculus has narrowed and hurried the mathematics curriculum leaving no time for appreciating or learning to use the mathematics that is taught before calculus. And use of mathematics has become essential in American life, requiring that the aims of grades 11–14 mathematics be broadened to include preparation for life.

Preparing for Calculus and Preparing for Life
Bernard L. Madison

Traditional college algebra courses are not working. That was the strong consensus of the participants in a recent Conference to Improve College Algebra, held at the U. S. Military Academy, February 7–10, 2002. This conclusion was based on the courses' outdated content, high FWD rates, and on the negative impact these courses have on student perceptions of mathematics. The large number of students enrolled in college algebra creates an urgency to transform these courses into ones that do work.

College Algebra: A Course in Crisis
Don Small

The traditional college algebra curriculum seems to assume that the course is a preparation for calculus. But surveys at many institutions have shown that only a minority of college algebra students go on to take calculus of any kind, and only a small fraction of those attempt a full-year calculus sequence. Thus, in recent years, some textbook authors have developed alternative curricula that address the future mathematical needs of the soft-sciences students.

Changes in College Algebra
Scott R. Herriott

The primary need for many students is to become quantitative literate citizens who are capable of interpreting and using information presented quantitatively. . . . In constructing and teaching a course designed to improve quantitative literacy, it is crucial that the goals and objectives be targeted at what students will most likely encounter outside of academia. That is, what skills and concepts are useful for informed citizens? For many mathematicians, this is a brand-new (and somewhat radical) idea.

One Approach to Quantitative Literacy: Understanding Our Quantitative World
Janet Andersen

7

Refocusing Precalculus: Challenges and Questions

Nancy Baxter Hastings
Dickinson College

Introduction

With this paper, I hope to initiate a dialogue about some of the challenges confronting refocusing pre-calculus. The paper articulates some of the forces for change in introductory mathematics courses at the collegiate level and how these concerns are being addressed. It describes the distinguishing features of alternative instructional materials and the pedagogical changes that the new materials have fostered in the teaching and learning environment. Many of these changes have been implemented in calculus; in fact, even the so-called traditional books reflect many of these new directions. Precalculus, however, lags behind.

What is the problem?

Introductory collegiate mathematics is in the midst of a revolution. Forces for change include:

- a national concern about the lack of quantitative literacy among exiting college students [9, 11, 12, 13]
- the need for improved education of prospective teachers to reflect the recommendations of the NCTM *Standards* and to allow prospective teachers to teach both the new subject matter and use the methods suggested [1, 8]
- the recognition that many introductory mathematics courses have failed to be appealing or accessible to vast numbers of students entering colleges and universities—especially women and minorities
- the concern that those teaching introductory mathematics do not take advantage of recent research findings in mathematics and science education [2, 3, 4, 6]
- the concern that many courses do not make effective use of technology [7, 10, 14]

In addition, new fields of interest in mathematics and new mathematical techniques used in allied disciplines have led educators to examine the content of introductory mathematics courses [5].

What is being done?

In response to these concerns, new instructional materials and new teaching approaches are being developed. For example, please see the texts described later in this volume in the section "Ideas and Projects that Work."

Characteristics of alternative instructional materials

In general, alternative curricular materials cover fewer topics, and the topics that are covered are introduced with more depth. Instead of introducing concepts strictly from a symbolic point of view, concepts are introduced from multiple viewpoints: verbal, graphical, numerical and symbolic. The new materials emphasize conceptual understanding and de-emphasize complicated pencil-and-paper manipulations.

Alternative materials typically use real-world applications to motivate mathematical ideas and to help students understand the relationship between these ideas and the world around them. Students model real-world data and use data analysis to determine the accuracy of the mathematical models. The integration of applications provides answers to questions students frequently ask: "Why do I need to know this?" "What good is this?"

The new instructional materials use projects to help deepen students' understanding of mathematical concepts and to emphasize inquiry. When undertaking a project, not only do students need to make decisions about how to solve the given problem, they also need to ask themselves: "What assumptions do I need to make?" "What issues do I need to pursue?" "What if . . . ?" In the process, students also develop their communication and writing skills.

Finally, alternative materials use technology to empower students to explore mathematical concepts, make conjectures and experiment, and tackle complex problems. Technology is also used to help students to form mental images associated with abstract mathematical ideas—for example, to form a mental image associated with the process of a function or with the end-run behavior of a function.

Changes in pedagogy

Alternative texts differ from traditional texts not only in content, but also in how they are used by both the instructor and the student. They help to foster changes in both the teaching and the learning environments.

Traditional mathematics books are written for instructors, not for students. Since being a good teacher is often equated to being a good lecturer, an instructor uses a traditional text to develop carefully honed lectures, which closely follow the text. The instructor's role in a lecture-driven setting is to tell students—in a clear and logical way—everything they are expected to know and to present examples of problems they are expected to be able to do—particularly on an exam. While the instructor lectures, students take copious notes, but occasionally their minds wander. When strong students stop paying attention, or "skip listen," they fall into I'll-just-have-to-figure-this-out-later mode, but when less able students skip listen, many decide that the material is hopelessly confusing and quit in frustration.

Homework in a traditional lecture-based class consists of mimicking examples students see in class or find in their book. Students study in isolation and compete with one another for grades. Students fear that discussing a problem might be viewed as plagiarism and helping each other might be detrimental to their own grades.

Alternative instructional materials, on the other hand, require students to read, write and talk about mathematical ideas. The materials are written for students to read. Moreover, as students explore new concepts, they think about mathematical ideas and make connections on their own, thereby developing a sense of ownership of new concepts.

The new materials foster interactive teaching formats, in which traditional lectures are replaced by mini-presentations, classroom discussions, and students working collaboratively. In this learning environment, students are actively involved in the learning process. The instructor becomes the guide or intellectual manager, mingling with the students as they work, posing probing questions, and encouraging students to share ideas with their peers and to listen carefully to what their peers have to say. The emphasis in the classroom shifts from the instructor teaching to the students learning. Students become responsible for learning and instructors become responsible for helping them learn how to learn.

The alternative materials seek to help students become confident, competent and creative problem solvers, rather than to help them learn problem-solving algorithms by rote. They seek to help students to utilize new ideas as well as to understand them. They seek to help students develop their mathematical communication skills. And maybe most importantly, they seek to make learning mathematics a more enjoyable experience.

A dozen challenges, plus one, confronting refocusing precalculus

As we said in the introduction, many of the changes described above have been implemented in the instructional materials for calculus. When the calculus reform movement got under way in the late 1980s and early 1990s, the mathematics teaching profession was energized and enthusiastic. Colleagues were excited about designing new materials and exploring ways to utilize emerging technologies, including both computer algebra systems and graphing calculators.

Changes in the course that feeds into calculus, precalculus, have followed much more slowly. New reform-based precalculus books are being published, and there are pockets of change, but there does not appear to be the wave of excitement that accompanied the changes in calculus. The good news is that a national movement to refocus precalculus and other courses below calculus is now getting under way.

The remainder of this paper articulates some challenges confronting the reform of precalculus and poses some associated questions. Some questions have easy answers. Some are hard. Many are addressed in the papers in this volume.

1. Serving two student populations: students for whom precalculus is a gateway course and those for whom it is a terminal course. Two possibly conflicting goals of precalculus are (1) to prepare students to continue their study of mathematics (as well as other quantitative disciplines) and (2) to provide students, especially those who choose not to continue, with a meaningful and positive learning experience.

> *Challenge*: We need to identify and meet the very different needs of students for whom precalculus is a gateway to studying higher level mathematics and those for whom it is a terminal course.

> *Questions*: Is it possible to meet the needs of both of these groups of students in the same course? If it is not, should we teach to the majority, for whom precalculus is the last math course they will ever take, or to the minority who plan to continue on? One possibility is to offer different courses for the various groups of students. But how would we place students in the appropriate course?

2. Serving two (other) student populations: high school students and post-secondary students. At the high school level, precalculus is often a capstone course for relatively strong students, whereas at the college level it is a low-level course for underprepared students.

> *Challenge:* We need to develop materials that meet the needs of both high school and post-secondary students, including non-traditional students who return to college after ten, twenty, or even thirty years.

> *Questions:* Is this feasible? Are the needs of high school and post-secondary students the same? Can the same materials be used effectively with both groups? Should separate materials be developed?

3. Positioning students to understand calculus. Since the term "precalculus" means "before calculus," it is reasonable to assume that one of the primary objectives of a precalculus course is to prepare students for calculus. Many traditional precalculus courses, however, simply reteach algebra and trigonometry. Moreover, by emphasizing symbolic manipulation, traditional courses help solidify students' misconception that the goal of mathematics is to use memorized routines to get the answers in the back of the book.

Students do not see any new mathematics in traditional courses, they are not exposed to thinking about mathematical ideas and they do not know what to expect in calculus. As a result, students don't understand why the course is called "precalculus," and when they take calculus, they find a huge disconnect between their precalculus and calculus courses. Consequently, many are lost from the start.

Challenge: Precalculus needs to position students to understand fundamental calculus concepts by providing them with a conceptual understanding of basic ideas, such as the limiting behavior of a function, the rate of change of a function, and the accumulation of quantities—without mentioning the terms limit, derivative, or definite integral.

Question: How can precalculus provide students with a firm basis for understanding fundamental concepts they will encounter in calculus?

4. Preparing students to do calculus.   The previous challenge claims that students need to understand what calculus is. However, students also need to be prepared to do calculus. There needs to be a balance. Too often we hear calculus students say: "I understand calculus. I just can't do it." In other words, they (think they) understand what a limit is or what a derivative is, but they don't have the necessary computational tools for evaluating a limit or finding the zeros of a derivative. As a result, many students become totally dependent on technology (see Challenge #8).

Challenge: Precalculus not only needs to position students to understand calculus concepts, but it also needs to equip students with the tools needed to do calculus.

Questions: What algebraic, geometric and trigonometric concepts do students need to acquire in order to do calculus? How do we test whether or not students have the appropriate skills before entering calculus? How do we help students succeed in calculus, especially when they have deficiencies and have already taken precalculus? Is an integrated precalculus/calculus I course, which provides a just-in-time review of precalculus concepts, an effective alternative?

5. Preparing students to use precalculus in other disciplines and in life.   For many students, precalculus is a terminal course, not because they don't like mathematics or didn't do well in the course, but because the course fulfills either the quantitative requirement for graduation or the mathematics requirement for their major. Students need to be able to transfer what they learn in these courses to the world outside the precalculus classroom. They need to feel that what they have learned is connected to their lives either because it interests them or because they feel that the concepts and skills they acquire will be useful in achieving their academic and career goals.

Challenge: Precalculus needs to prepare students not only to continue their study of mathematics, but also to use precalculus in other disciplines and in their lives beyond the classroom. Students need to understand why the ideas they are learning are important.

Questions: How can we help students feel that what they learn in precalculus is connected to their lives? How can applications be used to achieve this need? How can precalculus courses position students to succeed in what is to come, even if what is to come is not in another mathematics course? And in the long run, how can we prepare students to use mathematics in ways that we cannot foresee today?

6. Increasing students' self-esteem and helping them to develop confidence to do mathematics. Many students feel anxious about mathematics, especially if they are faced with an unfamiliar situation. Students can acquire self-esteem and develop confidence in their own abilities through peer support and as a result of achieving (even small) successes. This will not happen if they are simply mimicking problems in the book.

Challenge: We need to develop curricular materials and design learning environments that help increase students' self-esteem and help students develop confidence that they can do mathematics on their own.

Questions: What types of learning environments help students feel confident about taking risks, asking questions, and sharing ideas? What types of instructional activities help students think about mathematical ideas, explore new concepts, and tackle unfamiliar problems on their own? How can we help students develop a can-do attitude?

7. Making the study of precalculus an enjoyable experience.   Closely related to the previous two challenges is the fact that many students claim that they "hate math." It is just not their thing. They don't enjoy taking mathematics courses. They spend most of their time uncertain about why they are doing what they are doing and questioning why they need to do it anyway. They feel inept and are unable to utilize what they have learned in other courses. On the other hand, something about math turns us (as mathematicians) on. We talk animatedly. We scribble on napkins. We enthusiastically share our results.

Challenge: We need to develop curricular materials and design learning environments that make taking precalculus a positive learning experience for students.

Questions: How do we help our students share the excitement that we feel when we see a connection between things that we know or when we discover something new? And how can we do this, realizing that our students' interests, abilities, and inclinations may be quite different from ours?

8. Balancing the use of technology and the use of pencil and paper.   Technology is a powerful tool, and students quickly—frequently, too quickly—turn to it for help. However, in order to be able to interpret the responses provided by technology and determine whether or not the responses are reasonable, students need to know how to perform basic computations, do basic algebraic manipulations, and graph basic functions by hand. And they need to know when to use these skills and when to turn to technology for assistance.

Challenge: In addition to ensuring that students have acquired the basic computational, algebraic and graphing skills necessary to do calculus (see challenge #4), precalculus should help students learn to use technology wisely and appropriately.

Questions: What is the appropriate balance between the use of technology and the use of pencil and paper? How can alternative precalculus courses help students to think through a situation before turning to technology? How can we articulate questions so that students are not tempted to use technology as a crutch?

9. Using the outcomes of research in mathematics education.   The results of research about how students acquire mathematical ideas can help us develop effective curricular materials. [3, 4, 6]

Challenge: Research pertaining to how students acquire fundamental precalculus concepts needs to be undertaken. The results of this research should guide the development and refinement of curricular materials.

Questions: How do those who are developing curricular materials forge working relationships with colleagues who are undertaking research in mathematics education? What types of research might someone who is developing materials conduct? What types of data should they collect? What types of questions should they ask? How should they analyze the results?

10. Evaluating alternative courses.   The research described in Challenge #9 concerning how students learn is part of the formative component of the evaluation process, since the results are used to direct the development or revision of instructional materials. Other important parts of the formative component are analyzing student learning gains and measuring changes in student attitudes. The summative component of the evaluation process involves analyzing how well the new course or program meets its goals.

Challenge: Evaluation needs to be viewed as an essential part of the development process. All curriculum development projects should have clearly defined sets of objectives for the course and learning goals for the student, and they should have ways of measuring whether or not these goals are attained.

Questions: Again, what types of tools might we use to evaluate reform-based courses? What type of data do we want to collect? How can we compare students in a traditionally taught course to students in a course with a new focus? What instructional materials should be classified as alternative or refocused? What data has been collected that supports their effectiveness? For example, have the new materials helped increase the retention rate in precalculus courses where they are used? Have they had an impact on the continuation rate into calculus? Can students apply what they learn in other courses and in the client disciplines? Do students feel that what they learn is useful? Do students feel confident about tackling new problems? Are students able to tackle new problems? Did students enjoy the course? And so on.

11. Drumming up interest in precalculus at the post-secondary level.   Related to the observation that precalculus serves both the secondary and post-secondary student populations (see Challenge #2) is the fact that many college and university instructors view precalculus as a low-level course and the students as being mathematically disposable.

Challenge: We need to help colleagues to recognize the importance of precalculus in the curriculum and to help them change their attitudes not only about teaching precalculus, but also about the students who take precalculus.

Question: How? (See Challenge #13.)

12. Disseminating alternative pedagogical approaches and curricular materials.   As we mentioned earlier, the work of refocusing precalculus has already begun. In response to the calculus reform movement, innovative new application-driven precalculus texts have been developed that integrate precalculus and data analysis concepts and utilize hands-on, inquiry-based pedagogical approaches.

Challenge: We need to help spread the word about the existence and effectiveness of these new approaches and the rationale for change.

Question: How do we get the word out?

13. Defining the problem and convincing colleagues and administrators that changes need to be made.   The calculus reform movement was bolstered by the fact that, in the late 1980s, fifty-percent of the students who took calculus failed. Clearly change was necessary.

Challenge: We need to challenge our tradition-bound colleagues to rethink both what they teach in precalculus and the way they teach it. We need to challenge our administrators to support these efforts. But first of all, we need to convince our colleagues and administrators that change is necessary.

Questions: How do we clearly state what the problem is? How do we spread the word that there is a problem? How do we put forth simple, convincing arguments, supported by data, that change is necessary?

Summary

Refocusing precalculus is a formidable task. The student audience needs to be identified (see Challenges #1 and 2). The intended outcomes for students and the goals for precalculus need to be defined (see

Challenges #3-7). New instructional materials need to be developed and evaluated (see Challenges #8-10). A comprehensive dissemination program needs to be launched (see Challenges #11 and 12). And, most importantly, the current state of precalculus needs to be articulated, and administrators and colleagues need to join the movement (see Challenge #13).

References

1. Conference Board of the Mathematical Sciences, *The Mathematical Education of Teachers*, CBMS Issues in Mathematics Education, Volume 11, AMS/MAA, 2001.

2. Dubinsky, Ed and G. Harel (eds.), *The Concept of Function: Aspects of Epistemology and Pedagogy*, MAA Notes #25, The Mathematical Association of America, Washington, DC, 1992.

3. Dubinsky, Ed, Alan Schoenfeld and James Kaput (eds.), *Research in Collegiate Mathematics Education*, Volumes I–IV, AMS/MAA, 1994, 1996, 1998, and 2000.

4. Fisher, Naomi, Harvey Keynes and Philip Wagreich (eds.), *Changing the Culture: Mathematics Education in the Research Community*, AMS/MAA, 1995.

5. Ganter, Susan and William Barker (eds.), *A Collective Vision: Voices of the partner disciplines*, MAA Reports Series, Mathematical Association of America, Washington, DC, 2003.

6. Kaput, James and Ed Dubinsky (eds.), *Research Issues in Undergraduate Mathematics Learning: Preliminary Analyses and Results*, MAA Notes #33, The Mathematical Association of America, Washington, DC, 1994.

7. Leinbach, Carl et al. (eds.), *The Laboratory Approach to Teaching Calculus*, MAA Notes #20, The Mathematical Association of America, Washington, DC, 1991.

8. Leitzel, James R. C. (ed.), *A Call for Change: Recommendations for the Mathematical Preparation of Teachers*, The Mathematical Association of America, Washington, DC, 1991.

9. Paulos, John A., *Innumeracy: Mathematical Illiteracy and its Consequences*, Hill and Wang, New York, 1988.

10. Smith, David A. et al. (eds.), *Computers and Mathematics*, MAA Notes #9, The Mathematical Association of America, Washington, DC, 1988.

11. Steen, Lynn A. (ed.), *On the Shoulders of Giants: New Approaches to Numeracy*, National Academy Press, Washington, DC, 1990.

12. Steen, Lynn A. (ed.), *Mathematics and Democracy: The Case for Quantitative Literacy*, Princeton, NJ: National Council for Education and the Disciplines, Princeton, NJ, 2001.

13. Steen, Lynn A. and Bernie Madison (eds.), *Quantitative Literacy: Why Numeracy Matters for Schools and Colleges*, National Council on Education and the Disciplines, Princeton, NJ: National Council on Education and the Disciplines, 2003.

14. Zimmerman, Walter and Steve Cunningham (eds.), *Visualization in Teaching and Learning Mathematics*, MAA Notes #19, The Mathematical Association of America, Washington, DC, 1991.

8

Preparing Students for Calculus in the Twenty-First Century

Sheldon P. Gordon
Farmingdale State University of New York

Major changes have taken place in the mathematical education of students over the last decade. These changes have come about for a variety of reasons, including:

(1) the changing demographics of the students taking college-level mathematics,

(2) the growth of technology and what it can provide for the teaching and learning of mathematics,

(3) the changing mathematical needs among the people who use mathematics.

All of these factors have major implications for what we teach, and how we teach it, both at the precalculus level and in all other mathematics offerings.

The student population

We first look at the changes in the student population. Table 8.1 shows total collegiate enrollments, in millions, in various years since 1955 and indicates how dramatic this demographic change has been—an almost six-fold increase in forty years.

Year	1955	1965	1970	1975	1980	1985	1990	1995	2000
Enrollment (in millions)	2.66	5.92	8.58	11.19	12.10	12.25	13.82	14.95	15.31

Table 1. Source: Statistical Abstracts of the U.S., Digest of Educational Statistics

In recent years, it has become commonplace to read and hear the media describe any situation involving rapid growth as exponential, whether or not that is an appropriate model. Using the regression features available on all graphing calculators and spreadsheets, we find that the exponential regression function associated with this data is $C(t) = 3.418(1.0358)^t$, where $t = 0$ corresponds to the year 1950. The base, or growth factor, of 1.0358 indicates that collegiate enrollment has been growing at an annual rate of about 3.6% over this time period. The associated correlation coefficient, $r = 0.9089$, indicates a high degree of correlation. This function is shown in Figure 1 superimposed over the data points. We see that it is not a particularly good fit because of the concave down pattern in the data compared to the concave up growth pattern of the exponential function.

The pattern in the data actually suggests a power function of the form $y = At^p$, with $0 < p < 1$ as a more appropriate choice. Again using the regression features of a calculator, we find that a power function that fits this data well is $D(t) = 0.7719t^{0.7859}$, where $t = 0$ also corresponds to the year 1950. This

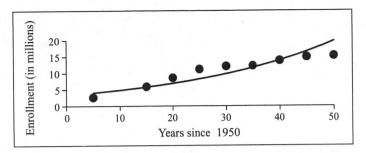

Figure 1. Exponential Model for Enrollment

function is shown in Figure 2 and we see that it is a much better match to the data than the exponential function in Figure 1. The corresponding correlation coefficient, $r = 0.9906$, indicates a very high level of correlation.

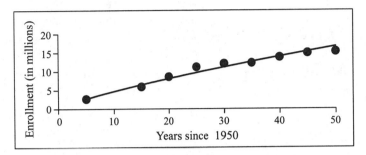

Figure 2. Power Function Model for Enrollment

The data in Table 8.2 shows the enrollment, in millions, in collegiate mathematics offerings (excluding statistics and computer science courses) since 1960.

Year	1960	1965	1970	1975	1980	1985	1990	1995	2000
Math Enrollment (in millions)	0.80	1.33	1.76	2.09	2.45	2.52	2.86	2.86	2.89

Table 2. Source: 2000 CBMS Study: Undergraduate Programs in the Mathematical Sciences

The concave down pattern in the data certainly suggests a power function model; we find that $N(t) = 0.1509t^{0.7898}$, where t is the number of years since 1950. The associated correlation coefficient is $r = 0.9752$. This power function for total mathematics enrollment, with power $p = 0.7898$, is growing slightly more rapidly than the power function for total collegiate enrollment where $p = 0.7859$. So, roughly the same proportion of students is taking mathematics courses, which can be interpreted as both good and bad news to the profession. However, if we just look at the two most recent entries from Tables 8.1 and 8.2, overall college enrollment is up 2.4% from 1995 to 2000, while math enrollment is up only 1%. So the short term trend is not a healthy situation for mathematics.

For comparison, the corresponding exponential regression function for this data is $M(t) = 0.8573 \times (1.029)^t$, where $t = 0$ corresponds to 1950. The growth factor 1.029 indicates an annual growth rate of about 2.9%. The corresponding correlation coefficient is $r = 0.9031$. Both the power function (solid) and the exponential function (dashed) are shown in Figure 3 and it is clear that the power function is a better fit to the pattern in the data.

Based on the two exponential models, we see that the growth in mathematics enrollment has proceeded considerably less rapidly (2.9%) than the growth in the overall college enrollment (3.6%). It is interesting to note that we come to the same conclusion about which process has grown faster; this is not always the case and often will depend on the model chosen.

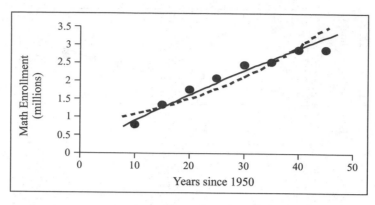

Figure 3.

Actually, the flattening in the data seen in the 1980–85 period (from $t = 30$ to $t = 35$) and again from 1990 to 2000 (from $t = 40$ to $t = 50$) suggests that a better fit might be achieved using a quartic polynomial. From a more sophisticated point of view, we should expect that this would be a much better fit because quartics are a five-parameter family of functions, while exponential and power functions are two-parameter families. The additional three parameters provide three extra degrees of freedom and so should lead to a significantly better fit to the data.

Next, Table 8.3 shows the total U.S. population, in millions, since 1950.

Year	1950	1960	1970	1980	1990	2000
Population	150.7	179.3	203.3	226.5	248.7	281.4

Table 3.

The associated exponential regression function for the U.S. population is $P(t) = 85.38e^{0.012t} = 85.38(1.0121)^t$, where t is the number of years since 1900. The growth factor 1.0121 indicates that the U.S. population has been growing at an annual rate of about 1.21% over the time period from 1950 to 2000. This is considerably slower than the rate of growth of either the entire college-level enrollment or the mathematics enrollment. The correlation coefficient $r = 0.9897$ indicates a very high level of positive correlation and, from the graph in Figure 4, we see that the exponential fit is quite good.

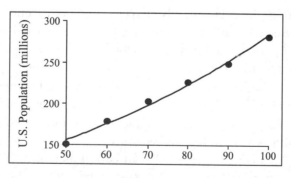

Figure 4.

There are several significant implications of the above information. Fifty, and even 30 years ago (when most of the current mathematics faculty were in school), the students coming to college represented a very small portion of the total U.S. population. From a traditional mathematical perspective, they were an elite group who had mastered a high level of proficiency in traditional high school mathematics, particularly algebraic manipulation. They entered college reasonably well prepared for the standard freshman course in calculus, which tended to have a strong algebraic focus.

More recently, as the cadre of college-bound students has increased dramatically, they can no longer be viewed as an elite group. Certainly, a comparable percentage of today's students are as good as the elite of the past, but these students likely attend the elite colleges. And, both today's elite students and the next tier of students have increasingly taken more sophisticated mathematics courses in high school. In 2004, about 225,000 students took the AP calculus exam; reportedly, about twice as many took AP calculus in high school, but did not take the AP exam. In addition, many other high school students took non-AP calculus—often polynomial calculus, though a growing number take International Baccalaureate courses and dual enrollment courses to get college credit for calculus at a local college. Together, this is greater than the number of students who take college calculus. Consequently, today's elite students are rarely seen in first year college calculus, let alone in precalculus or courses further down the collegiate mathematics sequence.

Furthermore, the number of students taking AP calculus has been increasing by about 8% for each of the last few years. This is considerably higher than the 2.9% rate of increase in college mathematics enrollment based on the exponential model. The implications of this fact are that collegiate calculus enrollments are not likely to increase much at all; if anything, they may continue the slight decline that has been observed over the last decade or so, as reported in the CBMS studies [1]. Consequently, we should expect, if anything, an increase in the proportion of students taking courses below the level of calculus in the foreseeable future.

Moreover, let's look at mathematics enrollments from a somewhat broader perspective. According to the 2000 Statistical Abstracts of the United States, 1,164,792 bachelor's degrees were awarded in 1996. Of these, only 13,143, or slightly over 1%, were in the mathematical sciences (which includes a large number in mathematics education). In the same year, 758 associate's degrees were awarded in mathematics out of a national total of 555,216 associate's degrees, which is on the order of one tenth of one percent. While we in the mathematics profession have a preoccupation with calculus in particular and the mathematics major in general, these offerings are only small potatoes at most colleges and universities. And by focusing on having these courses serve the needs of math majors, we tend to do a huge disservice to the mathematical needs of the overwhelming majority of the students we face.

Finally, an examination of the data in the CBMS survey [1] and other studies of mathematics enrollments in both high school and college show a dramatic drop-off from one year to the next and one course to the next. Historically, about 50% of the mathematics audience is lost each year in high school. The efforts of NCTM over the last decade to keep students enrolled in mathematics longer has improved these figures dramatically from algebra I to algebra II; the drop-off rate is now only about 15%. However, the 50% drop-off figure also applies to each semester in college. As several physicists have put it, "the half-life of math students is one semester." In an increasingly quantitative society, this should not be acceptable.

Technology and its implications for mathematics education

The student population that the majority of colleges face today consists predominantly of students who increasingly have not mastered traditional high school mathematics. In turn, ever greater proportions of students are being placed in remedial tracks designed to develop all the traditional algebraic skills that once were necessary for a traditional calculus course.

But, freshman calculus courses have been undergoing significant change in the last decade as a result of the calculus renewal movement. These reform calculus courses seek to achieve:

- a balance among graphical, numerical, and algebraic approaches (the Rule of Three),

- an emphasis on conceptual understanding rather than rote manipulation, and

- a focus on realistic applications from the point of view of mathematical modeling, often through an early introduction to differential equations.

Much of this is possible because of the availability of sophisticated technology, most commonly graphing calculators, although some schools make heavy use of computer software such as Derive, Maple or Mathematica with CAS (Computer Algebra System) capabilities. An analysis of the status of this movement is in [2] and [3]. A discussion of the challenges to be met in the forthcoming decade is in [4].

Technology has not stood still since the advent of the graphing calculator. The first generation of graphing calculators, such as the TI-81 and the TI-85, essentially provided the tools to implement the graphical aspect of the Rule of Three. The second generation, such as the TI-82, the TI-83, and the TI-86 and similar models from other manufacturers, provided additional tools to implement the numerical aspect through the use of lists, tables, and spreadsheet-like features. The newest generation, such as the TI-89 and the Casio CFX-9970G, now complete the triad by providing the CAS capability to perform algebraic operations such as FACTOR, EXPAND, SIMPLIFY, SOLVE, DIFFERENTIATE, and INTEGRATE at the push of a button. They can solve, in a fraction of a second, any purely manipulative problem that we would ever have expected our very best students to do. When CAS capability was available only on a computer, it could perhaps be ignored as being too inconvenient to require of all students, except possibly in some advanced courses with limited enrollment. But, given the availability and reasonable prices of these new hand-held tools, we must face the challenge of rethinking the content, as well as the long-term value to the students, of any mathematics course that continues to place the development of traditional manipulative skills as its raison d'être.

But, if the students in college-level math courses are using sophisticated technology to assist in learning and doing mathematics, the practitioners who actually apply mathematics in all quantitative fields are utilizing technology that is at least as powerful as what we have available in the classroom. And this trend will undoubtedly expand as the capabilities of technology grow and the array of problems encountered outside the classroom expands in their level of sophistication and complexity.

In reality, any routine operation that people use repeatedly has already been programmed. It therefore makes little sense to offer mathematics courses that focus primarily on making students into imperfect organic clones of a $150 graphing calculator with CAS capabilities! The students will never win the competition — they will never be as fast or as accurate as the machine. Instead, we should be focusing on the intellectual and applied aspects of the mathematics that the machines cannot do.

The challenge we face is to find a reasonable balance between the use of technology and the level of algebraic skill development that is essential for utilizing the technology wisely.

Changes in the mathematical needs of students

In a number of presentations—for instance, at the conference Confronting the Core Curriculum, which took place at West Point in 1994—Henry Pollak has made a strong issue of the changes that have taken place over the last half century in mathematics that is used in practice. Fifty years ago, virtually every problem was continuous and deterministic. Problems with a discrete or stochastic component were almost non-existent. Basically, algebraic methods and differential equations with closed form solutions ruled! Today, the tables have turned 180° — virtually every problem that arises is inherently discrete (in large part because of the digital age in which we live) and virtually every problem has some probabilistic component (there is always some uncertainty). But the mathematics curriculum, especially its first few years, has not changed appreciably to reflect the needs of the people who now use the mathematics.

The question we need to consider is: What should be the focus of mathematics education, especially at the precalculus level? I believe that the mathematical discussion at the beginning of this article is a large part of the answer. Very few people today, let alone in the future, will need to factor anything as complicated as $x^8 - y^8$. However, virtually any educated individual will need the ability:

1. to examine a set of data and recognize a behavioral pattern in it,
2. to assess how well a given functional model matches the data,

3. to recognize the limitations (often due to uncertainty) in the model,

4. to use the model to draw appropriate conclusions, and

5. to answer appropriate questions about the phenomenon being studied.

In turn, this process requires

- A deep understanding of the function concept, function notation, and the meaning of variable.

- A knowledge of different families of functions, including being able to distinguish between the different families graphically, numerically and algebraically.

- A knowledge of the behavior of the different families of functions depending on one or more parameters.

- The ability to select the appropriate tool, be it pencil-and-paper, graphing calculators, spreadsheets, or CAS system, to solve the equations that arise from using the models.

- The ability to interpret the mathematical results and to communicate these ideas to others.

All of these principles underlie the MAA's *CUPM Curriculum Guide 2004* [5], the NCTM *Standards* [6] and the AMATYC *Crossroads Standards* [7]. More extensive discussions of reform efforts in the courses that precede calculus are in [8], [9], and [10].

Let's see what these ideas mean in the context of precalculus courses and preparing students, not only for calculus, but also for courses in all other quantitative disciplines. Perhaps the best way to see this is from the perspective of some of our critics—the faculty in other disciplines who express recurring complaints about what mathematical skills and knowledge students bring with them to those other courses and the students themselves who vote with their feet to abandon mathematics in such relentless numbers.

In the past, the first mathematics course that appeared on the "radar screens" of the traditional quantitative disciplines (physics, chemistry, and engineering) was calculus. The introductory courses they offered were all calculus-based and so any course below calculus did not directly serve any of their needs. At most schools, these departments, especially physics and chemistry, now offer non-calculus-based versions of their introductory courses to much larger audiences than those who take the calculus-based courses. As a result, what students bring from precalculus and college algebra courses—and what they don't bring—is now a growing concern to the faculty in these other disciplines.

What then do students need to succeed in courses in other fields? They certainly need to know what a variable is, so that they can understand and use the formulas that arise. They need to know several fundamental classes of functions, most notably linear, exponential, logarithmic, and power functions. (Other than projectile motion, there are relatively few problems that lead to polynomials; can you think of any? Other than inverse proportions and inverse square laws, there are virtually no problems that lead to rational functions.)

Certainly, these are topics in standard precalculus and college algebra courses, but they tend to get buried in a much more extensive array of techniques for factoring polynomials and producing graphs of every possible type of rational function. Is this really necessary? Not for the other disciplines. What about for calculus? Let's see where our years-long development of rational expressions and rational functions ends up. In order to find closed-form solutions for a handful of differential equations, such as the logistic equation $y' = ay - by^2$, one usually applies the method of partial fractions. (Ironically, there are simple ways to avoid the use of partial fractions altogether using a clever substitution to transform the logistic differential equation into a simpler one that can be integrated easily without partial fractions; one can also utilize a CAS, if desired.)

But, to prepare for this, there is a heavy emphasis in traditional calculus II courses on integration using partial fractions—often all four exhaustive (and exhausting) cases. To prepare for this, calculus I courses often devote an inordinate amount of time to differentiating rational functions. To prepare for that, precalculus courses emphasize the behavior of all manner of rational functions and their graphs and occasionally even partial fraction decompositions. To prepare for that, college algebra courses emphasize the algebraic operations of adding, subtracting, multiplying, dividing, and reducing complex fractional expressions. Each of these is a hard algebraic technique that "separates the men from the boys". Is it any wonder that we see a 50% drop-off in mathematics enrollment with each subsequent course?

Is this what we want to do? If the techniques were so vital for success in subsequent courses (as was certainly true in the past) and if it was not possible to introduce what is needed on a timely basis in those subsequent courses to the small fraction of students who really need it, then a case can be made to include those topics. But these skills are no longer that important. Modern differential equations courses typically depend on computer software, including CAS systems to generate closed-form solutions, so that the qualitative behavior of solutions and their dependence on initial conditions has become a far more important aspect of those courses. Similarly, there is now a strong emphasis on mathematical modeling to demonstrate the power of differential equations to provide understanding of a wide variety of natural processes.

The reality is that our students will rarely, if ever, have to integrate those terrible differential equations by hand. So, do they really need all that algebraic preparation? And, if they don't need all of it, what else can and should we do with the resulting available time in all our courses from developmental algebra up through calculus?

The need for conceptual understanding

One of the most common complaints from faculty in other disciplines is that students do not know how to find the equation of a line. That is something we certainly teach, repeatedly, in every course in the curriculum. Just open any standard textbook from elementary algebra to precalculus and there are hundreds of problems that read: Find the equation of the line through the points $(1, 4)$ and $(5, 12)$. What more could the physicists, chemists, biologists, economists, etc. want?

Well, the problems that arise in their courses tend not to have just one digit, positive integers, for a start. The slope typically does not work out to be a one-digit integer or a simple fraction such as 1/2 or 1/3 at the worst. And, much more importantly, the faculty in the other disciplines tend not to give the students two simple points and tell them just to create the equation. They expect their students to make a connection between the mathematics and the context, so that the equation and its component terms provide insight into the situation. They also expect their students to use the equation to answer questions about the context. Shouldn't their students be able to do that based on what we teach them in traditional courses?

In a recent article by F.S. Gordon [11], the answer to that last question turns out to be a resounding No! As one part of an extensive study comparing student performance, success rates, and attitudes based on the type of precalculus course—reform with a modeling emphasis or traditional with an algebraic emphasis—the department posed a series of common questions of a purely algebraic nature on final exams for both precalculus groups. One of these common questions had a contextual flavor. The students were given values for the enrollment at a college in two different years and were asked to find the equation of the linear function through those points and to give an interpretation of the meaning of the slope of the line in the context. In both groups, virtually every student could calculate the slope and find the equation of the line. In the reform group, virtually every student could give a meaningful interpretation to the slope. But in the traditional group, only about one-third could give a meaningful interpretation! A third left that part of the question out altogether; a large number simply restated the formula for the slope in words—the change in y over the change in x—but did not interpret the value or the context.

As the author put it, "unless explicit attention is devoted to emphasizing the conceptual understanding of what the slope means, the majority of students are not able to create viable interpretations on their own. And, without that understanding, they are likely not able to apply the mathematics to realistic situations." She goes on to address the broader implications of this finding. "If students are unable to make their own connections with a concept as simple as that of the slope of a line (which they have undoubtedly encountered in previous mathematics courses), it is unlikely that they will be able to create meaningful interpretations and connections on their own for more sophisticated mathematical concepts."

We, and faculty in other disciplines, expect students to understand the significance of the base (growth or decay factor) in an exponential function. We expect them to comprehend what the parameters in a sinusoidal function tell about the phenomenon being modeled. We expect them to understand the significance of the derivative of a function and the significance of a definite integral. But, if students cannot create the connection between the slope of a line and its meaning in a context, it is clear that we should not expect them to create comparable connections of more sophisticated ideas on their own. It is our job to help them make those connections by emphasizing the meaning of the concepts, not just emphasizing the formulas to be memorized and applied by rote.

One of the main themes of the calculus reform movement is an increased emphasis on conceptual understanding of the fundamental mathematical ideas and methods, not just a focus on the development of manipulative skills. (Remember the old adage: You take calculus to learn algebra.) This same principle of stressing conceptual understanding must be applied in the courses below calculus as well. If nothing else, we want the students to be prepared for calculus intellectually, not just algebraically. If they have not developed the ability to understand mathematical concepts and to value the importance of that understanding before walking into a calculus class, they are not prepared for a modern course in calculus. Nor are they prepared for any associated quantitative course in any other discipline that uses calculus or precalculus or college algebra ideas.

In order to accomplish this, it is necessary to emphasize the importance of the concepts and this requires putting heavy emphasis on conceptual problems, as opposed to primarily computational problems. And this emphasis must be put both in homework assignments and on exams. Homework problems should not be only repetitions of worked examples in the text that serve as templates. Exam problems should not be only further repetitions of what the students have previously seen. When a final exam is just a compilation of problems from class tests with the numbers changed, and when the class tests are just a compilation of weekly quizzes, students are not being educated. They are being trained in the same way that Roy Rogers trained Trigger to answer such mathematical questions as: *How much is 2 plus 3?* The horse answered by tapping his hoof on command.

We owe it to our students to do much more for them—not just for calculus or for other courses, but to function effectively in a rapidly changing society where the one thing they can and should expect is more change over the course of their careers. Simply put, no one will pay our students $30,000 or more a year if all they can do is reproduce solutions to problems memorized in high school and college mathematics classes!

The need for realistic problems

Reform calculus courses also usually include more realistic, and hence more sophisticated, problems and applications than routine problems that tend to be highly artificial. This theme should also be carried over to college algebra and precalculus courses. Traditional algebra applications such as "Ann is 8 years older than Billy and in 5 years she will be twice what he was 4 years ago" are not in the least realistic. There is no way to pose such a problem without knowing the ages in advance, which makes the entire problem totally artificial.

But what then constitutes a realistic problem? Just as two points determine a line, two points also determine an exponential function or a power function; three points determine a quadratic function; and so forth. In any realistic context, one can find two data points—just open a newspaper, a magazine, a textbook in any other quantitative field, or a copy of the Statistical Abstracts of the U.S. or search the Web. Presuming that the process being studied follows a linear or an exponential or a power function pattern, ask the students to find the equation of that function and use the resulting equation to answer some predictive questions in context.

All the algebra that anyone could want, and then some, is imbedded in solving those questions. Because the functions are based on real values, not artificially concocted one-digit integer values, the parameters are almost certain to be unpleasant decimal values, so there is plenty of opportunity to practice one's skills. But that practice is done in a hopefully interesting context. We are not asking the students to solve equations for the sake of practice, but to answer questions that they can see make sense to ask. That makes an incredible difference in terms of convincing the students that they are learning something that is potentially valuable to them.

For instance, we created the exponential function $P(t) = 85.38(1.0121)^t$ to model the growth of the U.S. population, where t is the number of years since 1900. We could ask questions such as: Predict the U.S. population in 2005 using this model, or when will the U.S. population reach 350 million? The latter requires solving the equation $85.38(1.0121)^t = 350$, a rather more daunting request than solving something like $5(2^x) = 80$, but it is far more interesting and useful. Similarly, we created the power function $D(t) = 0.7345t^{0.8053}$ to model the growth in total college enrollment since $t = 0$ in 1950. We could then ask: Predict college enrollment in 2010, or when will there be 20 million people enrolled in college? The latter requires solving the equation $0.7345t^{0.8053} = 20$, which again is considerably more complicated, as well as considerably more meaningful, than the traditional type of problem such as solving $3x^4 = 48$.

The trigonometric functions can likewise be introduced using realistic situations. They serve as our primary mathematical models for periodic phenomena. For example, students can be asked to construct a sinusoidal function to model the temperature in a house where the furnace comes on when the temperature drops to 66°F and turns off when the temperature reaches 70°F, a cycle that repeats every 20 minutes. One possible result is

$$T = 68 + 2\sin\left(\frac{\pi}{10}t\right)$$

assuming that there is no phase shift.

As another example, students can be asked to model a person's blood pressure over time given readings of 120 over 80 and a pulse rate of 70. Again, each of these situations provides a wonderful opportunity to ask questions in context that go well beyond asking students to graph $y = 3\sin 4x$ or solving $6\sin 2x = 3$ in terms of both the level of interest and the level of algebraic manipulation involved.

Furthermore, every such realistic problem carries with it the opportunity to reinforce the fundamental mathematical concepts—the meaning of the slope of a line or the growth or decay rate of an exponential function or the vertical shift, amplitude, period, and frequency of a sinusoidal function, etc. It also gives an opportunity to discuss domain and range issues repeatedly—how far can you reasonably extrapolate from the data points? What are the limitations of the model?

Alternatively, when we assign a page full of exercises asking the students to solve a collection of 50 or 100 equations that all look the same with the numbers changed, we send a very different message. In reality, only a handful of the students ever bother to do more than a small number of these problems.

Other topics that should be emphasized

Another very common complaint from the other disciplines is that students do not have any understanding of or facility with exponents and logarithms. In partial response to this, the calculus reform projects have placed considerably more emphasis on exponential and logarithmic functions. They are no longer relegated to a chapter at the beginning of calculus II, but have been brought up front as some of the fundamental functions of mathematics. The same kind of emphasis is required in the courses preparatory to calculus, not just to prepare the students for the subsequent calculus experience, but perhaps even more importantly because these functions are so vital in all quantitative disciplines today.

In that regard, the treatment of these functions in college algebra and precalculus should not be just one stand-alone chapter and the functions never reappear. Instead, exponential functions and their properties, just like linear functions and their properties, should arise repeatedly in many different contexts throughout the course. If we want students to develop an appreciation for certain ideas, we have to give more emphasis to those concepts; if every topic or type of function receives equal attention, students do not learn what is important and, at best, make their own decisions of what they should learn for the long haul.

Furthermore, in the other disciplines, the various mathematical functions typically arise in the context of finding an appropriate function to model sets of data, just as they arose in the introduction to this article. There is a reason that these curve-fitting techniques are incorporated into all graphing calculators and spreadsheets such as Excel—they are the standard tools of today's practitioners, both in class and on the job. But, as I tried to demonstrate in the analysis of the various data sets earlier, it is not as simple a matter as just pushing a button to get an answer. Some very deep levels of understanding are essential. One has to know the behavioral characteristics of each family of function in order to make an intelligent selection of possible functions to use as models. There are some critical difficulties that can arise that are domain issues for these functions. For instance, the routines used by calculators and by spreadsheets to fit exponential, power, and logarithmic functions to data involve transformations of the data to plot $\log y$ versus x, or $\log x$ versus y, or $\log y$ versus $\log x$. (The first two are semi-log plots; the third is a log-log plot.) But, if any of the data entries is zero or negative, the logarithms are not defined. The error messages the systems give are not exactly self-explanatory; the person who pushes the button has to know the mathematics to understand the message and to know how to avoid the problems—say by performing a vertical or a horizontal shift in the data and then reversing the shift after the analysis is done.

These techniques and ideas are ideal ones to incorporate into college algebra and precalculus classes for a variety of reasons. First, they give the opportunity to reinforce the important characteristics about each family of function, so that the students see the ideas coming back again. Second, the students see how these functions arise in practical settings, which is a great motivation for topics that otherwise tend not to appear all that useful. Third, this gives us the opportunity to ask interesting, predictive questions in the contexts of the data, so that the students have even more occasions to practice their skills solving the resulting equations. Fourth, the students are being prepared for the specific kinds of applications that will arise in their other courses; in turn, this increases their level of appreciation for the mathematics course. That may not sound terribly important, but in the long run, it makes our courses far more important to the students. Instead of dropping out of mathematics, they are encouraged to continue to subsequent courses.

What can be removed

Clearly, there are many new topics and methods that I believe should be included in courses at this level. To do so, we have to find time. This means eliminating something.

Back when I studied trigonometry, we were all expected to know three fundamental laws: the Law of Sines, the Law of Cosines, and the Law of Tangents. These were not just theorems or formulas; they were *universal laws*! I've been asking mathematicians for the last several years about the Law of Tangents. So far, only one or two actually know it; a small percentage recall that there ever was such a law; and the overwhelming majority never heard of it. It is not that this law was repealed or that triangles stopped obeying it. The reality is that many topics, some of marginal significance, others that once were considered extremely important (otherwise the Law of Tangents would not have been called a law), have been removed from the curriculum in the past with seemingly minimal long-term impact.

We face the same decisions today. Some topics in the present syllabus have to be relinquished to make room for newer, more important topics. Over the last half century, as has been pointed out, the focus of mathematics in practice has changed dramatically and an incredible body of new mathematical ideas and techniques has been developed. In turn, we owe it to our students to at least acquaint them with some of these concepts—matrix algebra, probabilistic reasoning via simulation, recursion and difference equations, etc.—early in their mathematical experiences. Part of the need is to provide the students with a broader view of what mathematics is all about; more importantly, these are important mathematical techniques they will need for their other courses.

For example, in the 1930s, linear algebra was a graduate course; it gradually worked its way down to a junior-senior offering, then to a sophomore-level course, and today matrix methods and their applications are standard topics in modern high school mathematics. Once, the entire focus of algebra up through calculus was to prepare students for a traditional course in differential equations where they saw, for the first time, the power of the mathematics to create mathematical models and to formulate closed-form solutions. Today, the focus in most other disciplines is on difference equations instead of differential equations—they are conceptually easier, they are simpler to set up, and they are much easier to solve, numerically and graphically, with modern technology.

What then can be relinquished from precalculus and college algebra courses? I have already suggested downplaying the emphasis on extensive treatments of rational expressions and rational functions at all levels of the curriculum. Several other topics I would suggest eliminating are things like Descartes' rule of signs, the rational root theorem, and synthetic division. For a long time, when finding the roots of polynomials was a major undertaking, these were valuable tools of the mathematician and the practitioner. Today, many students have calculators with a FACTOR button. Every student has at his or her fingertips a calculator with a numerical root-finding routine built in, not to mention the ability to zoom in repeatedly either on the graph or in a table of values associated with any polynomial. Locating the real roots of a polynomial is no longer a challenge; it should not be a major emphasis in our courses. But understanding what the roots are and knowing how to use them intelligently is certainly still a critically important aspect of these courses.

In a similar vein, we have six trigonometric functions because, in the pre-technology centuries, it was much simpler to have tables available of all six possible ratios of the sides in right triangles to minimize hand computation. Today, computational issues are irrelevant. So, what valuable role do the secant, the cosecant, and the cotangent play? There are very few realistic problems that involve any of these functions and each of these problems can be solved quite easily by using only the sine, the cosine, and the tangent, along with several "new" identities:

$$1 + \tan^2 x = \frac{1}{\cos^2 x}$$

and

$$\frac{d(\tan x)}{dx} = \frac{1}{\cos^2 x}.$$

The Harvard calculus course, for instance, totally avoids the use of cotangent, secant, and cosecant without any loss; there is therefore little reason to bedevil students with them in precalculus and trigonometry courses. They take up an inordinate amount of time for virtually no gain. In fact, a number of other nations, including Russia, France, and Israel, apparently never mention these three functions in any of their mathematics courses and the mathematicians and scientists they produce never seem particularly handicapped by this loss.

Changes in pedagogy

So far, I have focused primarily on content issues. But giving any course incorporating the philosophies mentioned above necessitates some significant changes in pedagogy.

Let's first look at the dynamics when non-routine problems are assigned for homework. With purely algebraic manipulation problems, there is a clear correct answer. If a student raises his or her hand to question such a problem, it is almost always possible for an experienced instructor to glance at the problem and immediately anticipate the algebraic error that the student has made. One can just write out the solution on the board, point out the usual algebraic pitfall, and presume that the problem is finished.

This is not the case with conceptual problems. There may be an entire spectrum of possible, legitimate answers. There may be a variety of reasons that the student couldn't do the problem: a lack of understanding of what is being asked, some basic misinterpretations of fundamental ideas that are essential to the problem, or perhaps just thinking much too deeply into the situation. For example, I am reminded of one problem I gave on a test in calculus: *A cylindrical tank develops a leak at the bottom and the rate at which the liquid in the tank escapes is proportional to the height of liquid. Sketch a graph of the height of liquid as a function of time.* (Note that the assumption of the rate being proportional to the height is inaccurate.) Most of the students were able to solve this with little difficulty, but one student objected to the solution everyone else had—they had all assumed, as had I, that the tank was standing on end. He claimed, rightly, that no such tank would ever be so positioned; it would always be lying on its side, making the problem far, far more difficult. Incidentally, the graph he had drawn was correct for his interpretation.

As another example, a homework problem that I like to give very early in college algebra and precalculus involves a formula for the number of calories in a peanut butter sandwich: $C = 150 + 6P$, where C is the number of calories associated with P grams of peanut butter. I ask the students, among other things, to devise reasonable values for the domain and range of this function. The level of debate often grows quite heated as they come to grips with what domain and range really mean in a practical context. But these concepts take on a life of their own in the students' minds. It is no longer just a matter of looking for and avoiding places where you divide by zero or take the square root of a negative number!

When students are asked to do non-routine homework problems, the instructor must expect to devote considerably more class time to going over many of those problems than would be necessary if the emphasis is purely algebraic. The students have to be encouraged to ask questions—questions about the homework and questions about the concepts and methods being presented. The instructor has to be prepared for very different interpretations on the part of the students. This can lead to a very different classroom atmosphere, one where the instructor does less lecturing and the students assume a more active role in the learning process.

For instance, I recall one incident in college algebra when I was leading the class to compile a list of characteristics of cubic functions—how the number of turning points and the number of inflection points relate to the degree, the significance of the real and complex roots, etc. One student, who had an exceptionally poor opinion of her mathematical ability (a high school drop-out who had done miserably in all her previous algebra-oriented math courses) raised her hand to ask, somewhat tentatively: "*Is it true that every cubic is centered at its point of inflection?*" I asked her to explain what she had in mind, so the rest of the class could see what she was getting at. With her eyes half-closed as she envisioned the idea and with her hands moving in the air, she added: "*Well, if you start at the inflection point and move in both directions, don't you trace out the identical path both ways?*" This, from a student whose previous math teachers felt had no talent for mathematics!

There are some very different ways in which mathematical ability and talent can manifest itself. Traditionally, most of us have assessed students' ability purely on the basis of how quickly and accurately they could manipulate symbols. From that point of view, the best student any of us could ever have is a TI-89 calculator, as I said before. But mathematical ability means much more than that. At the professional level, we consider creativity and insight as the hallmarks of a good mathematician. The identical assessment should be placed on our students and we should provide them the opportunity to, and an environment in which they can, demonstrate those qualities.

Non-routine conceptual and realistic problems can also provide the opportunity to have the students work together in small groups using collaborative learning. Again, this changes the classroom dynamic considerably. For example, when developing the practical meaning of the parameters in a sinusoidal function, I hand out a set of data on historic high temperatures in Dallas every two weeks over the course of a year and ask the students to create a sinusoidal function based on the data. There have been occasions when I would assign this problem to groups of three or four students some 20 minutes before the end of class. Twenty-five or 30 minutes later, I would have to leave for another class while the students were all still sitting there, oblivious to the time, arguing over the problem and how to turn the numbers into amplitudes, vertical shifts, and phase shifts!

The fascinating thing about this project is that there are several different strategies that the students can develop for estimating the various parameters. One of the most memorable lines by a student in a written report based on this problem was: "*The next quantity to be determined is the frequency. This was deceptively simple.*" How often does a student in a precalculus course describe the frequency of a sine function as "*deceptively simple,*" particularly when the value he obtains for the frequency is .0172 (which is $2\pi/365$)?

I also recommend assigning individual or small group projects related to the mathematical content of a course. For example, I assign a project requiring students to find a set of data of interest to them and perform a complete analysis of it—finding the best linear, exponential, and power function to fit the data, and asking and answering pertinent questions (i.e., predictions) based on the context. Each student is required to write a formal project report. For instance, during a recent semester, a sample of the topics studied by the students in one of my precalculus classes included:

- The number of sexual harassment cases filed as a function of time.
- The likelihood of car crashes as a function of blood alcohol level.
- The growth of the prison population as a function of time.
- The time of high tide at a beach as a function of the day of the month.
- The amount of solid waste generated per person as a function of time.
- The time for water to come to a boil as a function of the volume of water.
- The size of the human cranium over time during the last three million years.
- The results of a serial dilution experiment in biology lab.
- The growth in the Dow-Jones average as a function of time.
- The Gini Index measuring the spread of rich versus poor in the population over time.
- The number of immigrants who entered the U.S. over time.
- The mean annual income as a function of the level of education.

In the process of writing such reports, the students must decide which variable is independent and which is dependent. They must come to grips with the practical meaning of domain and range as the limitations inherent in the model they are creating; again, it is not just avoiding division by zero, but rather a high level of mathematical judgment as they connect the mathematics to the real world. They must understand the practical meaning of the slope of a line, not just think of it as a ratio of the number of boxes in two directions. They must interpret the growth or decay rate of an exponential function in context.

Moreover, several of these data sets arose from topics the students encountered in other courses they were taking, particularly in the laboratory sciences. I cannot overstate the value to the students of seeing the direct link between the mathematical methods they were learning in their precalculus class and the use of these methods in their other courses. Even for those who found their data in a magazine article or a reference book, the ability to apply the mathematics to a topic of personal interest gave each student a feeling of ownership. And, once they have taken ownership that way, the battle to convince them that the mathematics is valuable, and worth working at, has been won.

Furthermore, one of the common themes that runs through the curricula discussions in all the quantitative disciplines is the need for students to develop writing and communication skills. Most of our courses are primarily service courses and we can help support this need in other fields by including a writing and communication component in our offerings. Other departments will certainly appreciate it. But more importantly, it also provides the opportunity for those students with strong verbal ability (and possibly relatively weak algebraic ability) to apply their forte to our subject. It also helps improve the writing skills of those who are weak in that area. (You may also want to find out what kind of support your school provides in the area of writing and direct some students to a writing lab, if they need help.)

Also, when students write about mathematics, it helps them develop a firmer grasp on the mathematical concepts and applications. At the same time, when they do not have a firm grasp on the ideas, this comes through loud and clear in a written report and allows the instructor to identify areas where the students are missing a concept or misinterpreting an idea. For instance, I recall one student in calculus II who consistently misused the word "interval" in a report; he clearly had no idea what an interval was, mathematically, and there was no way that I could possibly have observed that gap from tests.

Most importantly, though, the use of written project assignments or even the occasional written response to a homework or test problem makes the students internalize the meaning of the mathematics by having them create the verbal connections between the mathematics and their own understanding of it.

Acknowledgment: The work described in this article was supported by the Division of Undergraduate Education of the National Science Foundation under grants DUE-0089400 and DUE-0310123. However, the views expressed are not necessarily those of either the Foundation or the projects.

References

1. Lutzer, David J., James W. Maxwell, and Stephen B. Rodi, *2000 CBMS Survey: Undergraduate Programs in the Mathematical Sciences in the United States*, MAA Reports, Mathematical Association of America, Washington, DC, 2002.

2. Tucker, Alan and James Leitzel (eds.), *Assessing Calculus Reform Efforts*, MAA Reports, #6, Mathematical Association of America, Washington, DC, 1995.

3. Ganter, Susan, *Changing Calculus — A Report on Evaluation Efforts and National Impact from 1988–1998*, MAA Notes #56, Mathematical Association of America, Washington, DC, 2001.

4. Ganter, Susan (ed.), *Calculus Renewal: Issues for Undergraduate Mathematics Education in the Next Decade*, Plenum Publishing, New York, 2000.

5. Pollatsek, Harriet, et al, *Undergraduate Programs and Courses in the Mathematical Sciences: CUPM Curriculum Guide 2004*, Mathematical Association of America, Washington, DC, 2004.

6. *Principles and Standards for School Mathematics*, NCTM, Reston, VA, 2000.

7. Cohen, Don (ed.), *Crossroads in Mathematics: Standards for Introductory College Mathematics before Calculus*, AMATYC, Memphis, TN, 1995.

8. Solow, Anita (ed.), *Preparing for a New Calculus*, MAA Notes #36, Mathematical Association of America, Washington, DC, 1994.

9. Mays, Marilyn (ed.), *Crossroads in Mathematics: Programs Reflecting the Standards*, AMATYC, Memphis, TN, 1999.

10. Lenker, Susan (ed.), *Exemplary Programs in Introductory College Mathematics*, Annenberg/CPB Competition, MAA Notes #47, Mathematical Association of America, Washington, DC, 1998.

11. Gordon, Florence S., *What Does the Slope Mean?*, PRIMUS, vol. XI, 2001.

9

Preparing for Calculus and Preparing for Life

Bernard L. Madison
University of Arkansas

Introduction

Every teacher has been confronted with the question, "Where will I use this?" Very often in school and introductory college mathematics, the answer is, "In calculus." In fact, preparation for calculus has for several decades been a primary aim of grades 9–12 mathematics. Often this preparation extends to grades 13 or 14. Consequently, calculus has taken on extreme importance and extreme responsibility in US mathematics education. I will argue here that this importance and responsibility are inflated and serve neither calculus nor general education well. Neither students nor the general public see the value that calculus delivers and very few people ever have any use for calculus techniques. The focus on preparation for calculus has narrowed and hurried the mathematics curriculum leaving no time for appreciating or learning to use the mathematics that is taught before calculus. And use of mathematics has become essential in American life, requiring that the aims of grades11-14 mathematics be broadened to include preparation for life.

Calculus: An important role

Calculus is an enormously impressive body of knowledge, a major human intellectual achievement. Developed about 300 years ago, advances in science and engineering due to calculus are endless. The ideas of calculus have been expanded into the major area of mathematical research, analysis, and research results have proven useful in many fields, both within mathematics and in science and engineering. Calculus concepts — limit, derivative, integral, and approximation — are important in many areas of human activity, well beyond those of scientists and engineers. Interpretations of these concepts in terms of rates of change and accumulations of changing quantities are fundamental quantitative ideas that every person should know.

The power of calculus and its extensions are usually not apparent until one studies beyond the three-semester college calculus sequence. That sequence introduces the ideas of limit, derivative, integral, and approximation, first in one dimension and then in multiple dimensions in the third semester. Most of the time in this sequence is devoted to the methods of derivatives and integrals, and those methods are used almost exclusively in further study of mathematics, science or engineering. Consequently, full value of college calculus is realized by few except scientists and engineers, a group that constitutes less than 20% [9] of the baccalaureate degree recipients in US higher education. Understanding the concepts of calculus and how they are applicable to the world — needed by a broader population — is not accomplished by most calculus sequences, so the major general education value of calculus is missed. The proper role of the current calculus sequence is for educating future scientists and engineers. A proper role for calculus in general education requires courses based on concepts and their uses rather than on methodology.

Quantitative literacy: Need and responsibility

Over the past two decades, computers have transformed public discourse by generating piles of data and myriad analyses of these data. Ordinary citizens must deal with numbers and data every day. News media, advertisements, and government reports are filled with graphs and charts explaining medical reports, environmental issues, economic forecasts, educational results, and consumer protection. Most jobs use numbers and numerical tools. Investing, insurance, and taxes are immersed in data and rates of change. Political debates, laws, and court decisions abound in probabilistic decision making. Polling has become an influential and continuous activity. As has been said, the third R is no longer arithmetic. Understanding numbers and quantitative reasoning is a requirement for productive workers, responsible citizens, and discerning consumers. (See [4], [5], [10], [11], and [12].)

The term "quantitative literacy" (QL for short) is used to describe the quantitative reasoning capabilities required of citizens in today's information age. The demands of QL required in the information age have increased the need for postsecondary education in the US. Consequently, colleges and universities have a responsibility to see that levels of QL continue to rise as students move toward undergraduate degrees. This responsibility falls on all disciplines, but most heavily on mathematics and statistics. Mathematics is the only subject other than English that is required and tested in every grade K–10, and the geometry, algebra, trigonometry, and calculus (GATC) sequence contains some of the highest enrollment courses in all of school and college. Such a privileged place in the curriculum carries a heavy responsibility for general education, especially now that appropriate general education includes a high level of QL. Unfortunately, the GATC sequence is dominated by algebraic methodology in preparation for traditional calculus, and especially for traditional calculus that may be taught without using technology. Since QL is the ability to reason mathematically and quantitatively in contexts that arise in everyday activities, then the need for teaching mathematics in context seems clear. Contextual mathematics is rare in the GATC sequence. Examples of applications are often overly simplistic and contrived. There is little immediate relevance to students' everyday lives.

Currently, evidence points to low levels of QL, even among college graduates. In addition to many anecdotes, national and international surveys and examinations show low levels of QL among US adults. For example, in the 1992 National Adult Literacy Survey, more than half of college graduates placed at or below the literacy level partially characterized by being able to determine correct change using information on a menu but not being able to determine shipping and total costs on an order form for items in a catalog. Such persons will have many gaps in understanding quantitative issues that now confront citizens in everyday activities. Carnevale and Desrochers (see [11]) point to dangers of the current situation from both the workforce and citizenship views. First, from the workforce view, "we appear to have too many people who do not have enough basic mathematical literacy to make a decent living and many more people taking calculus than will ever actually use it on the job." Second, a more ominous warning, the "wall of ignorance between those who are mathematically and scientifically literate and those who aren't can threaten democratic cultures."

Because QL is a critical new challenge to US education, the National Council on Education and the Disciplines (NCED) at the Woodrow Wilson Foundation has initiated a national effort to improve QL education. The NCED effort has published *Mathematics and Democracy: The Case for Quantitative Literacy* [10] and commissioned several essays on various aspects of QL. This provided background for a national forum, *Quantitative Literacy: Why Numeracy Matters for Schools and Colleges*, at the National Academy of Sciences in December 2001. The commissioned papers, the proceedings of the forum, and recommendations for strengthened QL education are contained in two publications, [11] and [12].

Calculus: Domination and disappointment

The influence of calculus reaches down into the school curriculum and strongly affects mathematics as early as grades 8 and 9. Calculus has become the crowning achievement of successful school mathematics and the gateway in college to advanced study of science and mathematics. Courses in school and college carry the telling label of "precalculus," and the subtitle of the American Mathematical Association of Two-Year Colleges' *Crossroads in Mathematics* is "Standards for Introductory College Mathematics Before Calculus," [1]. Calculus is a destination benchmark and dominates mathematics in grades 11–14.

Enrollment patterns in college mathematics are very revealing about the lack of effectiveness and efficiency of grades 11–14 mathematics. In fall 2000 there were approximately 2.9 million enrollments [3] in two-year and four-year college mathematics courses. About one million were in remedial mathematics; about 1.1 million were in introductory college mathematics; 700,000 were in calculus-level courses; and 100,000 were in advanced mathematics courses. Since a similar pattern of enrollment has persisted for about two decades, one infers that about one of four students in college mathematics will ever get to any course in calculus, and about one in thirty will study mathematics beyond calculus. The 700,000 calculus-level enrollments include about 130,000 in what is referred to as "non-mainstream" calculus, mostly one-semester courses serving business majors. Excluding the 100,000 advanced enrollments and deducting the 130,000 "non-mainstream" enrollments leaves less than 600,000 of 2.8 million or about one in five in the "mainstream." Obviously, the main stream of college mathematics is not this "mainstream." The huge pool of students is in remedial and introductory courses, and most of this pool will never get to calculus of any kind. The largest pool — about 1.6 million, or more than half the total — is in courses whose content is mostly algebra.

This enrollment pattern indicates that the major purpose of college mathematics should be for general education of the three of four students, rather than preparation for calculus for the one of four. Unfortunately, courses dominated by algebra methods are not good general education courses.

The other revealing characteristic of college mathematics is the overlap in content with high school mathematics. The content of remedial mathematics courses is contained in school mathematics, as is most of the content of introductory college mathematics courses. Further, about half of US high schools offer courses in calculus. So about 60% of the enrollments in four-year college mathematics courses and 80% of those in two-year colleges are in courses whose content is covered in high school. If one includes the first course in calculus, the 60% becomes 77% and the 80% becomes 87%. There is little evidence that this repetition leads to enhanced understanding or better algebra skills. Quite to the contrary, the low success rates in college indicate that the repetition is counterproductive. Much of this repetition can be attributed to students not being able to demonstrate facility with algebra methods on college placement examinations. Because many college mathematics faculty believe such facility is necessary for entry into more advanced college mathematics, the students are enrolled in courses in algebra

Inefficiency and duplication are bad enough, but they are probably not the most negative effects of introductory college mathematics. Partly for efficiency, techniques of algebra and trigonometry have been gathered together into these courses. The logic of these techniques and their usefulness are not apparent to most students. Consequently, the material is uninspiring, and student interest is low. Success rates are low – often 50% or less – and many students develop antipathy toward mathematics. Calculus catches much of this distaste because "use in calculus" is often given as the justification for the unpopular content. Students who fall out and never get to calculus have fragmented algebra skills that they will likely never want to or be able to use. Students who do get to calculus will find more methodology and likely not realize the value of the power of calculus. Thus, calculus, which has been held responsible for years of preparatory study, often is either blamed by failing students for their failures or seen as a disappointment for not living up to its advanced billing.

Some college curricula — business and premedical, for examples — benefit from the GATC sequence because it filters out students and presumably selects those most likely to succeed. Mathematics and other college faculty recognize that mathematics is used as a filter, sometimes legitimately for acquiring necessary skills, but sometimes simply as a selection mechanism. Wide recognition of this use of mathematics as a fine filter strongly affects student and public attitudes toward mathematics, and makes it socially acceptable to be ignorant of mathematics. Now that ignorance of mathematics is incompatible with QL and that QL is essential for responsible living, it is no longer acceptable, social or otherwise, to be ignorant of mathematics. Therefore it is unacceptable to use critical education in mathematics as a filter. Higher levels of success in mathematics courses are required.

Calculus in high school

About a half-century ago, when the Advanced Placement (AP) Program of the College Board was describing its AP mathematics course, calculus was chosen as the topic. The belief was that the GAT part of the GATC sequence normally would be covered by the end of high school and the C would be the entry point for college. Even though AP calculus has grown to about 200,000 examinations now and about half of US high schools offer courses in calculus, algebra rather than calculus remains the dominant first college mathematics course. Nonetheless, calculus, the badge of achievement in high school mathematics, contributes to narrower and more hurried high school mathematics. Some believe that hurrying through GAT shortchanges both students' preparation for calculus and general education. Others believe that calculus in high school, including AP calculus, is not good preparation for subsequent college calculus courses. Comprehensive evidence to support these beliefs will be difficult to find, partly because the students in high school calculus are stronger students sifted out in high school in much the same way as students in calculus-level courses in college are sifted out by the GATC sequence. Various studies (e.g., [7]) have shown that successful AP calculus students do fare well in subsequent calculus courses in college, confirming the alignment of AP calculus standards and content with those of college calculus. Calculus in high school results from a compression of the GATC sequence which normally reaches into college, so criticism of high school calculus' effect on general education is also a criticism of college mathematics because that's what the high schools are emulating.

AP calculus enrollments are only a fraction of the calculus in high school. The annual 200,000 AP calculus examinations represent less than two-thirds of enrollments in courses labeled AP, and there are other calculus courses that are not labeled AP. Some of these courses are yearlong surveys of calculus techniques while others are partial year highlights courses. The Mathematical Association of America, [6] and [2], the National Council of Teachers of Mathematics [2], and the National Research Council [8] have recommended against such courses. Survey and highlights courses are not likely to contribute significantly to QL or to be good preparation for more rigorous college calculus courses.

If colleges were to provide gateways to advanced mathematics other than the engineering and science calculus sequence, then high schools and AP could also respond with alternatives. AP statistics, first offered in 1997, is an example of the AP Program creating a course not firmly entrenched in colleges. The phenomenal success of AP statistics (approximately 50,000 examinations in 2002) reflects the need for a more diversified offering of mathematical sciences courses at the transition from school to college. Modeling courses have several attractive features to consider: contextual use of mathematics; connections to other disciplines; and making connections among the strands of school mathematics.

Some directions for improvements

- The controversy over appropriate use of technology in calculus and in preparation for calculus needs to be resolved. Use of technology promises possibilities of relief from manipulation skills, better conceptual understanding, and more attention to authentic problems. Calculators and computers are ever-present in the world of students, including their school mathematics, yet there is not a clear consensus position among college mathematics faculty or departments.

- Many of the algebra and trigonometry methods that are used in calculus need to be incorporated into the calculus course and taught – or at least refreshed – when they are needed in the course. If the major justification for an algebra or trigonometry topic or method is its use in calculus, then it should be learned as part of that course rather than part of a course where the majority of students are not headed for calculus.

- High school mathematics should be faithful to balanced attention to the strands of number and operations, measurement, geometry, data analysis and probability, and algebra. Only if the standards are met in the broader curriculum should school mathematics venture into teaching calculus, and then that should be at the college level.

- Introductory college mathematics should emphasize the use of mathematics to solve realistic problems. In so doing, more algebra and trigonometry can be learned and the power of technology in promoting this learning can be realized. Wherever feasible, courses should open possibilities for study of additional mathematics and statistics.

References

1. American Mathematical Association of Two-Year Colleges (AMATYC), *Crossroads in Mathematics: Standards for College Mathematics Before Calculus*, Memphis, TN: AMATYC, 1995.

2. Dossey, John and Steen, Lynn A., Letter as Presidents of NCTM and MAA, 1985. Reprinted as Appendix B of the *Statement on Competencies in Mathematics Expected of Entering College Students*; Intersegmental Committee of the Academic Senates, California Community Colleges, California State University, and University of California; Sacramento, California, 1997.

3. Lutzer, David J., Maxwell, James W, and Rodi, Stephen B., *Statistical Abstract of Undergraduate Programs in the Mathematical Sciences in the United States*: *Fall 2000 CBMS Survey*. Providence, RI: American Mathematical Society, 2002.

4. Madison, Bernard L., "Quantitative Literacy: Everybody's Orphan," *Focus*, Vol. 21, No. 6, Washington, DC: Mathematical Association of America, 2001.

5. ——, "Quantitative Literacy: A Challenging Responsibility," *Notices of the AMS*, Vol. 49, No. 2, Providence, RI: American Mathematical Society, 2002.

6. Mathematical Association of America CUPM Panel, "Report of the CUPM Panel on Calculus Articulation: Problems in the Transition from High School Calculus to College Calculus," *The American Mathematical Monthly*, Vol. 94 (1987), pages 776–785.

7. Morgan, R. and Ramist, L., *Advanced Placement Students in College: An Investigation of Course Grades at 21 Colleges*, Unpublished Statistical Report No. SR-98-13. Princeton, NJ: Educational Testing Service, 1998.

8. National Research Council Committee on Programs for Advanced Study of Mathematics and Science in American High Schools, *Learning and Understanding: Improving Advanced Study of Mathematics and Science in U. S. High Schools*, Washington, DC: National Academy Press, 2002.

9. National Science Board, *Science and Engineering Indicators 2000*, Arlington, VA: National Science Foundation, 2000.

10. Steen, Lynn A. (Ed.), *Mathematics and Democracy: The Case for Quantitative Literacy*, Princeton, NJ: National Council on Education and the Disciplines, 2001.

11. Steen, Lynn A. and Madison, Bernard L. (Eds.), *Quantitative Literacy: Why Numeracy Matters for Schools and Colleges*, Princeton, NJ: National Council on Education and the Disciplines, 2003.

12. Steen, Lynn A. (Ed.), *Achieving Quantitative Literacy*, Washington, DC: The Mathematical Association of America, MAA Notes No. 62, 2004.

10

College Algebra: A Course in Crisis

Don Small
U.S. Military Academy

Introduction

This paper presents a case for transforming traditional college algebra from a failed program attempting to prepare students for calculus to one that enables students to address the needs of society, the workplace, and the quantitative aspects of disciplines. Characteristics of improved college algebra programs are described as well as the symbiotic relationship between a transformed college algebra and quantitative literacy.

Traditional college algebra

Traditional college algebra courses are *not working* [3]. That was the strong consensus of the participants in the Conference to Improve College Algebra, held at the U. S. Military Academy, February 7–10, 2002. This conclusion was based on the courses' outdated content, high FWD rates, and on the negative impact these courses have on student perceptions of mathematics. The large number of students enrolled in College Algebra creates an urgency to transform these courses into ones that *do work*.

Number of Students

College algebra has the largest enrollment (approximately 400,000 in fall 2000) of any college credit-bearing mathematics course [2]. (Another approximately 100,000 students are enrolled in combined college algebra/trigonometry courses.) This enrollment is approximately equal to the combined enrollment in all mainstream calculus courses, having increased from 73% in 1980. Precalculus has the second largest enrollment, which is about half that of college algebra. Almost all students are required to pass one of these courses, or a higher level mathematics course, as part of college distribution or major requirements. Some states, such as Texas and Louisiana, have legislated that students must pass a mathematics course at or above the level of college algebra before they are permitted to enter their third year of college. Thus college algebra and precalculus form the gateway to college mathematics for the large majority of students.

The combined enrollments in college algebra and precalculus increased 59% since 1980, while the enrollment in calculus I has remained relatively stable. This calls into question the traditional role of college algebra as a preparatory course for calculus. Commenting on this situation, Mercedes McGowen in her paper, "Redefining the First College-level Mathematics Course," which she presented at the Conference to Improve College Algebra, said:

A comparison of enrollments for all precalculus courses—particularly college algebra—with calculus I over the past twenty years indicates that the increasing enrollment in these courses has had little, if any, impact on the calculus I enrollments. There is a broad sense that the traditional college algebra course is not the appropriate course—particularly for those not going on in the hard sciences. In fact, it appears that large numbers of students do not have learning experiences in the college algebra courses which generate enthusiasm or enrollments in calculus I and a subsequent program of studies in mathematics or other math-intensive fields.

What then should be the role or roles of college algebra? If it is to be a stepping stone into calculus, then the traditional course must change as presently 10% or less of its students successfully complete calculus I. If college algebra is to be a service course in terms of satisfying distribution requirements, then it must change to emphasize applications across the disciplines. If it is to be a mathematics appreciation course, then it must change to emphasize student involvement in problem solving and discovery.

Content

The traditional content—factoring linear and quadratic polynomials, radicals, partial fractions, absolute values, inequalities, systems of equations, and so on—has its origins in the 1950s era when college algebra came into the curriculum as a capstone course for high school mathematics. This was long before the advent of graphing calculators or computer algebra systems and thus development of (hand) manipulation skills of algorithmic procedures was important for students going on to calculus. This thinking along with that era's emphasis on drill and symbol manipulation continues to shape traditional courses. Today, however, these courses are not successful in launching students into a standard calculus track. The largest cohort (approximately one-third) of students passing college algebra go into schools of business. However their need for elementary data analysis, modeling real-world problems, using technology, gaining small-group experience, and developing communication skills are not addressed in traditional college algebra courses. In spite of containing numerous exercises involving exponential functions, the traditional courses do not prepare social or life science majors to model growth situations or economic majors to model the multiplier effects of increased spending. Nor does the traditional content address citizenship or workplace needs such as the ability to interpret data, construct a budget, design a schedule, plan a multifaceted event, understand round off, or optimize a procedure. In the view of Arnold Packer, Chair of the SCANS 2000 Center at Johns Hopkins University's Institute for Policy Studies, "Interpreting data is more important than manipulation of algebraic skills that can be computerized."

FWD rates

The percentage of students who receive a grade of F or D or who withdraw from the course—are unacceptably high. Several studies place the FWD rate in the 40–60% range, although there are several schools in which this rate is considerably higher. In particular, one large urban community college system has a withdrawl rate of 50%. There are many contributing factors to the FWD rate—high school preparation, placement, content, attitude, pace of the course, pedagogy, out of school commitments, etc [1]. However, the fact remains that several alternative college algebra programs, which focus on real-world problem solving, have lowered the FWD rate by 15–25 percentage points. Students in these programs are drawn from the same pool and were subject to the same placement procedures as students in the traditional programs.

Attitudes

Negative attitudes, generated by high FWD rates, abstract content, and low expectations are pervasive among both faculty and students in traditional college algebra courses. These attitudes create additional barriers to the majority of students struggling to complete a course in which they see little relevance. Chris Arney, Dean of Science and Mathematics at St. Rose College, said:

> Traditional College Algebra is a boring, archaic, torturous course that does not help students solve problems or become better citizens. It turns off students and discourages them from seeking more mathematics learning.

Possibly more devastating is the fact that these attitudes influence life-long views of mathematics as College Algebra is the terminal mathematics course for the majority of students.

The urgency for creating alternative college algebra courses is heightened by the tremendous loss in student potential resulting from the traditional programs. Because of its gateway position in undergraduate programs, traditional college algebra courses block the academic opportunities and plans of approximately 200,000 students per semester. As educators, we can not accept this cost.

Vision for improved college algebra programs

The vision espoused at the Conference to Improve College Algebra was to create programs that empower all students to become competent and confident problem solvers. These real-world problem-based programs address the quantitative needs of other disciplines as well as those for citizenship and the workplace. The problem solving is to be understood in the sense of modeling as illustrated in the following diagram.

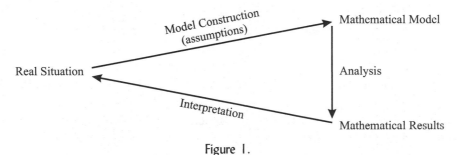

Figure 1.

The symbol manipulation type exercise that characterizes traditional courses only involves the right-hand side of the Modeling Process diagram. An example is to ask the student to factor the polynomial, $x^2 - x - 2$. The polynomial is the Mathematical Model and the factors, $x - 2$ and $x + 1$, are the Mathematical Result. This example is typical of the traditional program in that the model is provided and only the results are sought. In contrast, problem solving in an improved college algebra program begins and ends within a real-world setting. Students are expected to first create a mathematical model. This often involves a communication skill of transforming a written or verbal description into a mathematical description. The results are then obtained, often by means of technology. The final stage is to interpret the results in light of the real-world setting. This may result in modifying the model to obtain a more realistic result and/or "what-iffing" in order to gain a deeper understanding of the situation. The interpretation stage encourages conceptual abstraction that facilitates transferability to other (mathematically) similar situations. The soda can problem provides a nice illustration.

Soda can problem   Determine the dimensions of a 12 fluid ounce (355 ml) aluminum soda can in the shape of a closed cylinder that minimizes the amount of aluminum in the can. Use the following mathematical model:

Assume that the thickness of the aluminum is uniform, say one unit.
Let r be the radius of the can measured in centimeters.

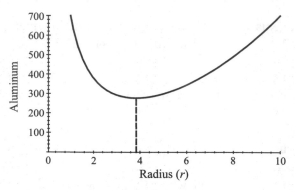

Figure 2. $\mathrm{alum}(r) = \frac{710}{r} + \pi r^2$

Let h be the height of the can measured in centimeters.
Conversion factor: 1 ml $= 1$ cm^3
Objective function: minimize amount of aluminum,

$$a(r, h) = 2\pi r h + 2\pi r^2$$

subject to the constraint equation: $355 = \pi r^2 h$.

Expressing h in terms of r in the constraint equation and then substituting for h in the objective function yields an *aluminum* function in terms of r : $\mathrm{alum}(r) = \frac{710}{r} + \pi r^2$.

We see from Figure 2, that the aluminum function achieves a minimum value when $r \approx 3.84$ cm and thus $h \approx 7.66$ cm. Interpretation of these results suggests the height of the can should be approximately equal to its width. Although this makes sense mathematically, it does not reflect the general shape of a commercial soda can. This calls for a rethinking of the model, starting with the assumption. An examination of a soda can suggest that the top and bottom are thicker than the sides. (Students are encouraged to go to a physics lab to measure the thicknesses of a commercial soda can.) Reworking the model with the assumption that the ends are twice as thick as the side yields a minimum for $r = 3.05$ cm and $h = 12.1$ cm. This is a reasonable approximation to the dimensions of a commercial soda can. Reflecting on the problem, students are asked to identify similar situations requiring the optimization of an objective function subject to constraints.

The *exercise* problem characteristic of traditional college algebra courses is contrasted with the *modeling* problem characteristic of improved college algebra courses as follows:

1. The *exercise* problem emphasizes algebraic manipulation while the *modeling* problem emphasizes conceptual understanding and realistic applications.

2. The *exercise* problem emphasizes solving in isolation while the *modeling* problem emphasizes solving in context.

3. The *exercise* problem lends itself to drill work while the *modeling* problem lends itself to inquiry.

4. The *exercise* problem remains in isolation while the *modeling* problem leads to conceptual abstraction and transferability.

Improved college algebra courses incorporate strong communication components—reading, writing, presenting, and listening. For instance, an objective of these courses is for students to be able to draw informed opinions from a news article containing data. The use of technology to enhance conceptual understanding as well as for computing is another strong component of these courses. For example, in the soda can problem, the value of the radius that minimizes the amount of aluminum was determined from the plot of the aluminum function. The pedagogy associated with improved college algebra courses recognizes that student experiences in constructing their own understanding are more important than coverage of topics. Thus lecturing is restricted to a minimum in order to maximize opportunities to engage students in activities and small group projects.

The major characteristics of an improved college algebra program include:

- Real-world problem-based: Introduce a topic through a real-world problem and then develop the mathematics necessary to solve the problem. Example: Schedule a multi-faceted process.

- Modeling: Transform a real-world problem into a mathematical construct by using power and exponential functions, systems of equations, graphs, and difference equations. Place primary emphasis on creation of a model and on interpretation of the results. Example: Model the stopping time versus speed data presented in a driver's manual by plotting the data and fitting a curve to the plot. The curve defines a stopping time function. Interpret how well this stopping time function models reality at small speeds. Revise the model, if necessary, to account for zero stopping time at zero speed. Use the resulting (revised) function to predict stopping times for speeds not given by the data. Revise the model to account for different road surfaces.

- Elementary data analysis: Display data, extract information from data, and extract knowledge from the information. Example: Nutrition labels on soda cans provide an example for extracting information from data. The label on a 12-ounce *Adirondack Ginger Ale* can lists 55 mg of sodium representing 2% of the recommended daily value based on a 2,000 calorie per day diet. In comparison the label on a 5.5-ounce Welch's Orange Juice can lists 15 mg of sodium representing 1% of the recommended daily value. What information concerning the recommended daily value of sodium can be extracted from this data?

- Communication: Emphasize communication skills as needed in society and the workplace as well as in academia—reading, writing, presenting, and listening. Example: Students learn how to read, understand, and critique news articles that include quantitative information and to make informed decisions based on the articles.

- Small group projects involving inquiry and inference: Provide experiences empowering students to become exploratory learners. Example: Analyze the soda preference of students by conducting a survey and comparing the results with data from the school's dining hall or a local fast food restaurant.

- Appropriate use of technology: Use technology to enhance conceptual understanding, visualization, and inquiry, as well as for computation. Example: Explore a model for paying off a credit card debt by changing the monthly payment, interest rate, size of debt, and so on. Plot the results to visually compare the different scenarios.

- Student-centered rather than instructor-centered pedagogy: Place the focus on student learning rather than on covering content. For example, maximize hands-on activities and minimize lecturing.

What you test is what you get—WYTIWYG or rather—what you test is what students focus on—WYTIWSFO. What does this say about assessment in real-world problem based courses designed to help students become competent and confident problem solvers? How can the seven aspects listed as characteristic of improved college algebra courses be assessed? These are very difficult questions to answer and answering them may be the biggest barrier to making curricular change. We can assign students to write an essay, but then how do we grade it? Do we grade grammar, spelling, number of words, and so on, or grade just on meaning? Do we feel competent to do this? Traditionally mathematics courses have addressed problem solving by focusing on well-defined, well-structured problems. However, many problems that arise in society or the workplace are ill defined and ill structured. For example, what are the three best predictors of success in college?

Traditional testing is not an adequate means of assessing student performance in an improved college algebra course. Traditional methods are teacher-centered in the sense of minimizing grading time, maximizing coverage, and focusing on well-defined, well-constructed skill type questions. In contrast, student-centered assessment needs to focus on process as well as results. It involves more subjectivity, more creativity, and more grading time than do the present traditional methods.

Collaboration with faculty in other disciplines and with representatives from the workplace is important to the improvement of college algebra and in on-going assessments of the programs. This collaboration is particularly important in the development or transformation of a course in order to ensure that content will align with student interests and needs. In addition, the collaboration establishes bridges to other disciplines that enhance opportunities for strengthening quantitative literacy throughout the academic program.

The past five years have seen the rise of a national movement to improve college algebra. This is evidenced by the large increase in the number of sessions at professional meetings devoted to improving college algebra and the number of college algebra workshops and conferences such as the Conference to Improve College Algebra held at West Point in 2002. The growth in the number of elementary modeling courses being offered as alternatives to college algebra as well as texts being written for improved college algebra courses are other indications of this national movement. Four samples of such texts are listed in the references [5]. An overview of each of these texts, by the author, appears later in this volume.

Improved college algebra: A base for quantitative literacy programs

Improved college algebra courses provide the focus, content, and interdisciplinary aspects on which to establish college-wide quantitative literacy programs. Lynn Steen describes "quantitative literacy," sometimes called "numeracy," as the quantitative reasoning capabilities required of citizens in today's information age [4]. Extended experience in problem solving in the modeling sense provides these quantitative reasoning capabilities. Thus a quantitative literacy program involves a two-step process.

1. Students understand the problem-solving process of transforming a problem situation into a mathematical description (model), solving, and then interpreting the results; and students develop confidence in their ability to apply this modeling process.
2. Students gain extensive practice in problem solving in a variety of situations.

With the emphasis on modeling, communication, and appropriate use of technology, improved college algebra courses provide the ideal opportunity for students to address the first step. Development of student self-confidence is facilitated by applying the problem-solving process in a variety of settings under the mentoring environment of a first year course. The second step, practice, needs to be addressed by all faculty so that students continue to hone their problem solving skills and strengthen their self-confidence as they progress through their academic careers. The interdisciplinary collaboration associated with improved college algebra courses facilitates this joint approach to developing students' quantitative capabilities.

An emphasis on quantitative literacy throughout the curriculum adds meaning and purpose to college algebra courses. In particular, attention to problem solving in the modeling sense throughout the curriculum serves as a laboratory for college algebra and thus extends these courses into programs. Thus the symbiotic relationship between college algebra and quantitative literacy is to the betterment of both programs.

Summary

As the primary college gateway course for thousands of students, college algebra has a great potential as a service course to address the quantitative needs of society, the workplace, and other disciplines. However traditional college algebra, based on a 1950s curriculum as part of the preparation sequence for calculus, does not fulfill this potential. To do so, college algebra needs to be transformed by refocusing both the content and pedagogy in order to develop competent and confident problem-solvers. The content needs to be real-world problem-based, to emphasize problem solving in the modeling sense, and to include elementary data analysis. Student-centered pedagogy involving development of communication skills, appropriate use of technology, and small group activities and projects should be designed to create student confidence and positive experiences. Student-centered rather than instructor-centered assessment procedures need to be developed.

The gateway function of college algebra means that transformed or improved courses provide a basis on which to develop a college-wide quantitative literacy program. The interdisciplinary collaboration that is important to the development and ongoing assessment of an improved college algebra course provides opportunities to link problem solving to the quantitative needs of other disciplines. In this sense, these disciplines provide laboratory experiences for college algebra students. Interdisciplinary collaboration is essential to realize the potential of the symbiotic relationship between college algebra and quantitative literacy programs.

Improved college algebra courses better serve the approximately 90% of students who do not enter into math-intensive programs as well as providing a more effective preparation for those going on to calculus I than do the traditional algorithmic type courses. Several improved college algebra courses have shown that FWD rates can be significantly lowered and positive student attitudes obtained. More importantly these courses have demonstrated, in a variety of schools, ways in which college algebra can be transformed from extracting an unacceptable cost to providing a valuable asset in educational programs.

References

1. Herriott, Scott R., "Changes in College Algebra," paper in this volume.

2. McGowen, Mercedes A., "Redefining the First College-level Mathematics Course," paper presented at the U.S. Military Academy Conference to Improve College Algebra, 2002.

3. Small, Don, "An Urgent Call to Improve Traditional College Algebra Programs," MAA *Focus*, May/June 2002 issue. (Summary of the Conference to Improve College Algebra held at the U.S. Military Academy, February 7–10, 2002.)

4. Steen, Lynn A., *Mathematics and Democracy: The Case for Quantitative Literacy*, the Woodrow Wilson National Fellowship Foundation, Washington, DC, 2001.

5. Samples of texts written for improved college algebra programs:

   • Crauder, Bruce, Benny Evans and Alan Noell, *Functions and Change: A Modeling Alternative to College Algebra*, 2nd edition, Houghton-Mifflin, Boston, 2003.

   • Herriott, Scott R., *College Algebra through Functions and Models*, Brooks-Cole, Pacific Grove, CA, 2005.

   • Kime, Linda A., Judy Clark, and Beverly Michaels, *Explorations in College Algebra*, 3rd ed., Wiley, New York, 2005.

   • Small, Don, *Contemporary College Algebra: Data, Functions, Modeling*, 5th ed. updated, McGraw-Hill, New York, 2005.

11

Changes in College Algebra

Scott R. Herriott
Maharishi University of Management

Introduction

The high rate of students' failure in the college algebra course could be a problem originating in admissions policies, placement activities, curriculum design, or instruction. This paper focuses on the curriculum as a partial solution to this problem, in light of the needs of the types of students who tend to enroll in the course.

The traditional college algebra curriculum seems to assume that the course is a preparation for calculus. However, surveys at many institutions have shown that only a minority of college algebra students go on to take calculus of any kind, and only a small fraction of those attempt a full-year calculus sequence. In the mid-1990's, the National Science Foundation sponsored a colloquium to examine the nature and purpose of the college algebra course [17,18], and in recent years, some textbook authors have developed alternative curricula that address the future mathematical needs of the soft-sciences students. This paper contrasts these new curricula with the traditional college algebra course and identifies the related issues of national and local educational policy.

The DWF problem

The 1997 MAA panel discussion on college algebra reform had one participant who was certainly an unusual guest at such meetings. He was a university president, and not himself a mathematician. Raymond Hicks, of Grambling State University, told the audience of his administration's concern about the high failure rate of college algebra students. It is not only bad public relations for the math department, he reminded the attendees, but it is demoralizing to the students.

The college algebra course has a reputation nationally for failing an unusually high percentage of students. The percentage of students who earn D, W or F grades—the DWF rate—may be 40–50% on average across colleges, and some institutions report that it is as high as 90%. A study conducted for a two-year institution in Georgia found DWF rates of only 25–31% for the freshman courses in English composition, psychology and government. The DWF rate was 47% among the 4,400 students in the mathematical modeling course that takes the place of college algebra there, despite the fact that the distributions of mathematical requirements implied by the majors declared by students in these four freshman courses were nearly identical.

In business and industry, a product defect rate of 40–50% would quickly cause some vice-president to lose his or her job, but the DWF problem of the college algebra course has persisted in American higher education. As President Hicks told his audience at the MAA meetings, this problem demands a solution.

In reality, the problem is likely to have its origin in more than one stage of the educational process. It could be a problem of **placement,** especially at institutions that do not use placement tests.

But it can also be a problem at schools that use either high-school exit exams or standardized college placement tests that do not mesh well, either with the high school curriculum or with the college's curriculum. Furthermore, many students want to get out of the requirement to take college algebra, so there can be a problem at schools whose computer systems cannot enforce prerequisites or block prohibited requests to change a registration. The placement issue is more serious at schools whose admission is open or nearly so. An open admissions policy can promote a general attitude among the faculty and administration that all students deserve the right to try, but that not many are really suited to succeed. Thus, the institution's faculty may pretend that failure is the fault of the student.

The DWF problem may also have its origin in **curriculum design**. Some college algebra courses are designed as a preparation for calculus on the assumption that most, if not all, students will need to take calculus for their majors. At such an institution, if the course is attended by students headed toward majors that do not require calculus, there will probably be a higher failure rate among the students in the less mathematically intensive majors. Another two-year college in Georgia found this in a study of the DWF rate among students of various majors who were taking college algebra. The overall DWF rate was 49% among more than 2,300 students. The largest groups of majors (business, education, and undeclared) had DWF rates very close to this average. But the 108 majors in communication arts had a DWF rate of 59%. There were smaller groups in other majors, such as sports management, English, building construction, interior design, hotel administration, sociology, child development, and recreation, where students had DWF rates in excess of 60%.

Even a well-designed curriculum suited to the composition of the class can be poorly implemented by the instructor, so **instruction** is one possible cause, especially at schools that hire current or former high school teachers as college algebra instructors. The college algebra course *should not* be just a louder version of high school algebra [22]. It is significant that in the study of the two-year college students in Georgia, the group of students in the college algebra class that had far and away the lowest DWF rate (17%) was that of high school students taking the course for college credit. The typical student takes college algebra in college precisely because he or she did poorly in algebra II in high school. Typically across the U.S., this student will be motivated by examples from business, health sciences, social sciences, and education, but less so by examples from the sciences or the contrived examples from personal life that tend to appear in high school texts.

As a footnote to the problem of instruction, we should remind ourselves that assessment is a closely related issue. If our examinations do not mesh with the course's content, its type of applications, and its level of abstraction, then students may still appear to do poorly.

In summary, placement is probably a major contributor to the problem, but its solution centers greatly on an institution coming to grips with its own mission and defining its responsibility to students. We may hope that instruction is not the core of the problem, though faculty development certainly deserves attention at all institutions. In this paper, we focus on the curriculum as the center of attention in reforming or renewing the college algebra course. The basic principle of curriculum design is that the curiculum should be suited to the needs of the student, so we now consider the types of students who take college algebra.

Who takes college algebra?

Dunbar and Herriott [10] cite data on the intended majors of college algebra students at the University of Nebraska at Lincoln and at ten colleges and universities in Illinois. It is not unusual for an institution to have 40% of its college algebra students headed toward a business major, and at urban locations it may be over 50%. Students from the life and allied health sciences tend to comprise 20% of the class, and those in the social sciences about 15%. Education and humanities majors may be another 10%, and the remaining 10–15% will be planning a major in the sciences, math, engineering, or computer science.

Thus, the vast majority of students in college algebra are headed toward the managerial, social or life/health sciences. It is the mathematics needed in these subjects and the needs of these students that should guide the curriculum for college algebra.

Curricular models for college algebra

The curricula, taught under the name college algebra, can be dramatically different at different institutions, yet overall there is a fairly consistent pattern in the type of student who takes the course. Many of the students have not fared well in prior mathematics courses and have a fear of mathematics.

In this section, we consider several distinct models for a college algebra course, as expressed in the content, organization, and pedagogy of textbooks. In doing so, we will see an emphasis on the extreme examples of each type. Certainly, there are curricular models that mix elements of these, but characterizing the extremes helps us understand the choices that the curriculum designers have made.

The *sine qua non* of a college algebra course is the solution of equations, as one would expect of a field in which the fundamental theorem concerns the existence of a solution to an important class of equations. However, this still leaves much room for the achievement of other mathematical and educational objectives, as these examples show.

Traditional college algebra

Purpose and clientele   The traditional college algebra textbook covers the key concepts of high school algebra in the context of studying functions and their graphs. This context and the selection of functional forms and other topics for coverage is designed to prepare the student for calculus. Indeed, the preparation for calculus may well be a guiding feature of this curriculum. As Sobel and Lerner [26] point out in their introduction to the instructor's annotated edition: "Since calculus is a subject numerous students study after this course, special emphasis is given to the preparation for the study of calculus. Thus, one of the *major objectives* of this book is *to help the student make a comfortable transition from elementary mathematics to calculus*[italics in original]." The text by Sullivan and Sullivan does not make that assumption [28, p. xii], but its content still reflects a preparation for calculus.

Other examples of textbooks along this model are Larson and Hostetler [19], Lial, Hornsby and Schneider [20], Ruud and Shell [24], Bittinger et al. [3], Cohen [5], Dugopolski [9], and Stewart, Redlin and Watson [27].

Content   The course covers the analysis of linear, polynomial, exponential and logarithmic, radical, and rational functions through the study of their graphs and the algebraic methods of solving their corresponding equations. Systems of equations are solved by matrix inversion. Systems of linear inequalities are studied as a context for linear programming problems, which are solved by the graphical method. Sequences are covered near the end of the course as a way to introduce the concept of limit. The text often includes a unit on conic sections, and it may include a chapter on elementary probability theory.

The solution of polynomial equations is based on the method of factorization. The study of rational expressions frequently includes partial fraction decomposition, which is necessary in calculus for the integration of rational expressions. The complex number system is likely to be discussed in the context of solving polynomial equations. Systems of linear equations are solved first by substitution and then by matrix methods.

Organization   The first thing a student sees when opening the textbook, on the inside cover, is a page or two of equations in small type with lots of exponents and parentheses. In addition to serving as a handy reference, this probably brings back memories of the student's high school algebra experience. Those memories are reinforced in the first chapter, which as often as not, is given the demeaning label Chapter R rather than Chapter 1, where the student receives a quick review of the properties of the real number system, exponential notation, operations on polynomials, factorization of polynomials, rational and radical equations, and perhaps complex numbers as a way to get started in the course.

The first substantive part of the text is usually a pair of chapters dealing with functions and graphs and with linear and nonlinear equations, in either order. Following those in fairly consistent order are chapters on polynomial and rational functions, exponential functions, systems of equations and matrices, conic sections, sequences and series, and sometimes finishing with permutations and combinations or probability.

Pedagogy   Applications may appear in the text to motivate the student's interest, but the vast majority of the homework exercises concern symbolic manipulation, not the formulation of an equation. Thus, problem-solving and critical thinking skills are not emphasized, nor are communication skills.

The applications tend to be drawn from the hard sciences or geometry rather than the soft sciences. Interdisciplinary applications are limited, and applications tend to be shallow and quick. One textbook listed a problem concerning the sum of the digits in the year that the transcontinental railroad was completed as an application in the field of "transportation."

The curriculum teaches students to work with functions under their numerical, graphical and symbolic representations, the Rule of Three, but it does not emphasize the verbal representation of functional relationships nor the process of modeling functions from written descriptions.

Pedagogy tends to emphasize the traditional lecture and makes little use of technology.

College algebra with trigonometry

This curriculum is the traditional college algebra course together with the study of trigonometric functions (cf. Aufman, Barker, Nation, [2]). Many textbooks teach precisely this curriculum under the title "precalculus." Including the trig, this course is clearly designed as a preparation for calculus. Its review of high school algebra and of the traditional college algebra's functions and graphs is accelerated in order to fit in the trigonometry.

At the University of Nebraska at Lincoln, the stated prerequisite for calculus I is either trigonometry or college algebra and trigonometry. In contrast, the prerequisite to the one-term calculus for the managerial and social sciences is college algebra. Not surprisingly, an unpublished study of mathematics enrollments at UNL found that of all the students who enrolled in college algebra and trigonometry in 1993-94, 59% took calculus I within the next five semesters. Only 5% of the students in the college algebra and trigonometry course took the soft calculus course for business and the social sciences. These data show that college algebra with trigonometry is genuinely pre-calculus.

Models of reformed college algebra

Purpose and clientele   Efforts at reforming the college algebra course in the last 5–10 years have the same spirit as the earlier movement to reform the teaching of calculus. As with the calculus movement the following hold for the college algebra movement:

- The DWF rates indicate a serious problem. They show that the course is not fulfilling an educational purpose.
- The course has been used as a filter, when it should be used as a pump.

A tenet of the reform movement is that the course should be designed to serve the needs of the students who take it, so mathematics faculty must understand which other departments in their university or in their transfer schools expect the students to have taken this course. And they must understand specifically which aspects of college algebra are used in the various majors that require it.

College algebra is frequently required of students who major in business, the social sciences, biology and health sciences, and education. Many state licensing boards require nurses to take college algebra. The students who take algebra at college typically did poorly in the subject in high school, or as adults have not seen the subject for many years, and the content and organization of the reformed college algebra curriculum is designed for these students.

Content   The curricular themes of college algebra reform are expressed well in the AMATYC *Crossroads* standard [6]. The main goals for **intellectual development** concern transferable problem-solving skills such as exploration, modeling, inductive and deductive reasoning, and tenacity as well as technical skills of a general nature including communication and the use of technology. In **content**, the theme of reform is that problem solving is the heart of doing mathematics and that students gain the power to solve *meaningful* problems through in-depth study of specific mathematical topics. But depth does not necessarily refer to theoretical depth, such as study of the complex number system along with quadratic equations. It can mean an extensive exposure to a range of applications of a single functional form, so the student learns to see the pattern that identifies a linear relationship, an exponential relationship, and so on.

The orientation toward meaningful applications is a key theme in the content of the reformed curriculum. In the Algebra Initiative Colloquium, Davis [8] commented, "One could argue that no concept of algebra should be taught unless it can be motivated by a problem that is likely to be part of the students' experience in the near future." Many topics in the traditional curriculum would have difficulty passing this test.

The concept of function is essential to modeling relationships among real-world phenomena. Reform emphasizes the Rule of Four—understanding functions verbally, numerically, graphically, and symbolically—along with the ability to transform functional relationships from one representation to another, which is the essence of mathematical modeling. However, it is possible to define a function contextually, as a relationship between variables x and y, rather than abstractly as a rule of transformation f.

The emphasis on ideas in context, rather than ideas in the abstract, seems to characterize the main efforts at reform. In the Algebra Initiative Colloquium, Artin [1] commented:

> For an undergraduate course, the most important thing the students should come out with is a familiarity with some examples—some basic structures on which they can build their understanding. That is more important than theory.

> We should be ruthless in asking: 'Is it important for the average student in the class to learn this material?' If not, throw it out.

Given that the average community college student is 27 years old, Pollack [22] had a particularly salient observation, "An algebra experience centered around the usefulness of the subject may succeed where previous attempts at rote learning did not, especially with an older student."

Davis, Artin and Pollack all reinforce the notion that the design of the college algebra curriculum should reflect a keen understanding of who the students are and what their needs are. In the reformed curriculum, each topic is evaluated according to its usefulness to the student and for the transferable skills that their study provides.

Organization   When opening the textbook of a reformed course, on the inside cover the student sees a page or two listing the areas of application of the material covered in the book, not a mass of equations in small type with exponents and parentheses. This sends a signal to the student that the textbook will have practical value.

From the reformed perspective, the traditional textbook's initial Chapter R (named without even the dignity of a chapter number) that covers equations, exponents, factoring, rational and radical expressions, and complex numbers as a review is a sure formula to instill mathematics shock in the typical college algebra student. In contrast, the first part of the reformed curriculum recognizes that the college algebra student may have been out of school for a while or did poorly with algebra in the last encounter. The initial reformed chapter covers linear functions through the study of linear equations and their graphs, in various forms corresponding to the various natural language descriptions of linear relationships. Significantly, systems of linear equations and inequalities follows this *immediately*. Linear systems are likely to be solved by substitution and Gaussian elimination but not by matrix methods. Matrix algebra, in the reformed view, belongs in a linear algebra course.

Following the study of linear functions, the reformed curriculum treats exponential functions, not polynomials. There are several reasons for this organization of the curriculum, centering on the ease of transition from linear to exponential functions. These functions each have two parameters, one of which is the vertical intercept. They each describe growth or decline, though in different ways. And the solution of an exponential equation by logarithms results in a linear equation that the student knows how to solve. Exponential equations may be studied in depth through applications of the various forms (standard, base two and natural base) that correspond to the various natural language descriptions of exponential relationships. Variations on the exponential theme may include vertical shifts (Newton's model of heating and cooling) and the reciprocal of that, the logistic function.

Following the study of exponential relationships comes polynomial functions, with emphasis on the quadratic. Complex numbers are typically introduced in a limited manner. Factorization is treated as a property of the polynomial, but not given much emphasis as a solution procedure for polynomial equations. The curriculum may emphasize the formulation of quadratic equations from real-world descriptions or from graphs or tables of data. The solution of the equation by the quadratic formula is treated as a useful, generic technique.

Sequences, arithmetic and geometric, may appear optionally in the curriculum through the study of the general first-order difference equation, which subsumes linear and exponential functions defined over the whole numbers and thereby helps integrate the curriculum [12, 14, 25].

Pedagogy   The reformed curriculum of college algebra tends to treat the *formulation* of an equation as an intellectual skill no less important than the *solution* of the equation. That is a significant departure from the traditional curriculum. Faculty schooled in the traditional curriculum probably reply, "That's not algebra. Algebra is about the solution of equations." True enough, but the purpose of the curriculum is to do more than teach algebraic manipulations. The AMATYC *Crossroads* standard describes a broad range of intellectual skills that should be developed in the subcalculus curriculum. The solution of equations is only one step in the more general objective of *problem solving*. Kenschaft [15] reports the experience of a mathematics alum who wrote in a survey, "Business is one 'word problem' after another." The task of taking a verbal representation and formulating it symbolically is an intellectual skill of high order and one that students will need in their careers. But it is harder to teach than the manipulation of algebraic expressions.

A consistent pedagogical theme in the reformed courses is this transformation of one representation into another. Transforming a verbal, numerical, graphical or symbolic representation of a relationship between variables into one of the other forms is the essence of mathematical modeling. Some curricula, such as Herriott [13] place relatively more emphasis on the verbal-to-symbolic transformation, reading a description and writing an equation. Others such as Gordon, et al. [12], Kime and Clark [16], and Crauder, Evans and Noell [7] and Small [25] give special attention to the numerical-to-symbolic transformation, fitting equations to data (see papers by the authors, later in this volume, for overviews of these texts). Rockswold, Hornsby and Lial [23] introduce this idea in their adaptation of a traditional curricular organization.

The mere use of technology does not tend to distinguish a reformed course from the traditional curriculum, because even traditional courses are using graphing calculators. But the reformed courses tend to use technology, including spreadsheet software, more intensively with applications.

Does reformed college algebra still prepare students for calculus? It is often said, "Student's don't fail at the calculus. They fail at the algebra." It is a reasonable guess that about 10% of the college algebra students nationally will go on to attempt the first semester of a year-long calculus sequence, and about 35% will take a one-semester course in calculus for the social, managerial and life sciences [10]. The qualities that students need to succeed in either of these courses are an understanding of functions and graphs, the ability to solve equations, and confidence in their ability to learn math. The reformed curricula tend to give a thorough graphical treatment of functions, including a discussion of end behavior and turning points. They give plenty of practice in solving equations. And their content and organization are designed to encourage the student's confidence in the ability to work with abstractions. It is a reasonable guess that such students will be equally prepared to succeed in a calculus course that emphasizes applications in the soft sciences. But students who attempt the full-year calculus sequence will need trigonometry, which is not usually in the reformed algebra curriculum.

College algebra as a general education course

In states such as Texas, Louisiana, and Georgia, college algebra or a near substitute is required of all students as general education. There may not be a single curricular design that serves well the future needs of the English and fine arts majors and those in the managerial, social and life/health sciences. Thus, adaptations of the college algebra curriculum to suit a general education objective tend to be radical departures from the traditional curriculum.

Small [25] developed his *Contemporary College Algebra* explicitly for those institutions where the college algebra course has the broader objective of developing quantitative literacy (see paper later in this volume for an overview of this text). Topics in his curriculum include the display and interpretation of data, linear equations and inequalities, and linear programming (graphical). Functions are studied in all four representations, and modifications of a collection of elementary functions (linear, exponential, power, quadratic) are obtained using shifts and rescalings and using the algebra of functions. Through case studies of applied problems students learn the skill of modeling, and they develop confidence as problem solvers in meaningful situations. Technology enhances the study of data and graphs, and small group work and written assignments develop communication skills. As one example, the distance it takes a vehicle to stop from various initial speeds is presented first as data in a table from a state driving manual. Students plot the data and fit a curve to the data plot. The resulting function is used to extend the data in the driver's manual.

Small group projects can be an important element of the pedagogy. These projects involve an inquiry aspect that involves students in real life activities such as pricing materials in a store, writing a business letter, interviewing a bank official, conducting a survey, and searching the Internet. Projects conclude with a written report that involves group reflections. The business world places a high priority on group work, so these projects develop students' readiness for the workplace.

At other institutions, the general education curriculum includes a wider range of subjects studied less deeply including the real number system, graphing and solving linear equations, geometric calculations, elementary probability and statistics, and consumer mathematics.

Local and national policy issues

The preceding survey of changes in the curriculum raises several issues that mathematics faculty must resolve at each college and university, and it poses a few questions that should be considered in national policy making.

Algebra as mathematicians know it vs. algebra as graduates use it

It should be easy to argue that the content of the fundamental theorem in any branch of mathematics belongs in the college-level curriculum of the subject. This would suggest that the study of polynomials should get main billing in the course, and indeed it does in the traditional curriculum. But algebra as *mathematicians know it* is not necessarily the algebra that college graduates (of all majors) will need and that they should learn in college. It is very difficult for a devoted mathematician to abandon the most beautiful ideas of his or her field and teach what some perceive to be a watered-down curriculum to students from across campus. That is why topics such as partial-fraction decompositions, complex numbers, the algebra of functions, and inverse matrices find their way into the textbooks and why words such as *group* and *field* can slip from the lips of the faculty teaching this course. The desire to share the extraordinary beauty of higher mathematics comes from deep in the heart of the mathematician.

That is why it takes some work to keep a focus on the goal of helping the student up the next step of their own career path. Examples—applications—may indeed be more important than theory. It may be more important for the student to come out of the course knowing how to fit verbal descriptions to a variety of forms of the linear equation. It may be more important to have a good facility with exponential functions than a knowledge of polynomials. It may be more important to understand a multivariate linear function than to understand a univariate rational expression.

Every mathematics department should develop a clear sense of which academic departments they are serving through the course, which topics their students will need to understand, and what mathematical skills students will need.

Admission and placement vs. instruction and assessment

If it has been used deliberately as a filter, then it may be timely to reconsider the assumptions of that policy, especially at public institutions. Restrictive admissions policies may be politically difficult for some state institutions to defend, and failing grades may seem to be a more objective way to discriminate between those students who deserve higher education and those who do not. But the implicit contract that an open admissions policy creates between the student and the institution is much more ambiguous than the restricted admissions policy of an institution that defines its clientele clearly, admits according to standards appropriate to that clientele, and allocates resources to whatever educational functions are necessary to support the success of admitted students. An ambiguous, implicit contract could become the focal point of complaints by dissatisfied parents and students.

The remarks of the former president of Grambling State University to the MAA community in 1997, paraphrased at the beginning of this paper, suggest that there may be a declining tolerance for high DWF rates. In this age of consumer activism, is it far-fetched to imagine that a group of parents or flunk-outs would organize a class-action lawsuit against a public university for recovery of the tuition they paid—plus damages? How would an academic institution, or a mathematics department chair, explain to a judge the reasons for the DWF rate at their institution?

Science policy and precalculus vs. subcalculus mathematics

If less than a majority of college algebra students go on to take calculus of any kind, then it seems more appropriate to refer to this curriculum as subcalculus rather than precalculus.

Should the NSF support subcalculus mathematics education? Only a minority of college algebra students typically go on to take a semester of calculus, but this does not imply that the reformed college algebra course is terminal mathematics, nor that it should lose its significance in the preparation of a scientifically literate citizenry. A *majority* of the college algebra students will major in subjects that require the study of statistics—business, psychology, the life and health sciences, some social sciences, and even some education programs. Statistics is required in these majors because it is fundamental to understanding phenomena in nature and evaluating the claims of scientific theories.

The American system of higher education should cultivate a scientifically literate electorate as well as scientifically literate policy makers at all levels of government. These people are rarely hard science majors and are far more likely to have been college algebra students than calculus students. Data collected by the U.S. Department of Education show that only 10% of the college graduates in 1997–98 had hard-science specializations [4], yet around 800,000 students take college algebra each year [21].

The specialization of college algebra for different majors

Large universities may run 20 or 30 sections of college algebra each year. This presents an opportunity, at the least, to specialize various sections of the course for different majors and at the extreme to create different courses that focus on the needs of different types of students. Monmouth University in New Jersey did just that in 2000–01 [11]. Their single course had a large number of education majors, and many other students were taking it to fulfill a general education requirement, along with the usual mix of students in business and the social and natural sciences. The math department split the course into four distinct curricula:

- Mathematical modeling for the social sciences took about 250 students, leading to statistics for most students and to quantitative analysis for business for some.

- Mathematical modeling for the biological sciences split off about 60 students and leads to statistics.

- Foundations of elementary math took nearly 100 education majors and was terminal.

- College algebra remained as a calculus-preparatory course for majors in the hard sciences and computer science and enrolled about 50 students.

The courses with 50–60 students seemed to be about as small as they could be and still allow enough sections that students would have some flexibility for scheduling. The social sciences course and the biological sciences course had to be made interchangeable as prerequisites to their respective majors, because students tended to switch majors between these fields. However, the course for education majors was not interchangeable with those.

Staffing is an issue to consider when subdividing the college algebra course. At Monmouth, the full-time mathematics faculty found it quite interesting to teach the applications of math in the social and biological sciences. Monmouth uses many adjuncts, so the ones with applied math backgrounds were selected for the specialized courses. One point of concern was how well the specialized courses would be taught by adjuncts who were also full-time high school teachers, as they may not be as comfortable with applications drawn from college-level majors and may have taught in the traditional manner for many years.

Conclusions

Change in college algebra must become a priority of the mathematics community. The high DWF rate of college algebra, as compared to other courses taken by large numbers of freshmen, is—or should be—an embarrassment to the chairs of mathematics departments.

The basic premise of contemporary changes in college algebra is that the curriculum should be guided by the future mathematical needs of the students. Research on the intended majors of college algebra students suggests that the vast majority are headed toward business or the life/health or social sciences. Most college registrars can give a mathematics department chair a report showing the distribution of students' intended majors in the college algebra course, so this type of market research is not difficult to obtain.

However, a curriculum oriented toward the soft-science student is decidedly different from the traditional college algebra course. The higher-order thinking skills required by the soft-science students are developed by formulating algebraic models of relationships between variables, not by performing clever manipulations of algebraic expressions. A review of textbooks in the soft sciences will show that the mathematical functions used by students in those fields are predominantly linear and exponential, not polynomial.

It is difficult for any mathematician to restrain the desire to share with students the extraordinary beauty at the depth of algebra. For many, it is almost painful to give short shrift to the fundamental theorem of the field in a college-level course. However, in our efforts at curriculum and instruction in college algebra, we must attend first to the needs of our students and reserve for the mathematics majors our unbridled enthusiasm for higher mathematics. Our goal in college algebra should be to develop the students' confidence in their ability to work with mathematical concepts and expressions. A curriculum oriented toward the needs of the students' future majors is more likely to inspire their interest and active participation than any other. With such a curriculum, college algebra can be a pump rather than a filter.

Acknowledgements: I am grateful for the contributions of Della Bell, Bonnie Gold, and Donald Small to this article.

References

1. Artin, Michael, "Algebra at the College Level," in *The Algebra Initiative Colloquium, Volume 2*, C. Lacampagne, W. Blair and J. Kaput (eds.). U.S. Government Printing Office document SE 056 573, Washington, DC, 1995, p. 72.

2. Aufman, Richard N., Vernon C. Barker, and Richard D. Nation, *College Algebra and Trigonometry*, Houghton-Mifflin, Boston, 1997.

3. Bittinger, Marvin L., Judith A. Beecher, David Ellenbogen, and Judith A. Penna, *College Algebra: Graphs and Models*, Addison-Wesley, Reading, MA, 1997.

4. *Chronicle of Higher Education*, "Earned Degrees Conferred, 1997–98," http://chronicle.com/weekly/almanac/2001/nation/0102501.htm, Oct. 1, 2001.

5. Cohen, David, *College Algebra*, 4th edition, West, Minneapolis, 1996.

6. Cohen, Don (ed.), *Crossroads in Mathematics: Standards for Introductory College Mathematics before Calculus*, American Mathematical Association of Two-Year Colleges, Memphis, TN, 1995.

7. Crauder, Bruce, Benny Evans, and Alan Noell, *Functions and Change: A Modeling Alternative to College Algebra*, 2nd edition, Houghton-Mifflin, Boston, 2003.

8. Davis, Paul, 1995. "Algebra, Jobs and Motivation," in C. Lacampagne, W. Blair and J. Kaput (eds.) *The Algebra Initiative Colloquium, Volume 2*, Office of Educational Research and Development, Washington, DC, p. 150.

9. Dugopolski, Mark, *College Algebra*, Addison-Wesley, Reading, MA, 1995.

10. Dunbar, Steven R. and Scott R. Herriott, "Renewing the College Algebra Course: Toward a Curriculum Suited to the Future Mathematical Needs of the College Algebra Student," unpublished manuscript, 2001.

11. Gold, Bonnie, "An Alternative to the One-Size-Fits-All Precalculus/College Algebra Course," in this volume.

12. Gordon, Sheldon P., Florence Gordon, Alan C. Tucker, and Martha J. Siegel, *Functioning in the Real World*, 2nd edition, Addison-Wesley, Reading, MA, 2004.

13. Herriott, Scott R., *College Algebra Through Functions and Models*, Brooks-Cole, Pacific Grove, CA, 2005.

14. Kalman, Dan, *Elementary Mathematical Models*, Mathematical Association of America, Washington, DC, 1997.

15. Kenschaft, Patricia C., "455 mathematics majors: What have they done since?" *College Mathematics Journal* 31 (2000), pp. 193–196.

16. Kime, Linda A., Judy Clark, and Beverly Michaels, *Explorations in College Algebra*, 3rd edition, Wiley, New York, 2005.

17. Lacampagne, Carole B., William Blair, and Jim Kaput (eds.), *The Algebra Initiative Colloquium, Volume 1*, U.S. Government Printing Office document SE 056 572, Washington, DC, 1995.

18. Lacampagne, Carole B., William Blair, and Jim Kaput (eds.), *The Algebra Initiative Colloquium, Volume 2*, U.S. Government Printing Office document SE 056 573, Washington, DC, 1995.

19. Larson, Roland E., and Robert P. Hostetler, *College Algebra*, 4th edition, Houghton Mifflin, Boston, 1997.

20. Lial, Margaret L., E. John Hornsby, Jr., and David I. Schneider, *College Algebra*, 7th edition, Addison-Wesley, Reading, MA, 1997.

21. Loftsgaarden, Don O., Donald C. Rung, and Ann E. Watkins, *Statistical Abstract of Undergraduate Programs in the Mathematical Sciences in the United States: Fall 1995 CBMS Survey*, Mathematical Association of America, 1997.

22. Pollack, Henry, "Algebra and the Technical Workforce," in C. Lacampagne, W. Blair and J. Kaput (eds.) *The Algebra Initiative Colloquium. Volume 1*, Office of Educational Research and Development, Washington, DC, 1995, p. 89.

23. Rockswold, Gary K., John Hornsby, and Margaret L. Lial, *College Algebra Through Modeling and Visualization*, Addison-Wesley, Reading, MA, 1999.

24. Ruud, Warren L. and Terry L. Shell, *College Algebra*, Worth Publishers, New York, 1997.

25. Small, Donald, *Contemporary College Algebra*, 5th edition, Wiley, New York, 2004.

26. Sobel, Max A. and Norbert Lerner (1995), *College Algebra*, 4th edition, *Annotated Instructor's Edition*, Prentice-Hall, Englewood Cliffs, NJ, page v.

27. Stewart, James, Lothar Redlin, and Saleem Watson, *College Algebra*, 2nd edition, Brooks-Cole, Pacific Grove, CA, 1996.

28. Sullivan, Michael and Michael Sullivan III, *College Algebra Enhanced with Graphing Utilities*, Prentice-Hall, Upper Saddle River, NJ, 1996.

12

One Approach to Quantitative Literacy: Understanding our Quantitative World

Janet Andersen
Hope College

Overview and philosophy of the course

The calculus reform movement sparked numerous conversations about pedagogy and curricula on campuses throughout the country. However, it is only recently that the national conversation has spread to courses below calculus. These include courses that potentially lead to calculus, such as college algebra and precalculus, as well as courses that students take primarily to fulfill general education requirements. Part of the confusion is that many courses serve one purpose in theory yet serve another purpose in practice. For example, many topics in college algebra and precalculus textbooks are included because they are necessary background for standard topics in differential or integral calculus. Yet the percentage of students enrolled in college algebra or precalculus courses who eventually take two semesters of calculus is quite small. So, in practice, college algebra and precalculus are primarily serving as general education courses or as preparation for courses outside of mathematics. However, a course designed for one purpose (preparation for calculus) is not necessarily well designed for another purpose (general education credit or preparation for courses outside of mathematics). One response to this situation is the creation of additional mathematics courses and implementation of effective advising procedures so that students take the mathematics course that is most appropriate for their needs.

The primary need for many students is to become quantitative literate citizens who are capable of interpreting and using information presented quantitatively. As stated on page 2 in the book *Mathematics and Democracy: The Case for Quantitative Literacy*, "Quantitatively literate citizens need to know more than formulas and equations. They need a predisposition to look at the world through mathematical eyes, to see the benefits (and risks) of thinking quantitatively about commonplace issues, and to approach complex problems with confidence in the value of careful reasoning."

In constructing and teaching a course designed to improve quantitative literacy, it is crucial that the goals and objectives be targeted at what students will most likely encounter *outside* of academia. For many mathematicians, this is a brand-new (and somewhat radical) idea. Typically, every course in the mathematics curriculum is designed to prepare students to take another course whether it is another mathematics course or a course in another discipline. Yet, many students do not take another course in mathematics and therefore have been drilled and tested on skills and concepts that they often will never use or encounter in the future. There are many students who take a particular mathematics course because it is a cognate for their major, such as social science majors taking statistics. In these cases, it is important that the mathematics department engage in ongoing conversations with the other departments to ensure that the mathematical preparation adequately addresses their needs. However, there are many students who are in majors that do not require a particular mathematics course, or who are undecided about a major when they enroll in a mathematics course. These students are typically taking a mathematics course to fulfill a general education requirement and this is likely to be the last mathematics course they will take. Once the pattern of assuming that mathematics courses should fit the needs of both general education and preparation for future mathematics courses is broken, we are left with two basic approaches for general education mathematics courses:

- What skills and concepts are useful for informed citizens?
- What great ideas of mathematics should all educated people know?

Note that these questions do not imply the same approaches to the answer. The great ideas of mathematics such as chaos theory, graph theory, the concept of infinity, and so on, are not the same as the skills and concepts needed to correctly interpret quantitative information found in common media such as *Newsweek* and *The New York Times*. While it would be affirming if the general populace recognized that fuzzy math is an actual branch of research mathematics rather than a glib slogan used during a political campaign, it would be more affirming if the general populace was able to correctly interpret commonly encountered statistical and graphical information. This is not to say that all general education mathematics courses should be focused on the idea of "What skills and concepts are useful for informed citizens?" There is certainly a strong and valid argument for giving non-mathematicians a glimpse of the beauty and creativity of mathematics and developing a sense of math appreciation. The appeal of this approach is demonstrated by the popularity of textbooks such as *The Heart of Mathematics* (Burger and Starbird) and the intersection of art and mathematics in areas such as Escher's paintings, the movie *A Beautiful Mind*, and the Broadway play *The Proof*. However, we cannot ignore the need to improve the quantitative literacy of the general American populace.

There are several projects that have concentrated on helping students to correctly interpret and understand commonly encountered quantitative information. *For All Practical Purposes* (COMAP) shows students new mathematics, such as voting theory, while demonstrating that mathematics can be used to interpret and inform life outside of academics. *Essentials of Using and Understanding Mathematics: A Quantitative Reasoning Approach* (Bennett and Briggs) focuses on how to interpret mathematical information commonly found in the media. *How Do You Know? Using Math to Make Decisions* (Hirst) is a workbook covering four modules (trigonometry and measurement, finance, statistics and linear programming) recognized with the Innovative Programs Using Technology (INPUT) award. Many statistics and finite mathematics courses devote considerable time to helping students interpret statistical or financial information that they will find useful as informed citizens. The National Council on Education and the Disciplines at the Woodrow Wilson Foundation has designated Quantitative Literacy as one of its four major emphases in literacy. Information can be found at http://www.Woodrow.org/nced/quantitative_literacy.html. The National Science Foundation funded several "Mathematics Across the Curriculum" projects, many of which addressed the needs of quantitative literacy (e.g., Dartmouth College and the University of Nevada at Reno).

At Hope College, we looked at several of these projects. However, we wanted something that forced students to actively engage in discussing mathematics and was connected to targeted general education science courses. We applied for and received partial support from the National Science Foundation (DUE 9652784), to develop a mathematics course designed around the question "What skills and concepts are useful for informed citizens?" Mathematics in Public Discourse was first piloted in spring 1998 with one section and quickly became a popular mathematics option at Hope College with approximately 270 students (out of a student body of 3,000) enrolled every subsequent year. Thus, about 35–40% of all students at Hope take this course.

Understanding Our Quantitative World is intended for students who do not have a specific mathematics course requirement (such as statistics or calculus) for their major. In practice, this means that the main audience for this course is humanities and fine arts majors. Most of the students in the course have taken two to four years of mathematics in high school and no other mathematics course in college. However, the knowledge base and skills base of the students varies widely from students who have taken a high school calculus course to students who cannot correctly solve simple algebraic equations or combine fractions. Almost all of the students enrolled in the course dislike mathematics at the beginning of the course.

The goals of the course are for students:

- to realize that mathematics is a useful tool for interpreting information;
- to see mathematics as a way of viewing the world that goes far beyond memorizing formulas; and
- to become comfortable using and interpreting mathematics so that they will voluntarily use it as a tool outside of academics.

The course is structured so that there is little lecturing. Rather, students are assigned readings from the text materials developed under the NSF grant. They read the material and answer a series of questions (some of which are computational and some of which require interpretation and written explanation) before the material is covered in class. The rationale for having students read the materials is to emphasize the importance of being a self-learner. For that same reason, the group activities are taken from public resources such as newspapers, magazines, and the World Wide Web. This demonstrates to students that they can use mathematics as a tool in interpreting the world they encounter. The course is organized so that students do group activities approximately 75% of the class time with the other 25% devoted to clarification and summary of the material, review, and exams. Students are required to have a TI-83 graphing calculator.

The readings cover the following topics (see [1]):

Functions. Four representations of functions (symbolic, graphical, tabular, and verbal) are emphasized. Specialized vocabulary (such as domain and range) is introduced. Examples include the stock market, population of the U.S., and the cost of internet services. Group activities include cell phone rates and credit card bills.

Graphical representations of functions. Correct interpretation of graphical information is emphasized, particularly with regards to shape and labels. The concepts of increasing/decreasing and concavity are introduced. Instruction on using the calculator to construct graphs is included. Group activities focus on analyzing a variety of graphs from magazines, newspaper, and non-mathematical textbooks.

Applications of graphs. The connections between and the meaning of the graphs of $f(x)$, $f(x + a)$, $f(x) + a$, $f(ax)$, and $af(x)$ is emphasized. This is introduced via the context of a motion detector graph of time versus distance. Group activities include working with a motion detector and converting baby weight charts from English units to metric units.

Displaying data. The emphasis in this section is on visual display of data. Histograms, scatterplots, and xy-trend line graphs are included. Students receive instruction on using the calculator to graph data in each of these formats. Group activities include looking at arm span versus height and data given from the American Film Association on best movies.

Describing data. Concepts underlying one variable statistics are emphasized. This includes ideas of center (i.e. median and mean) and ideas of spread (i.e., standard deviation and quartiles). The emphasis is on the difference between the median and the mean, particularly with skewed data. Normal distributions are also introduced. Instruction on using the calculator to compute one-variable statistics is included. Group activities include salary versus winning percentage of basketball teams and looking at house prices.

Multivariable functions and contour diagrams. Commonly occurring multivariable functions (such as computing the payment on a car loan) and commonly occurring contour maps (such as weather and topological maps) are emphasized. Treating a multivariable function as a single variable function by holding all but one input constant is also included. This allows the students to connect some of the ideas in this section with those encountered earlier in the text. Group activities include a contour map of Mount Rainer and looking at car loans.

Linear functions. The emphasis is on translating a situation with a constant rate of change into the mathematical concept of a line. There is also an emphasis on the concept that only two pieces of information—a starting point and a rate of change—are necessary to determine a line. This section ends by showing that proportional changes (such as unit conversions) can be thought of as linear functions. Group activities include working with a motion detector and looking at an electric bill.

Regression and correlation. Students are introduced to the concept of using regression analysis to determine if two variables exhibit a linear relationship. Calculator instructions are included. Other types of regression (e.g. exponential) are introduced in later sections. Group activities include Olympic race data and atmospheric carbon dioxide data.

Exponential functions. The concept of an exponential function is introduced via the idea of doubling. Exponential functions are contrasted with linear functions. In particular, the idea of a constant rate of change versus a constant growth factor is emphasized. This section also explores vertical and horizontal shifts of exponential functions, connecting with the ideas introduced in the section on applications of graphs. Group activities include a cooling experiment and looking at prices of DVDs.

Logarithmic functions. Logarithms are emphasized as functions that compute the magnitude of a number. Only base 10 logarithms are used. Properties of logarithms and using logarithms to solve simple exponential equations are included. Group activities include sound decibels and verifying Bedford's law on the occurrence of numbers in print.

Periodic functions. Periodic functions are introduced as a way of modeling cyclic behavior. The behavior of a clock and a swing are used to motivate the concepts. Sine and cosine are defined only in terms of the circular definitions. The concepts of amplitude and period are related to the ideas of shifting functions introduced earlier in the text. Group activities include an experiment with sound waves and looking at the seasonal change in the hours of daylight.

Power functions. Power functions are the last type of function covered in the text and are introduced graphically. Behavior of polynomials with even and odd positive integer exponents is contrasted. Positive rational exponents are also included. Group activities include Kepler's law of planetary motion and looking at the wingspan of birds.

Probability. The basic concepts of counting and determining simple probabilities are introduced. Systemic ways of listing (or counting) all possible outcomes are emphasized. Multi-stage experiments and expected value are included. Group activities include codes for garage door openers and roulette.

Random samples. This section explores how to set up a random sample and why this is desirable. The concepts of variability, bias, and confidence intervals are included. Group activities include looking at phone-in surveys and simulating a capture-recapture experiment.

Each of these readings is a single unit on the topic. The goal is to give students an intuitive sense of the mathematical concept so they can adequately interpret (rather than necessarily create) mathematics. In addition to the readings, we have also written four to eight group activities for each section, of which we typically assign two to four.

Even though the course was designed with the premise that this is the last mathematics course a student will take, we have found that this course functions equally well as an introduction to other courses. In fact, about half of the students enrolled in Understanding Our Quantitative World take additional mathematics courses. Understanding Our Quantitative World has always been a pre- or co-requisite for two of our general education science courses, Populations in a Changing Environment and The Atmosphere and Environmental Change. Students in these science courses see and use many of the mathematical skills repeated in another context. This is in response to the recommendation by the MAA that quantitative literacy should occur throughout the curriculum and not be isolated to one or two mathematics courses. In fact, these two general education science courses hold students accountable for more mathematics than do the introductory courses for the biology or geology major. This is due to extensive conversations among faculty in biology, geology and mathematics. Knowing exactly what mathematics is covered in the Understanding Our Quantitative World course, plus knowing the topics covered in the science course and ways that mathematics occurred naturally in these contexts, enabled us to create courses whose content fit together. Some of the topics, such as multivariable functions and the emphasis on the behavior of various types of functions, were included because they would be applied in the science courses. Several of the group activities developed for the math course, such as the capture-recapture experiment and looking at levels of carbon dioxide in the atmosphere, appear again in the science courses. This spiral approach of having students see similar material in multiple courses and contexts increases the expertise and comfort with which students successfully use and interpret mathematics.

Another unexpected benefit of this course is that it also functions as a precursor to our calculus with review sequence. [Note: Calculus with Review is a year-long sequence that incorporates precalculus topics just in time while covering slightly more calculus than is typically found in the first semester calculus course.] One option for students at Hope College enrolled in the elementary education program is to complete a mathematics minor for elementary teaching. For this minor, we require students to master mathematical content up to differential calculus, in addition to taking courses designed for elementary teachers. While we have some students with strong mathematical skills enrolled in this minor, we also have several with weaker mathematical backgrounds. For these students, we have found that having them take Understanding Our Quantitative World followed by Calculus with Review I & II works well. Part of the reason for this is that it builds on the spiral approach to content that has been shown to be effective in K–12 mathematics curriculum design. (See, for example, the paper by Rebecca Walker, in this volume, and the evaluation of the Core Plus mathematics project, http://www.wmich.edu/cpmp/evaluation.html.) Most of our students have seen at least some material on functions in high school, even if their understanding of these concepts is weak and flawed. They then see functions in the Mathematics in Public Discourse class included in a context and discussed in a group setting with their peers. By connecting the mathematical concepts to topics taken from common sources such as newspapers, magazines, and the internet, students gain an intuitive understanding of the type of behavior described by various types of functions. Functions and their properties are also covered in the Calculus with Review courses, this time in a more mathematical, algebraic context. Notice that this is different from the college-algebra-leads-to-precalculus-leads-to-calculus approach because the students are seeing the topics presented differently in the courses and in a context that automatically addresses the question of why do we need to know this? In Understanding Our Quantitative World, students see that the mathematics allows them to interpret information they will encounter in their lives as citizens. The emphasis is on developing an intuitive understanding and interpretive ability. In the Calculus with Review I & II courses, students see that the same mathematics is necessary for understanding the calculus. This time, the emphasis is on algebraic manipulation and computation. Seeing the material in multiple contexts and courses leads to better understanding and retention of the material.

Sample section: Periodic functions

One of the topics covered in Understanding Our Quantitative World (our general education course) is periodic functions, a topic found in all precalculus courses. Yet our approach to this topic is different from the traditional precalculus approach. Our goal is for students to be critical consumers of mathematics (rather than primarily creators of mathematics). While we want students to become facile with formulas, we consider it more important that students can appropriately analyze and interpret quantitative information. Therefore, our approach is for students to develop an intuitive understanding of what a periodic function is, what type of behavior it describes, and how to interpret information given in the context of periodic functions. The focus is on the word "periodic"–we only allude to the connection to the triangular definitions in a footnote.

Each section starts with a paragraph giving some motivation for the topic, describing the content of the section and relating it to previous sections in the text. Periodic functions are motivated by considering the cyclic behavior of daily temperatures (over the year), the position of the second hand on a clock, and the position of a child on a swing.

Periodic functions are introduced as functions whose outputs repeat at regular intervals. The first examples are graphical. The definition of period is given early in the text. The text describes how to determine if a function is periodic and if so, to determine the period. Function notation is used. Some examples connect the students to the definition of a function given at the beginning of the text. After this introduction to the concept of periodic functions, the section develops the circular definitions of sine and cosine by looking at the behavior of a six inch second-hand on a clock. First, the vertical position of the hand (from the center of the clock) is graphed over the course of two minutes. Next, the horizontal position of the hand is graphed. Why these are periodic functions as well as the similarities and differences of the two graphs are stressed. Figures 1 and 2, taken from the text, illustrate this approach.

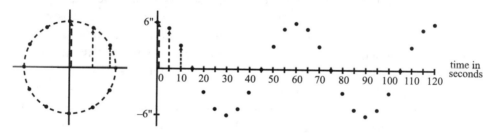

Figure 1. A graphical representation of the vertical position function

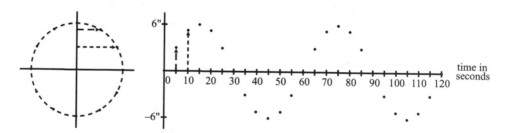

Figure 2. A graphical representation of the horizontal position function

The section goes on to define amplitude, period, the sine function, and the cosine function. The explanations are rooted in two examples: the clock example and a swing example introduced when explaining amplitude and period. Both of these are familiar examples that help students relate the mathematical concepts to their own experience. The emphasis throughout the section is on a graphical understanding.

A similar approach to introducing periodic functions is certainly feasible in a precalculus or a trigonometry course. Giving students familiar examples and beginning with a graphical understanding helps students understand the concept of periodic functions in a way that is not feasible when we restrict our teaching to mathematical symbols. In a precalculus or trigonometry course, it is necessary to also include significant practice in using and interpreting the symbolic forms of these functions. But only emphasizing the symbolic forms (the traditional approach to precalculus) is likely to produce students who blindly memorize formulas, but have little to no understanding of the mathematical concepts.

Students in our general education course are expected to read the section and complete the reading questions BEFORE coming to class. This approach, while at first uncomfortable for many students, helps combat the common misconception that mathematics is something that is only understandable to the experts. It is empowering when students realize that they do have the capability to read and understand mathematics when it is presented in a familiar context. Reading questions include both computational and conceptual questions. For example, in the section on periodic functions, students are asked to contrast the difference in behavior between exponential and periodic behavior, to determine if a graph does or does not represent a periodic function, and to find simple periodic models for representing the motion of a child in a swing or a seat on a Ferris wheel.

Students typically do 15 to 20 reading questions per section. About half of the questions ask students to *write* sentences in response to questions such as "explain why this is true," "give an example," "what would happen if." The other questions ask students to perform a mathematical computation in the context of a graph, a table, a physical description, or mathematical symbols. The approach in our general education course differs from a precalculus course by putting a greater emphasis on writing and a reduced emphasis on symbolic manipulations.

When students come to class, the instructor summarizes the section and emphasizes the major points. The instructor also spends time answering homework questions and clearing up misconceptions. Having students work on group projects takes the remainder of the class time and typically all of the following class period. The group projects involve longer applications. Two to four group assignments are typically assigned per section. One of our group activities is Sound. In this activity, students first look at sound waves produced by tuning forks. They are then led to derive the relationship between the period of the sound wave and the frequency of the note. Next, they produce unknown notes by blowing across a bottle and are asked, by looking at the sound wave, to figure out what note was produced. Finally, we bring in musical instruments and/or ask students to sing to see what happens.

This group assignment is appealing to most students. It allows them to make connections between mathematics and music—two areas that many students assume are unrelated. There are always musically talented students in the class who enjoy showing off their expertise. One of the instructors brings in his keyboard to demonstrate the graphs produced with playing chords and other combinations of multiple notes. This group assignment relies on technology. We have purchased a classroom set of CBLs (Calculator Based Laboratories) and microphone probes. Students use their TI-83 calculators to collect the data and draw the graphs. The necessary software for the TI-83 is a program called Sound provided on the TI-83 website.

Other group activities for the unit on periodic functions include Biorhythms, Daylight, and Current. The biorhythm activity gives students a generic formula for computing physical, emotional, and intellectual biorhythms. They have to modify these formulas using their birth date and graph them for the following two weeks. This is, again, an activity that is appealing to most students who enjoy seeing when they are supposed to peak in the next two weeks. The daylight activity gives students a chart of times the sun rose and set in a nearby city every ten days for a year. They use this data to create a graph of hours of daylight versus day of the year. Using the graph, they are asked questions such as "which day had the most daylight," "when is the amount of daylight changing the most rapidly," "how does this correspond to your experience of the seasons." The current activity uses the CBL and a light intensity probe. Students measure the light intensity of a fluorescent light bulb (over a short time period). Students are astounded to see that the light they thought was steady is actually alternating.

These activities are adaptable to a precalculus course. When we use them in precalculus, we typically add more questions asking students to compute or use a symbolic formula. The advantage of such activities is that they help students understand that mathematics has a variety of applications and is prevalent throughout their world.

Evaluation

We have solicited information from students regarding Mathematics in Public Discourse through pre- and post-attitude surveys, course evaluations, and focus groups conducted with students at least one year after completing the course. We have found that Understanding Our Quantitative World is having a slow, positive influence on student attitudes towards mathematics. In contrast, when general education students are placed in traditional mathematics course (such as college algebra or precalculus), attitudes towards mathematics are often noticeably more negative at the end of the course. Typical student comments regarding the course include:

> I liked that we applied it to real life, so it sort of felt like there was at least a purpose to learning the stuff.

> I liked learning how to keep an open eye for misleading information, biased data, and manipulative statistics.

> I liked to see the many different ways math is viewed in today's society.

> This class was great! I think that I learned more in this short semester than I ever did in any high school courses.

The focus groups found that, even a year or more after taking the course, students could still recall particular activities and at least a vague understanding of its connections with the mathematics. The course did seem to have a long-term effect on student perceptions of the usefulness of mathematics.

The biggest difficulty with the course is attendance. The heavy reliance on group activities makes it problematic when students are absent. To address this, we have created an attendance policy that helps students realize the direct connection between coming to class and doing well. We call this attendance policy our "C– guarantee." In the syllabus and on the first day of class, we announce to students that anyone who attends every class session and completes all of the homework, will be guaranteed a grade of C– or higher regardless of test scores. We find that this is a policy that rarely needs to be implemented since it is unusual that students who are attending class and completing assignments fail to pass the exams. However, having such a policy gives some students, particularly those that have been unsuccessful in previous mathematics courses, a clear means to success.

We have found that about 40% of the students who take this course also take an additional mathematics course. Most of the time this is a statistics course. This course is also taken by several pre-service elementary teachers (in addition to the mathematics course for elementary teachers) to strengthen their mathematical background and experience a different form of pedagogy in a mathematics classroom

The course works best when there is a clear match between the objectives of the course and the reasons that students have for taking the course. When students are taking a course that is likely to be their last formal mathematical experience, the best option may be a course focused on issues of quantitative literacy and attitude rather than a course that is pre to a subject they may never study.

References

1. Andersen, J. and T. Swanson, *Understanding Our Quantitative World*, Mathematical Association of America, Washington, DC, 2004. https://enterprise.maa.org/ecomtpro/Timssnet/products/TNT_products.cfm.

Theme 2. The Transition from High School to College

At the collegiate level, precalculus is frequently categorized as a "remedial course" and too often it is a terminal mathematics course for students who enter college not prepared to take calculus. On the other hand, at the high school level, precalculus seeks to prepare the best and the brightest to study calculus. In both cases, enrollments are at record levels. Not only are the audiences different, at the high school level, as a result of implementation of the *Curriculum and Evaluation Standards*, published by the National Council of Teachers of Mathematics (NCTM), the way precalculus is taught and what is taught is changing. In this section Zalman Usiskin, an invited speaker at the conference *Rethinking the Preparation for Calculus*, gives an overview of what is happening in the high schools and comments about the transition from high school to college. Daniel Teague discusses precalculus reform at the high school level, and Eric Robinson and John Maceli give a perspective on the *Standards* vision as it relates to the preparation of school students, particularly secondary students, for calculus.

Zalman Usiskin observes that enrollments in college preparatory courses are at historically high levels, enrollments of high school students in calculus, whether advanced placement or not, are also at record levels, and on all long-term national measures of performance, mean scores have increased steadily over the past twenty years. Yet, in colleges enrollments in remedial courses are also at record levels and the number of mathematics majors has declined. Using data from the National Assessment of Educational Progress, SAT and ACT college entrance exams, and the Advanced Placement program of the College Board, Zalman discusses and analyzes these phenomena and offers some recommendations for action.

> I have been asked to present a "high school overview" with some comments about the transition from high school to college. I am interpreting the first part of this mission to mean an overview of the mathematics curriculum in schools, instruction in classrooms, and performance among students today in the United States. ...Any overview is an oversimplification. The word "precalculus" itself can refer either to the student's mathematical experience before taking a calculus course, or to the specific concepts felt to be prerequisites for calculus. Students, teachers, the mathematics curriculum itself, the broader school context, and the surrounding community vary widely. Furthermore, we lack some important data that might help us to understand what is going on. In particular, we lack data on what works and what does not. But we do have some trend data and some specifics that may help inform discussion of these issues.
>
> *High School Overview and the Transition to College*
> Zalman Usiskin

Daniel Teague, who is on the faculty at the North Carolina School of Science and Mathematics, observes that in many secondary schools precalculus reform has preceded calculus reform. Daniel shares some of his insights from 15 years of teaching a reformed precalculus course, focusing on two issues: "Successful students in a reformed precalculus curriculum will have new attitudes and abilities that can be utilized successfully in the teaching of calculus, and reformed precalculus curricula can play an important role in preparing students for the rigor of theoretical mathematics." Daniel notes that "both of these issues have implications in the way the mathematical community views and accepts precalculus reform." He observes that the transition from high school precalculus to post-secondary calculus is difficult for many reasons, but one important reason is that the post-secondary teacher doesn't know the skills, understandings, and expectations of their new students.

> If you try to teach my students with the mistaken belief that they know the mathematics I knew at their age, you will miss a great opportunity. My students know more mathematics than I did, but it is not the same mathematics; and I believe they know it differently. They have a different vision of mathematics that would be helpful in learning calculus if it were tapped.
>
> *Precalculus Reform: A High School Perspective*
> Daniel J. Teague

Eric Robinson and John Maceli describe "changes in the school mathematics curriculum that would or could result from full implementation of a NCTM *Standards* vision, concentrating on changes that affect the preparation of school students for calculus." They consider the questions: "Should the primary driver of the K–12 curriculum, especially in the upper grades, be preparation for calculus? Don't we just need to examine the topic differences that there are in *Standards*-based school mathematics programs in order to understand the changes in students' preparation for calculus? How does the content in *Standards*-based school mathematics education lay a foundation for calculus? What other things influence students' preparation for calculus?"

> It is fair to say that as of this writing most mathematics programs in our nation's schools are neither completely traditional (in the sense of what was typical in the early 1980s) nor completely *Standards*-based (in terms of the changes suggested in the NCTM documents). Massive change does not occur quickly. Even if it did, there has been barely enough time since the publication of the visionary 1989 documents mentioned above for students to be entering calculus courses with a complete K–12 *Standards*-based education. Nonetheless, our major task here is to describe changes in the school mathematics curriculum that would or could result from full implementation of a *Standards* vision, concentrating on changes that affect the preparation of school students for calculus. To that end we will structure this article by asking questions and then providing answers with a *Standards*-based K–12 point of view.
>
> *The Influence of Current Efforts to Improve School Mathematics on the Preparation for Calculus*
> Eric Robinson and John Maceli

13

High School Overview and the Transition to College

Zalman Usiskin
The University of Chicago

Editor's note: This paper is the text of an invited address given by Zalman Usiskin, in October 2001, at the conference Rethinking the Preparation for Calculus. (Please see the paper about the conference, by Jack Narayan and Darren Narayan, in the introduction to this volume.)

Introduction

I have been asked to present a "high school overview" with some comments about the transition from high school to college. I am interpreting the first part of this mission to mean an overview of the mathematics curriculum in schools, instruction in classrooms, and performance among students today in the United States. Six years ago at the conference "Preparing for a New Calculus" I concentrated mainly on the curricular changes moving into grades 7–12 [13] and could only begin to speculate on the implications of those changes. Now we have some rather consistent information about changes that have occurred in student exposure to and learning of mathematics.

Any overview is an oversimplification. The word "precalculus" itself can refer either to the student's mathematical experience before taking a calculus course, or to the specific concepts felt to be prerequisites for calculus. Students, teachers, the mathematics curriculum itself, the broader school context, and the surrounding community vary widely. Furthermore, we lack some important data that might help us to understand what is going on. In particular, we lack data on what works and what does not. But we do have some trend data and some specifics that may help inform discussion of these issues.

Trend data on student performance

A major source of trend data on student performance in mathematics is from the National Assessment of Educational Progress (NAEP). Two performance trends are currently being maintained. The first, the *long-term trend*, also known as the *longitudinal study*, involves random samples of students age 9, 13, and 17 and is administered separately from the other mathematics assessments of NAEP. The trend dates from 1973, and since 1978 the test has used exactly the same items. On this assessment, students are not allowed to use calculators, and the items tested are those considered important in 1973 (such as paper-and-pencil computation skills, direct application of measurement formulas in geometric settings, and the use of mathematics in daily-living skills involving time and money). On these tests, performance of 9-year-olds increased significantly between 1973 and 1996, by perhaps a grade level. The increase was monotone with a particular jump between 1986 and 1990. Performance of 13-year-olds also increased significantly from 1973 to 1996, by perhaps a half grade. In slight contrast, performance of 17-year-olds was steady through the 1990s, a little higher than in the 1970s and 1980s, but not significantly higher [2, p. 54]. These data can be interpreted as showing that, on the average, the basic arithmetic skills of students have not decreased in the past quarter century.

Why would the greater increases at lower grades on the NAEP not be reflected at higher grades four years later? It could be that the test, being a test of rather basic skills, is not sensitive to the mathematics students encounter at the higher grades. Or it could be that middle schools are not taking advantage of the increased knowledge of younger students, and high schools are not taking advantage of the increased knowledge of their entering students.

A second trend, the *short-term trend*, is part of the regular National Assessment administered to 4th graders, 8th graders, and 12th graders in 1990, 1992, 1996, and 2000. The short-term trend was begun to determine possible effects of the NCTM Curriculum and Evaluation Standards [9]. Calculators (4-function at grade 4, scientific at grades 8 and 12) are allowed on some sections of the test and are distributed to students who take the test. As with the long-term trend, the short-term trend indicates greater increases at lower grades than at higher grades. Grade 4 mean scores increased steadily through the 1990s, from 213 in 1990 to 228 in 2000. The gain of 15 scaled points by 4th graders is almost matched by a similar steady gain of 12 points, from 263 to 275, by 8th graders in the same decade. At 12th grade, the performance is also generally positive. Although the mean scaled score of 12th graders decreased from 304 to 301 from 1996 to 2000, both scores are significantly higher than the mean score of 294 in the year 1990 [1, p. 24].

The positive directions of these trends agree with trends on the two quite different college entrance tests that are widely taken in the U.S., the SATs and the ACTs. Mean scores on the mathematics portion of the SAT-I, the basic SAT test taken by about 1.3 million seniors, increased from 501 in 1990 to 514 for last year's senior class [3]. Mean scores on the ACT test, taken by over one million seniors in 2000, increased from 19.9 to 20.7 in the same period [17, p. 267]. Unlike the NAEP tests, students taking the SAT and ACT do not constitute random samples of U.S. 12th graders. However, the number of students taking the ACT test has increased 30% since 1990, which normally would lead to a lower mean score as it did for the SAT in the 1960s, when there was a similar increase in numbers of students taking the test [16]. Thus the increase in mean scores on the ACT can be considered as an underestimate of an actual increase in mathematics performance by a population comparable with that of 1990.

Variability within the United States

When I speak of the extraordinary variability in our country, I do not mean the obvious variability that must exist among individual students. Nor do I mean differences between special schools such as the North Carolina School of Science and Mathematics and typical public schools. I am referring to the variability among larger jurisdictions such as cities and states.

The performance of 8th grade students on the NAEP provides one example of this variability. (I focus here on 8th grade scores because there are no state scores collected for 12th graders.) In 2000, mean state scores for 8th graders ranged from 254 for Mississippi to 283 for North Dakota and Ohio. (Minnesota and a few other states had higher mean scores, but they did not meet all the sampling guidelines, so I am not referring to them.) Of course, as the jurisdiction gets smaller, the variability of mean scores grows. For instance, the mean 8th grade score in Washington, DC is 234.

Some items on the NAEP are given at more than one grade level so that the NAEP scaled scores for different grades can be reasonably viewed as being on the same scale. That means that the typical 8th grader in Washington, DC is at about the same level in mathematics as the typical 4th grader in the rest of the country. The historical difference of 80 points between the grade 4 and grade 12 means comes to about 10 points a year. This suggests that the typical student in North Dakota is 3 years ahead of the typical student in Mississippi at 8th grade.

These differences have major implications for policy decisions. A "one-size fits all" policy decision for the nation is easy to defend with arguments of democracy and equal opportunity for all, but even stretch socks have their limitations. Grouping mathematics students by prior performance exists in most schools despite many stories of improper and biased allocation of students to groups, and despite cogent arguments against grouping. And mathematics teachers favor grouping even in an individual school. In a write-in response to a question in the November 1998 NCTM *Mathematics Education Dialogues* (reported in the April 1999 issue of that newsletter) regarding when it might be appropriate to group, only 22 mathematics teachers out of 360 preferred no grouping of students at any time in grades K–12.

Differences in performance through the U.S. are reflected in startling differences of graduation rates. In some public schools, virtually 100% of students expect to go to college, while in others fewer than half the students graduate. However, precise graduation rates are a matter of debate. Because of student mobility, because some students take more than 4 years to graduate, because some students never enroll in 9th grade and so are lost from the system, and because of the questionable equivalence of the GED to any high school diploma, estimates of graduation rates in the U.S. vary depending on how the rate is calculated.

One set of data, collected by the National Center for Education Statistics, compares the number of high school graduates as a percentage of 9th grade enrollment 4 years earlier. With this definition, graduation rates for entire states in 1997–98 ranged from 53% in South Carolina to 89% in North Dakota [17, p. 265]. In some cities fewer than half the students graduate from high school with their age cohort. My own city of Chicago is one of these. Using this definition, the graduation rate for the entire nation in 1998 was 68%, a decline from 73% 10 years earlier.

A second set of data, collected annually by the U.S. Census Bureau, results from asking adults how far they have gone in school. The data for 20–24 year-olds may be taken as a reasonably current measure of high school graduation rates. This percent in 2000 was 85.5%, compared to 84.1% in 1990. For adults age 25–54 in 2000, 88% reported having finished high school and about 29% reported having a bachelor's degree, and these percents are relatively constant within that entire age range. These percents have been steadily increasing at least since 1940 [12].

Where fewer students graduate from high school, fewer go on to college. And where fewer go on to college, it is more difficult to teach a college-preparatory curriculum. Consequently, these differences in student graduation rates have major implications for the high school mathematics teacher and the high school mathematics curriculum.

Specifically, calculus plays completely different roles in the schools near the two endpoints of the range of college-intending students. Schools in which almost all students go to college are likely to have advanced placement calculus courses and high percents of students taking precalculus or calculus courses by the time they graduate. Calculus is perceived by students and teachers as a first-year college subject that is important for a diverse collection of fields of study and that might be appropriate for any well-prepared student. But in schools in which small numbers of students graduate from high school, calculus is rarely taught and enrollments in precalculus courses are low. Calculus is viewed as a course well beyond high school, seldom taken even by college freshmen, a mysterious course that is hard and taken only by those particularly interested in mathematics or science.

Another circumstance that affects student views towards calculus is that more than a quarter (364,000 out of 1.3 million who took the SAT) of students who are going to college this year are first generation college students [3]. These students are likely to view calculus as more arcane than other students whose parents went to college and know many people who studied calculus there.

Curriculum

What courses do students take in high school? The high school student typically takes four major subjects. Most four-year high schools recommend four years of English, three years of mathematics, three years of science, three years of social studies, and three or four years of a foreign language. This totals 16–17 major classes, and students who wish to get into the better colleges and universities usually take 4 years of mathematics, science, and languages; consequently they either take 5 major subjects most years or go to summer school. Students are also encouraged to participate in sports or other activities, such as band. Furthermore, more than two-thirds of students taking the NAEP 12th grade test reported having a part-time job, 56% reported working more than 10 hours a week, and 27% reported working more than 20 hours a week. Even ignoring family, a social life, and the world in general, many things vie for the time and attention of a typical high school student.

It is quite difficult to determine what mathematics U.S. students are taking in high school. NAEP questionnaires have historically asked students for the highest mathematics course taken. From this we can roughly determine the percents of students who have enrolled in each high school mathematics course. These percents have been rising steadily for 30 years [various NAEP reports cited in **5**, p. 15]. From the reports of the 12th graders in 1996, 92% had taken or were enrolled in first-year algebra, 79% geometry, 63% a second year of algebra, and 13% precalculus or calculus. The datum on precalculus or calculus enrollments disagrees with TIMSS data for enrollments collected a year earlier in which 22% of 12th-grade students were classified as taking or having taken precalculus or calculus [8, p. 19].

On the 2000 NAEP, the course-taking question was changed from prior years. Students were asked to identify the course they had taken in each of grades 8 through 12. From their responses, 94% report taking a year of algebra, 88% report taking a year of geometry, and 80% report taking a second year of algebra. That last statistic seems quite high to me, and I wonder if some students are not reporting a second year of first-year algebra as if it were what we might call "advanced algebra." Responding to the same question, 37% reported taking precalculus, 18% reported taking calculus, and 18% reported taking a course in statistics [1, p. 169]. Examining individual student responses, NAEP evaluators found that 50% of students reported taking either trigonometry, precalculus, statistics, discrete mathematics, or calculus before or during 12th grade.

These percentages gathered from the NAEP seem high, but they are not as high as those found from reports of students taking the SAT. The SAT asks students to report how many years of high school mathematics they have taken, but some students take courses over two years and others start high school mathematics well before high school. SAT-takers are not a random sample of college-intending students since weaker students often do not take them, and in non-SAT states only the best students take the SAT. But SAT data can indicate whether the NAEP data are on track. Last year, 24% of seniors taking the SAT reported taking calculus before graduation, and 45% reported taking precalculus. These numbers are up from 19% for calculus and 32% for precalculus ten years ago.

The most reliable data come from various studies of transcripts done by the U.S. Department of Education. These data tend to support these high percents. Table 13.1 shows that 10.7% of students in 1982 took mathematics beyond a second year of algebra and trigonometry, while 27.0% took such mathematics in 1998, almost triple the number.

The cluster of statistics on enrollments may be rife with overestimates but yields a robust conclusion: students are taking more mathematics now than they did ten years ago and far more than they did twenty years ago.

What content is covered in these high school mathematics courses? To answer this question, it helps to distinguish the traditional high school mathematics curriculum from others. The traditional high school mathematics curriculum emphasizes algebra and functions, de-emphasizes Euclidean geometry except as an important vehicle for learning about proof, and ignores statistics almost completely. From pre-algebra in the middle school through second-year algebra, much time is spent having students do algebraic manipulations of the type that are used in calculus, such as solving linear and quadratic equations and systems, and factoring and performing operations on polynomials. Functions are introduced starting in the second year of algebra and are a main focus of study through the remainder of the precalculus experience. Almost

Year	Level I alg/trig, trig, stat, anal. geom.	Level II precalculus Intro to anal.	Level III calculus anal.geom./calc
1982	15.5	4.8	5.9
1987	12.9	9.0	7.6
1990	12.9	10.4	7.2
1992	16.4	10.9	10.7
1994	16.3	11.6	10.2
1998	14.4	15.2	11.8

Source: National Center for Education Statistics, *The Condition of Education 2000*, pp. 66, 157, 216.

Table 1. Coursetaking in Advanced Mathematics (from various transcript studies)

all the content of the traditional high school mathematics curriculum is designed to prepare students for calculus.

Not only are greater percents of students taking geometry and second-year algebra than ever before but also these percents may be greater than the percent that will go on to college. The increased enrollments in these courses have an obvious effect on what teachers feel their students can successfully learn. Teachers can adjust either by easing the course somewhat or by broadening its content to be of interest, or they can keep the course the same and fail a significant number of students. Most teachers adjust the course, even if only slightly. So I believe less time is being spent on complicated algebraic manipulations involving rational expressions or radicals than used to be the case, even though this material is in the texts from which they teach. This also may be due to the influence of the NCTM *Curriculum and Evaluation Standards* [9].

The newer curricula, those that follow the guidelines of the NCTM Standards [9, 10] are clearly influenced by these increased enrollments. They are designed not only to prepare students for any of the first mathematics courses students might take in college—calculus, statistics, computer science, or finite mathematics—but also to appeal to students who may not continue their education. They give significant attention to mathematical modeling and statistics, de-emphasize the proof aspect of Euclidean geometry while concentrating on properties of measures, and discuss algorithms and discrete functions. They tend to downplay abstract work with polynomials, rational expressions, and radicals (e.g., see Core-Plus [4] and IMP [7]).

I have written elsewhere of the many similarities between the roles that calculus plays in the college curriculum and the roles that algebra plays in the high school curriculum [14]. Then perhaps it is not surprising that a common rhetoric is found in both discussions of calculus reform and reform of the high school curriculum. Broadening the mathematics experience by using real-world data and modeling, utilizing the latest technology, involving students more in their own learning, and downplaying manipulative aspects and proof are some of the commonalities. Accordingly, the same tension that exists at the college level between traditional and reform calculus exists at the high school level between traditional and newer curricula.

It would be nice to know the percent of students who are encountering the various curricula available to them, but these data are not systematically collected by any agency and are considered confidential by publishers. I estimate that at least 70% of students in the country are in a traditional curriculum. Probably a higher percent of the stronger (honors) students are in a traditional curriculum. These students are now regularly taking first-year algebra in 8th grade. The report of the NAEP 12th graders in 2000 that 29% had taken first-year algebra before 9th grade is in reasonable agreement with earlier reports of 20–26%.

If a significant percent of students are in the NSF-sponsored integrated curricula, it is not being captured by NAEP data. On the NAEP, fewer than 5% of students reported using an integrated or sequential curriculum in any year. This percent seems too low to me; perhaps we need a better descriptor than "integrated" or "sequential" (the New York state term) for these curricula. There are hybrids of old and new, like the University of Chicago School Mathematics Project (UCSMP) curriculum and some other materials, that have much of the spirit of the NCTM Standards while keeping the traditional course names and sequence. I think that about 20% of students encounter such hybrids, sometimes because their teachers supplement what is in one set of materials with what is in another set. Most teachers would like to teach everything, old and new, but there is not time. We and all other curriculum developers are constantly being asked what can be omitted in our books without hurting the student.

Calculators are used in all curricula. On the 2000 NAEP, 69% of 12th graders reported using a calculator every day, while only 10% reported never or hardly ever using a calculator [1, p. 159]. My recollection from other studies is that at least 50% of 12th grade students are using graphing calculators. On the NAEP, mean scores of students increased significantly with frequency of calculator use. Twelfth-grade students who reported using a calculator every day had a mean score of 309; those who never or hardly ever used a calculator had a mean score of 279. No causal connection can be inferred because when there is a choice, teachers are more inclined to allow better students to use calculators. The data do seem to indicate that, on the whole, calculator use did not harm students on the NAEP.

Advanced placement calculus

To segue from this overview into a discussion of the transition from high school to college, it seems appropriate to offer some data about the most visible college program in high school, the Advanced Placement (AP) program of the College Board. The AP calculus course has a significant impact on the mathematics programs of many high schools. It provides a road map to teachers indicating what their students should know to be prepared for calculus as well as providing a destination for the top students in a school. In 1999, 127,744 students took the AB Calculus exam (covering about 2/3 of a typical year-long calculus) and 30,724 took the BC Calculus exam (covering a full-year course), totaling about 5% of the graduating seniors in the country. This number constitutes about a 10% increase from 1998 and a 29% increase since 1995. Of those who took the AB test, 64% scored 3 or higher (considered a satisfactory score). On the BC test, 79% scored 3 or higher [3].

Calculus remains the course of choice in the mathematical sciences for those taking AP exams. In 1999, the AP exam in statistics was taken by 25,240 students, and 18,837 students took one of the two AP exams in computer science.

Above it was noted that 18% of all 12th graders on last year's NAEP and 24% of SAT-takers last year reported being enrolled in a calculus course. This suggests that only about one in three students enrolled in calculus in high school went on to take the AP exam. (An article in *USA Today* [6] uses the same estimate.) The majority seem to be taking the course in high school in order to show colleges that they are taking difficult courses and to increase their chances of getting a high grade in college calculus. That is, if they ever get to calculus. The data collected 15 years ago by Waits and Demana [15] may still apply today. They found that only about 28% of the 1721 freshmen who entered Ohio State University in 1986 with *five* or more years of college-preparatory mathematics were ready for calculus. For Ohio State, readiness for calculus was determined by a placement test that had "remained essentially unchanged for the past twenty-five years" [15, p. 11].

Calculus in high school? — Yes.

Data like that collected by Waits and Demana [15] are often used by college mathematics departments to discourage students from studying calculus before college. The argument is that the teaching of calculus in high schools is poor, often done by individuals unqualified to teach the subject, and results in students learning concepts in wrong ways. I do not understand the argument for two reasons. First, the environment in which calculus is taught is far better in high schools than in colleges. The teachers care about teaching and have often taught the course for many years. They know their students personally, having seen many of them in their courses in prior years, and they care about them. The teachers are typically the best-qualified mathematically and among the most experienced and able teachers in the school. Also, the students know and help each other in a familiar and comfortable setting. Second, when the same argument is (in my opinion, wrongly) offered to discourage the teaching of algebra before high school, it is found to be riddled with holes. Students need to have studied some algebra before high school in order to be successful in the typical high school algebra course. There is simply too much material to digest to expect a student to progress from never having worked with variables to the study of linear systems, quadratic equations, radicals, and rational expressions in a single year. To meet this problem, traditional textbooks now introduce significant amounts of algebra in the course preceding algebra. Likewise, some of the NSF curricula handle this problem by spreading introductory algebra over two or three years. Perhaps the most significant reason that calculus is so difficult for many students is that we try to teach it from scratch in a single year.

Technology has blurred the traditional boundaries between algebra and calculus. For instance, problems of maxima and minima can be solved by successive approximations to values of functions. Graphs of functions can be examined to estimate points of inflection and radii of curvature. Values of integrals can be estimated by entering a function into a graphing calculator and giving it appropriate commands. This makes some concepts and problems of calculus quite accessible well before formal treatments of these ideas are given.

Likely the best approach to an area of mathematics (such as calculus, algebra, functions, geometry, statistics, etc.) is to develop its concepts over many years. Then, when the time has come, that area should be studied in some detail. With this approach, all of the ideas that were introduced separately over preceding years can be studied in detail, reintroduced in a more formal manner, and seen as related to each other logically.

For calculus, this means students should encounter *multiple times* the topics of inequality, distance and areas on the coordinate plane, area, rate and rate of change, infinity, sequence, function, limit, max-min problems, and summation as part of their experiences with algebra, geometry, functions, statistics, and discrete mathematics. It also suggests that students have at least one introduction to derivatives and integrals. This is done in the UCSMP curriculum [11]. However, because calculus is not the only area of college-level mathematics for which students need some background, we found that this approach requires an extra year between second-year algebra and calculus devoted not only to these precalculus concepts but also to statistics and discrete mathematics.

Unfortunately, the push to get students to calculus is so strong that, except for the one-year geometry course, most schools ignore mathematics that is not viewed as precalculus. I believe that K–8 curricula should be designed so that algebra is taken by most students in 8th grade not because it makes it possible for them to take a calculus course in high school, but because it provides an extra year to prepare them for calculus and the other mathematics they are likely to encounter in college. Likewise, in an optimal calculus experience students should revisit some algebra, geometry, probability, and statistics from earlier years, and also be introduced to concepts of differential equations, complex variables, and algebraic structures that they might encounter in later years. College students' lack of exposure to the latter topics in early undergraduate mathematics courses is surely one of the reasons they have difficulty with them in later undergraduate mathematics.

So, who should take calculus in high school? Either those students whose K–6 experience is solid enough to prepare them to begin concentrated study of algebra in 7th grade, or those students who have studied the equivalent of an algebra course in 8th grade and are willing to put in the extra work in grades 9–11 needed to ensure success in their calculus experience in 12th grade.

Teachers and tests

If average performance is going up, and if students are taking more courses, why do students throughout the nation perform so poorly on college placement tests? Many factors may contribute. As had to be the case with Ohio State's 25-year-old test in 1986, placement tests often ignore much of the newer content and many of the newer ways of teaching that high school teachers are trying to incorporate: working with a variety of representations of functions, including graphical and tabular approaches, modeling and otherwise dealing with data; and using the technology of calculators and computers as a helpful tool in solving problems. There exist no review classes for placement tests as there do for the SAT and ACT. Often there are no sample exams nor a detailed syllabus given to students from which to study. High school teachers cannot prepare their students for a particular exam because their students go on to a variety of institutions. And these placement tests are often taken under conditions that are far from optimal. They are given to students who may be away from home for the first time, who may have just had a medical exam or come from long waits for ID cards, and who may have stayed up late the night before talking to other freshmen in their dorms.

Still, the large numbers of students taking remedial (*pre*-precalculus) courses suggest that many students do not know enough of the mathematics they need even for precalculus courses. How can this be, given that mean mathematics scores have risen? At the same time that the percent of the population going on to college has increased, college mathematics requirements have themselves increased. Some institutions require a certain competence in mathematics of all their students, and almost all institutions have seen mathematics requirements increase in fields such as psychology, business, the biological sciences, and the social sciences where requirements were once minimal. Students are taking these exams who, in previous decades, would not have had to take them or would not have attended college.

Most high school mathematics teachers realize that many of their students will study statistics and computer science in college, and that many will enter the health professions (currently the most popular major reported by students taking the SAT) or become business majors. All these areas require calculus, but not necessarily a mathematician's calculus. Students are well aware that computers and calculators are universally available outside the classroom, but that tests may not allow them to use this technology. Faced with a diverse group of students coming in who will be going out to a diverse set of institutions or to various workplace settings, high school mathematics departments and their teachers are faced with difficult decisions regarding what curricula and technology to use.

These decisions are made more difficult because the various high-stakes external tests have quite different goals. The SAT and ACT college entrance exams are themselves quite different from each other, but both are used as a criterion for college admission. Each of these also differs dramatically from the college placement tests given to students after they have been accepted by an institution. In turn, all these tests differ from the increasing number of state tests now being given to determine how a school is performing relative to other schools in the state, and which in a few places are being used to judge teacher performance or to determine whether a student can graduate.

Each of these tests has its own idiosyncrasies, content, and technology requirements. The SATs and ACTs allow graphing calculators. Many college placement tests disallow all calculators. Some college placement tests emphasize symbolic manipulation; others don't. Few touch upon statistics or geometry. Parts or all of these tests may conflict with the high school teacher's or the school mathematics department's own tests over the curriculum they are using, tests employed to determine student grades. Yet the teacher is under pressure to ensure that students perform well on all of these tests. Thus the high school teacher is beleaguered, faced with pressures from professional organizations and from a variety of testing agencies to get students to perform at high levels with and without calculators, on traditional content and newer content. They are expected to do this with a large population of students whose parents never studied the courses and thus have no idea why they are needed.

Calculus in college and the mathematics major

That a course or set of courses is entitled "precalculus" is a sign of the importance of calculus itself. High school and college mathematics faculties both tend to view calculus as the first college course in mathematics. But, for students, calculus is very much the culmination of five or more years of study. For their entire high school experience, and for any precalculus courses they take in college, much of what they study is justified by "You will need this for calculus." Many students believe that when they study calculus, they will finally learn what mathematics is all about, how mathematics explains the universe, and how to solve all sorts of problems.

This is too high a burden to place on any single course. Traditional calculus may have mathematical sanctity, but it rarely yields the epiphany students hope for. Students may have unreasonable expectations for their calculus course, but for most students, calculus does not bring even a small measure of power; it is just another mathematics course. I recognize that the reform calculus movement is motivated in part by a desire to attack this problem, but I wish to take the argument beyond calculus. This may be pushing the envelope since I have been asked to speak about precalculus, but one cannot examine preparation for calculus without examining all of the roles of calculus itself.

We need to examine why the number of mathematics majors in the country is declining even though high schools are turning out record numbers of students who have been successful in calculus. Calculus has lost its role as the first course for mathematics majors and cannot have that role for students who take it in high school. College mathematics departments need a broad-based post-calculus course for committed and potential mathematics majors to induce them into majoring in the mathematical sciences. Such a course should not assume that all smart people believe mathematics is beautiful, important, and useful, or that all mathematics majors become either teachers or research mathematicians. Discussions *about* mathematics should be part of the course's agenda.

Summary

Data from the National Assessment of Educational Progress and from SAT and ACT college entrance examinations establish that high school students know more mathematics than their counterparts at any time in the past 30 years. This is likely due to a significant increase in enrollments in college-preparatory mathematics and in calculus by high school students. For students formerly at the lower end of the performance range, these increases in enrollments have been encouraged by recommendations of "mathematics for all" from organizations like NCTM and the existence of reform curricula with a broader agenda than the preparation for calculus. Because we do not have data on classroom use of materials, we do not know whether or not the raising of performance is due to the influence of the NCTM *Standards* and these reform curricula. However, the use of calculator technology does not seem to have hurt performance and may have enhanced it.

Despite this rather uniformly positive picture, the number of students in remedial (*pre*-precalculus) courses in colleges is at record levels. While some of this enrollment increase is due to the increasing mathematics requirements for college students, some of this enrollment increase is likely played by mismatches between the mathematics taught in high school and the mathematics tested on college placement exams, reflecting differences between the content of precalculus mathematics taught in high school and taught in colleges. In particular, high school courses tend to place less emphasis on paper-and-pencil manipulative algebraic skills than their college counterparts, and high school courses influenced by reform efforts cover a broader range of content than the standard prerequisites for calculus.

For many students, the college mathematics placement test carries with it extraordinarily high stakes. It can keep a student from calculus for one or two years, thereby greatly influencing the majors possible to the student. Yet these tests receive little publicity and are accountable to no one. A national effort involving all the mathematical sciences is needed to bring these tests in line with high school curricula and teaching practices so that they can reward, rather than penalize, today's students for the wider range of knowledge and techniques which they bring to their overall study of mathematics.

At the same time, we need to encourage, rather than discourage, the teaching of calculus concepts before college, for we should apply what we have learned about successful approaches to algebra to the teaching of calculus. We also need to examine the role that calculus has played in the decisions of students to become mathematics majors, and possibly create post-calculus courses that will serve to encourage more students to study more college mathematics.

References

1. Braswell, James S., Anthony D. Lutkus, Wendy S. Grigg, Shari L. Santapau, Brenda Tay-Lim, and Matthew Johnson, *The Nation's Report Card: Mathematics 2000*, National Center for Education Statistics, Office of Educational Research and Improvement, U.S. Department of Education, Washington, DC, 2001.

2. Campbell, Jay R., Kristin Voelkl, and Patricia Donahue. *NAEP 1996 Trends in Academic Progress*, National Center for Education Statistics, Washington, DC, 1997.

3. College Board, Press release August 28, 2001. http://www.collegeboard.com (September 2001).

4. Core-Plus Mathematics Project, *Contemporary Mathematics in Context: A Unified Approach,* Courses 1–3, Everyday Learning, Chicago, IL, 1997.

5. Dossey, John, and Zalman Usiskin. *Mathematics Education in the United States 2000*, National Council of Teachers of Mathematics, Reston, VA, 2000.

6. Henry, Tamara, Advanced Placement faces crossroads, *USA Today*, September 26, 2001, p. 6D.

7. Interactive Mathematics Program (IMP), *Introduction and Implementation Strategies for the Interactive Mathematics Program*, Key Curriculum Press, Berkeley, CA, 1998.

8. Mullis, Ina V. S., Michael O. Martin, Albert E. Beaton, Eugenio J. Gonzalez, Dana L. Kelly, Teresa A. Smith, *Mathematics Achievement in the Primary School Years*, TIMSS International Study Center, Boston, MA, 1997.

9. National Council of Teachers of Mathematics (NCTM), *Curriculum and Evaluation Standards for School Mathematics*, NCTM, Reston, VA, 1989.

10. ——, *Principles and Standards for School Mathematics*, NCTM, Reston, VA, 2000.

11. University of Chicago School Mathematics Project (UCSMP), *Transition Mathematics* (2002), *Algebra* (2002), *Geometry* (2002), *Advanced Algebra* (2002), *Functions, Statistics, and Trigonometry* (1997), *Precalculus and Discrete Mathematics* (1998).

12. U.S. Census Bureau. http://www.census.gov/population (September 2001).

13. Usiskin, Zalman, Recent Developments in Secondary School Mathematics, and Their Implications, In Anita Solow (Ed.), *Preparing for a New Calculus*, MAA Notes 36, The Mathematical Association of America, Washington, DC, 1994.

14. ——, Algebra and Calculus for All? UCSMP Newsletter No. 18 (Winter 1996); also in the *Illinois Mathematics Teacher* (September 1996) and the *Journal of Mathematics and Science: Collaborative Explorations* (Spring 1999).

15. Waits, Bert K., and Franklin Demana, Is Three Years Enough? *Mathematics Teacher* 81:1 (January 1988), 11–15.

16. Wirtz, Willard, et al., On Further Examination, *Report of the Advisory Panel on the Scholastic Aptitude Test Score Decline*, College Entrance Examination Board, New York, 1977.

17. *The World Almanac and Book of Facts 2001*, World Almanac Books, New York, 2000.

14

Precalculus Reform: A High School Perspective

Daniel J. Teague

The North Carolina School of Science and Mathematics

In many secondary schools, precalculus reform has preceded calculus reform. Consequently, those of us in the high schools have considerable experience with the capabilities of successful precalculus students in a reformed curriculum and their performance the next year in calculus. From a personal perspective, two important issues are brought forward for consideration.

- Successful students in a reformed precalculus curriculum will have new attitudes and abilities that can be utilized successfully in the teaching of calculus.
- Reformed precalculus curricula can play an important role in preparing students for the rigor of theoretical mathematics.

Both of these issues have implications in the way the mathematical community views and accepts precalculus reform.

Introduction

At the post-secondary level, precalculus reform is a consequence of successful calculus reform. To the extent that calculus reform has been successful at a university, the modifications in the content and pedagogy of calculus argue for similar modifications in the preparatory precalculus course. However, at the high school level, and particularly at the North Carolina School of Science and Mathematics (NCSSM) where I teach, precalculus reform came first and has been the center of our curriculum for over a decade.

NCSSM opened in 1980 as the nation's first public boarding school for academically talented 11th and 12th grade students who show promise of exceptional development and/or special interest in science and mathematics. The approximately 585 juniors and seniors attending the public high school live in six residence halls on the Durham campus.

The mathematics department at NCSSM first revised its precalculus curriculum in 1986. As a consequence of this reversal in order, we have a lot of experience with the consequences of precalculus reform, particularly as it relates to success in calculus.

At the high school level, mathematics education reform has focused on the vast majority of students who are not preparing for calculus. The National Science Foundation has funded five major projects that restructure all of secondary mathematics, including precalculus. The web address for each project is presented in the appendix.

After reviewing many of these revisions in the structure and content of secondary mathematics, Tom Berger and Harvey Keynes in their article *Everybody Counts/Everybody Else* [2], expressed concern that students who will be our future mathematicians and engineers may not be sufficiently well prepared by the reformed curricula for those vocations. Can a reformed precalculus course prepare students for a career as a "professional user of mathematics"? In this paper, I will give some background on my department's movement into precalculus reform and present a personal view of its effects on the teaching of calculus.

Precalculus reform at NCSSM

All of the students taking precalculus at NCSSM are in preparation for calculus, and most aspire to careers in mathematically intensive professions. Since its first graduating class of 140 students in 1982, approximately 75% of the school's graduates have gone on to earn university degrees in mathematics, engineering, science, or computer science.

The core courses of the mathematics program, precalculus and calculus, are taken by 80% of the student body. Because we offer a modeling approach to learning mathematics, students are able to engage in the creative aspects of mathematics early in their mathematical development. The creative aspects of problem-solving engender a concomitant task commitment which, in turn, enhances and encourages greater creativity. Working in groups on investigative problems, the students also learn how to develop convincing arguments to support their conjectures as they defend their ideas with other members of their work group. During these investigations, students leave behind the role of "lesson-learner and exercise-doer," as described by Renzulli [4], to become creative producers of knowledge. This is the context within which I will address the issues of influencing the mathematics community.

Calculus reform at NCSSM

Students who have consistently engaged in creating mathematical solutions to challenging and intriguing problems have a very different view of mathematics than those fed an unrelenting diet of computations and formulas. Once we began teaching our reformed precalculus course, both students and faculty quickly became dissatisfied with the standard Advanced Placement Calculus course we offered. In precalculus, we had given students a voice; we were interested in their solutions, not just whether they could mimic our solution. By returning to an algorithmic calculus course, we had taken their voice away, and they rightfully resented it. In precalculus, we had also presented a vision of mathematics as a creative, investigative and challenging subject, where careful, reflective thought was prized above quick memory. Our calculus reform was forced on us by this dissatisfaction with the difference between our precalculus and calculus instruction.

Our experiences over the last 15 years suggest that a mathematics program based on challenging, open-ended problems, mathematical modeling in groups, using mathematical techniques and procedures in new and varied contexts, and activity-based lessons can give students a dramatic head start in mathematics.

Influencing the mathematics community

What insight does our 15 years of teaching a reformed precalculus course give us in influencing the university mathematics community as it begins to contemplate the necessity of precalculus reform? What aspect or result of a new precalculus course will make the mathematical community take note of the changes and support this effort? One simple answer is to deliver mathematically capable students to university calculus classes. So an essential question is, what makes a student ready for calculus? What mathematical skills are necessary to be ready for calculus and what habits of mind are sufficient? What role do the students' beliefs about mathematics and themselves as mathematicians play?

The transition from high school precalculus to post-secondary calculus is difficult for many reasons, but one important reason is that the post-secondary teacher doesn't know the skills, understandings, and expectations of their new students. If you try to teach my students with the mistaken belief that they know the mathematics I knew at their age, you will miss a great opportunity. My students know more mathematics than I did, but it is not the same mathematics; and I believe they know it differently. They have a different vision of mathematics that would be helpful in learning calculus if it were tapped. One example will illustrate some of the capabilities of my students that I did not have entering calculus.

The CO_2 Concentration Problem

In this problem, we will fit a combination model to the data from the Mauna Loa observatory describing the amount of CO_2 in the atmosphere as a function of time. A portion of the data is given below:

Year	Jan	Feb	Mar	Apr	May	June	July	Aug	Sept	Oct	Nov	Dec	Avg
1968	322.3	322.9	323.6	324.7	325.3	325.2	323.9	321.8	320.0	319.9	320.0	322.4	322.8
1969	323.6	324.2	325.3	326.3	327.0	326.2	325.4	323.2	321.9	321.3	322.3	323.7	324.2
1970	324.6	325.6	326.6	327.8	327.8	327.5	326.3	324.7	323.1	323.1	324.0	325.1	325.5
1971	326.1	326.6	327.2	327.9	329.2	328.8	327.5	325.7	323.6	323.8	325.1	326.3	326.5
1972	326.9	327.8	328.0	329.9	330.3	329.2	328.1	326.4	325.9	325.3	326.6	327.7	327.6

The actual table includes both the monthly concentration and the average for each year from 1958-1989. A graph of the full data set has the basic shape shown below. Find a model to describe the concentration of CO_2.

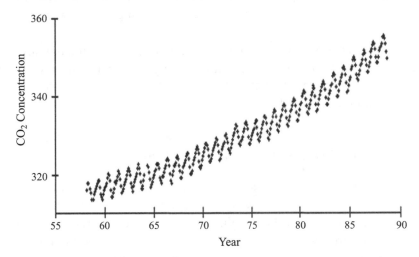

Students use the year and average CO_2 to fit an exponential model by first performing a vertical translation of around 300 ppm, then do a semi-log re-expression. The data is classically exponential, with a model similar to this:

$$F(t) = 1.65e^{0.03923t} + 299.5.$$

By subtracting the exponential model from the data to generate a trigonometric residual plot, the monthly fluctuations can be modeled. Using the maximal and minimal values from this residual graph, and the fact that the period is one year, the monthly variations can be modeled by the function

$$S(t) = 3.5\left(2\pi\left(t - \frac{1}{24}\right)\right) - 0.5.$$

Putting these two functions together creates a very accurate model for the data

$$C(t) = F(t) + S(t) = 1.656e^{0.03923t} + 3.5\sin\left(2\pi\left(t - \frac{1}{24}\right)\right) + 299.$$

The CO_2 concentration can be modeled surprisingly well with this model, a simple sum of two standard functions from precalculus.

Students can now give interpretations of the various parameters in the model and what they say about the increase in CO_2 concentration. (Modified from [1])

What is of value in these abilities to a teacher of calculus? The technical ability required in the problem, the knowledge of data analysis and transformations of functions, the willingness to think through the problem, to see the pieces in the whole, and to use known techniques in new combinations are all valuable. Why, exactly, is a student versed only in simplifying expressions like $\sqrt{abc^3}\,\sqrt[3]{a^4bc^2} = ac^2\,\sqrt[6]{a^5b^5c}$, as I was, better prepared to understand the concepts of calculus than the student who can develop a model for CO_2 concentration? And if it is true that they are, is it something inherent in the subject matter of calculus, the way we teach calculus, or what we expect to see when we test students in calculus that makes it so?

Often, students' experiences in mathematics classrooms convince them that being talented in mathematics means being able to give the expected answer quickly and with little thought. Their experiences in mathematics have convinced them that creative responses are counter-productive and just slow the class down. Mathematical creativity takes time. Creative students must be willing to persevere when the ideas do not come easily, to stay with the problem, to refine their solutions and improve upon their less creative work. Students must be working on problems that are sufficiently rich to allow for extended work on them and sufficiently interesting and engaging that they are willing to give the problems their time and intellectual energy.

The reformed precalculus courses seek proficiencies in both symbolic manipulation, which is essential for the performance of calculus, and modeling, which is important for understanding the nature of and applicability of calculus. The issue is the balance between these aspects of calculus readiness. I believe we all agree that all of only one is a bad idea.

My students are not me, your students are not you

I think about my preparation in mathematics and how it differs from my students' preparation. I have created a Venn diagram of the knowledge and understandings I believe I had when I began calculus and those I believe my students have when they begin calculus. The comparison is, of course, unfair, since what I knew is hampered by my memory (although it is interesting to see the residue of my high school mathematics career) while what my students know is inflated by optimism. Nevertheless, the comparison is striking. What would your diagram look like?

When you look at these overlapping ellipses, what do you see? If you look at what used to be in the knowledge set but is now missing, you will be disappointed. Look at what they can't do that I could. How can they be sucessful in a mathematics related field if they can't do X as well as I could? (Of course, some Xs are more important than others.)

If you look at what was never in my knowledge set but is in my students', you will be pleased. Look at what they can do that I could never have attempted. OK, they can't do X, but they can run circles around me on Y and Z. (And some Ys and Zs are more important than some Xs.)

If your school's placement test in mathematics contains only questions from the Xs, you will believe the students are poorly prepared and you will teach accordingly. More importantly, you will never know

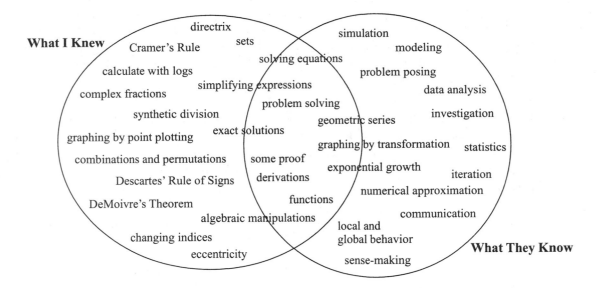

what abilities they truly have if you don't give them problems that allow them to show you what they can do. You will never know about *Y* and *Z* and, consequently, you will not be able to take advantage of these abilities in teaching them mathematics.

One essential aspect of influencing the mathematics community is to explain clearly what new capabilities and habits of mind students from a reformed course are expected to have. What do they know that we didn't know? If they don't have what you want, do they have something you can use, and perhaps use even more successfully than the algebraic skills of old? I think my students are better prepared for calculus than I was because they understand the geometry of functions much better than I did, and they have come to expect mathematics to make sense—and further, they expect to understand it. This expectation is an important, but seldom mentioned aspect of understanding the importance of proof.

Proof and sense-making

The purpose of proof and deductive argument should be made clear to students preparing for significant utilization of mathematics. Ken Ross, in the "Second Report of the MAA Task Force on the NCTM Standards" said, "It should be emphasized that the foundation of mathematics is reasoning. While science verifies through observation, mathematics verifies through logical reasoning... If reasoning ability is not developed in the student, then mathematics simply becomes a matter of following a set of procedures and mimicking examples without thought as to why they make sense" [5]. The AMS Association Resource Group, chaired by Roger Howe added, "The most important thing to emphasize about mathematical reasoning is that it exists—more, that it is the heart of the subject, that mathematics is a coherent subject, and that mathematical reasoning is what makes it so. ... Mathematics should simply be taught as a subject where things make sense and where you can figure out why they are the way they are." [3]

Can a reformed precalculus course support this view of mathematics as sense-making and simultaneously encourage the students to develop convincing arguments and proof? Based on our experience with a reformed precalculus at NCSSM, the answer is yes.

All of my high school friends and I believed mathematics was a list of things to do, each with an appropriate technique. I was considered good at mathematics because my list was longer than most. I thought a proof was another part of the list. I don't believe my students have the same misunderstanding of the nature of mathematics that I did.

Students have not been historically, nor are they now, focused on proof in the precalculus course. It is primarily a development of the elementary functions and their rules of operations. Proof in precalculus is primarily proof by derivation. Students derive results like the sum of a geometric series, the law of cosines, sum and difference formulas for sine and cosine, and the rules of logarithms. Most of these comprise a rewriting of expressions via algebraic manipulation. The level of manipulation taught in the course is often geared to being able to accomplish these derivations successfully.

It seems to me the first requirement of proof in mathematics is an expectation on the student's part that the mathematics should make sense and be supportable by argument. Do you expect it to make sense? This is the first requirement of understanding a proof. A proof is an effort at sense-making.

Looking at precalculus as a way of making sense of the world around you by modeling natural activities with the elementary functions sets the stage for proof as sense-making in a way that the formal, and often sterile, proof-based geometry course could never do, particularly when the focus of the proofs was on the format.

I received an e-mail from a student now in a PhD program in electrical engineering at Michigan saying, in part,

> Math Modeling occasionally makes a big difference in terms of understanding how to pose problems as well as contributing to that nebulous thing called "mathematical maturity" that is important for developing proofs and devising other mathematical tricks.

This is common sentiment from students as they move into a rigorous treatment of mathematics. Seeing mathematics as a body of knowledge, with connections and an underlying structure, rather than as a string of techniques is an excellent way to develop the meaning of and need for proof. There is often a sense that theory and modeling are mutually exclusive approaches to teaching mathematics. The mathematics department at NCSSM believes that they can coexist happily in the same classroom and that each supports the other.

Conclusion

Precalculus reform is just getting underway at the post-secondary level. The experience of high schools teaching from a variety of non-traditional precalculus texts can be very useful to university mathematics departments as they contemplate changes in their precalculus programs. Two-year colleges have also made significant recommendations for changes in the approach to teaching precalculus as a preparation for calculus. Communication among these groups can provide important insight as the process continues. Our experience with students in calculus suggests that successful students in a reformed Precalculus course have the essential skills for success in calculus. They also have an understanding of the nature of mathematics and a vision of mathematics as sense-making that, if fully utilized, can be very useful in the teaching of calculus.

References

1. Barrett, Gloria, et al., *Contemporary Precalculus Through Applications*, Everyday Learning, 1999.

2. Berger, Tom, and Harvey Kenyes, "Everybody Counts/Everybody Else," *CBMS Issues in Mathematics Education, Volume 5, Changing the Culture: Mathematics Education in the Research Community*, AMS, Providence, RI, 1995.

3. Howe, Roger, "Reports of AMS Association Resource Group," *Notices of the AMS*, Volume 45, Number 2, 1998.

4. Renzulli, J.S., & Reis, S.M., *The schoolwide enrichment model: A comprehensive plan for educational excellence*, Creative Learning Press, Mansfield Center, CT, 1985.

5. Ross, Kenneth, "Second Report from the Task Force," *MAA Online*, http://www.maa.org/past/maanctcm3.html, June 17, 1997.

Appendix

The five NSF curriculum projects with a description of the project taken from the included web address are described below:

1. Application Reform in Secondary Education(ARISE): *Mathematics: Modeling Our World* (Southwestern Educational Publishing)

   Mathematics: Modeling Our World is founded on the principle that mathematics is a necessary tool for understanding the physical and social worlds in which we live. This is not the same as saying that mathematics can be applied. Rather, important questions about the "real world" come first and serve to motivate the development of the mathematics. Thus the contextual questions "drive" the mathematics. As students discover a variety of ways to solve a problem, they not only learn mathematics and content in other curriculum areas, but they also learn how to reason mathematically, organize and analyze data, make predictions, prepare and present reports, and revise their predictions based on new information. Using technology and group work, students explore situations that offer a wide variety of mathematics concepts. Mathematical modeling is a central focus of the curriculum. In the modeling process, they identify key features of the context being studied, build a simple model, test it against various criteria, and modify the model in an effort to improve its description of the real content.

   For more information, go to: http://www.comap.com/highschool/projects/arise.html

2. Core-Plus Mathematics Project (CPMP): *Contemporary Mathematics in Context* (Everyday Learning Corporation)

   Contemporary Mathematics in Context is a four-year curriculum that replaces the traditional Algebra-Geometry-Advanced Algebra/Trigonometry-Precalculus sequence. Each course features interwoven strands of algebra and function, statistics and probability, geometry and trigonometry, and discrete mathematics. The first three courses in the series provide a common core of broadly useful mathematics for all students. They were developed to prepare students for success in college, in careers, and in daily life in contemporary society. Course 4 continues the preparation of students for college mathematics. It formalizes and extends important mathematical ideas drawn from all four strands, with a focus on the mathematics needed to be successful in college mathematics and statistics courses.

   For more information, go to: http://www.wmich.edu/~coreplus/

3. Interactive Mathematics Project (IMP): *Interactive Mathematics Program* (Key Curriculum Press)

   The Interactive Mathematics Program (IMP) has created a four-year program of problem-based mathematics that replaces the traditional Algebra I-Geometry-Algebra II/Trigonometry-Precalculus sequence and that is designed to exemplify the curriculum reform called for in the Curriculum and Evaluation Standards of the National Council of Teachers of Mathematics (NCTM). The IMP curriculum integrates traditional material with additional topics recommended by the NCTM Standards, such as statistics, probability, curve fitting, and matrix algebra. IMP units are generally structured around a complex central problem. Although each unit has a specific mathematical focus, other topics are brought in as needed to solve the central problem, rather than narrowly restricting the mathematical content. Ideas that are developed in one unit are usually revisited and deepened in one or more later units.

   For more information, go to: http://www.mathimp.org/

4. Math *Connections* Project: *MATH Connections: A Secondary Mathematics Core Curriculum Initiative* (It's About Time Publishing)

   MATH *Connections* is a unified approach that blends traditional mathematical topics around a common thematic thread. Using the NCTM *Standards* as a guideline, MATH *Connections* blends algebra, geometry, probability, statistics, trigonometry and discrete mathematics into a meaningful package that is interesting and accessible to *all* students. The text materials are designed to provide students with mathematical experiences that excite their curiosity, stimulate their imagination and challenge their skills. All the while, our primary concern is the conceptual development of the learner while focusing on these goals: mathematics as problem solving, mathematics as communication, mathematics as reasoning, and mathematics as making connections.

For more information, go to: http://www.mathconnections.com/

5. Systematic Initiative for Montana Mathematics and Science Project (SIMMS): *Integrated Mathematics: A Modeling Approach Using Technology* (Pearson Custom Publishing)

An integrated mathematics program consists of topics chosen from a wide variety of mathematical fields. It emphasizes the relationships among topics within mathematics as well as between mathematics and other disciplines. SIMMS IM materials are designed to replace all currently offered secondary mathematics courses, with the possible exception of Advanced Placement Calculus, and build on recent middle school reform initiatives. SIMMS IM modules were written by high school mathematics teachers.

For more information, go to: http://www.montana.edu/~wwwsimms/

15

The Influence of Current Efforts to Improve School Mathematics on the Preparation for Calculus

Eric Robinson and **John Maceli**
Ithaca College   *Ithaca College*

Introduction

In this paper, we will describe some of the changes in K–12 education that affect the mathematical preparation of students entering colleges and universities and who pursue a study of mathematics that includes calculus. Although ideas and research can be traced back further, we will take the year 1989 as a starting point, when two significant publications appeared that served as catalysts for many individuals engaged in efforts to improve school mathematics education. They are: *Everybody Counts: A Report to the Nation on the Future of Mathematics Education* [14], published by the Mathematical Sciences Education Board (MSEB), the *Curriculum and Evaluation Standards for School Mathematics* [10], published by the National Council of Teachers of Mathematics (NCTM). Some other influential publications followed including: *Reshaping School Mathematics* [15], from MSEB in 1991, *Professional Standards for Teaching Mathematics* [11], from NCTM in 1991, and *Assessment Standards for School Mathematics* [9], from NCTM in 1995. A planned update that also combined the three aforementioned NCTM documents into a single NCTM publication appeared in the spring of 2000 and is titled *Principles and Standards for School Mathematics* (PSSM) [12].

In general vision, the documents mentioned above are largely consistent with each other[1] and serve curriculum developers, teacher educators, policy makers and others with a general vision of school mathematics education as well as a challenge to make that vision a reality. This collection of documents should provide the interested reader with many details of that vision and rationale for the suggested changes in school mathematics that are not mentioned here. See [1], [8], and [13], also. For brevity in this article, the terms *Standards* and *Standards-based* refers to the common vision set forth in these documents, particularly as described in the documents from NCTM.[2] The term "traditional mathematics program" will refer to school curricula that predominated in schools prior to 1989.

The *Standards* vision suggests changes in every aspect of the school mathematics classroom in terms of content, pedagogy, student expectations, the use of technology, and student assessment. Because the suggested changes were so comprehensive, beginning in the early 1990s, the National Science Foundation funded a number of comprehensive multi-year curriculum projects at each of the elementary, middle and

---

[1] The update of the original three NCTM Standards documents, the PSSM, differs from its predecessors in some aspects but not in overall vision.

[2] See also [19].

secondary school levels intended to make the *Standards* vision a reality. These projects are listed in the appendix of this paper. Each project was developed over at least five years with input from practicing schoolteachers, mathematicians, teacher educators, and educational researchers. Each of these curricula realized the *Standards* vision in a somewhat different way. While there are other curricula commercially available that embody aspects of the *Standards* vision, this group of curricula is most familiar to the authors of this paper. In particular, the secondary curricula in the collection serve as a basis for many of the examples, comments, and conclusions below.

It is fair to say that as of this writing, most mathematics programs in our nation's schools are neither completely traditional (in the sense of what was typical in the early 1980s) nor completely *Standards-based* (in terms of the changes suggested in the above documents). Massive change does not occur quickly. Even if it did, there has been barely enough time since the publication of the visionary 1989 documents mentioned above for students to be entering calculus courses with a complete K–12 *Standards-based* education. Nonetheless, our major task here is to describe changes in the school mathematics curriculum that would or could result from full implementation of a *Standards* vision, concentrating on changes that affect the preparation of school students for calculus. To that end we will structure this article by asking questions and then providing answers with a *Standards-based* K–12 point of view.

Should the primary driver of the K–12 curriculum, especially in the upper grades, be preparation for calculus?

There is an argument that traditional school mathematics programs function in a way that the answer to the above question would be *yes*—especially if calculus is seen as *the* first college-level mathematics course. The scheme is something like the following: all students take at least two years of high school mathematics (usually algebra followed by geometry) and then enter courses designed to culminate in calculus (usually algebra II followed by precalculus). For those accelerated students who begin algebra in the eighth grade, there is "room" before graduation to take an AP calculus course. (There is increasing movement in some states to shift the study of algebra into the eighth grade for all students.[3] Presumably, then, the study of calculus would be accessible for all who succeeded in the calculus track.) Students move along this calculus track for as long as they can—or until they lose interest and are not required to take any more school mathematics. Students who don't succeed at some point along the way in the calculus track are deemed less capable. They either quit taking mathematics courses or are tracked into other mathematics courses that are seen as less difficult and that do not have calculus as the ultimate goal. Many parents, administrators, and possibly some readers of this article, see success in the study of school mathematics as synonymous with success in the calculus track.

However, even when students received passing grades, there is a body of evidence suggesting that, in reality, students in traditional mathematics programs were *not* being well prepared for calculus. See for example results from the Third International Mathematics and Science Study (TIMSS) begun in 1995 that suggest that even our best students rated poorly when compared with their international counterparts, [20], [7]. Studies such as these suggest that U.S. students have trouble retaining mathematical knowledge. Indeed, we repeat some of the same material at the same level over and over again in multiple grades. These same studies show our students have difficulty with multi-step and non-routine problems where they have to synthesize knowledge. That is, they have trouble transferring knowledge to new situations.

There are other indicators that suggest we need to look at the effectiveness of a college or university precalculus course if the goal is to prepare students to do well in calculus. In a study by Ruddock [16], significantly fewer students (about 20% fewer) at the University of Texas at Austin achieved a grade of

---

[3] According to Zalman Usiskin, the idea of shifting algebra into the eighth grade in the U.S. originated over fifty years ago. The influence of calculus (including its placement as a first college course and later as an AP option in some high schools) is one among several factors involved in this shift—but an important one.

B or better in both calculus I and calculus II if they preceded their study of calculus with a university precalculus course, than if they began their collegiate study of mathematics with calculus I. With regard to the effectiveness of prerequisite courses, see [17], also.

If the goal of traditional school mathematics programs was to prepare students for calculus and the college remedial classes follow suit, the set of statistics mentioned in this volume by Bernie Madison (see "Preparing for Calculus and Preparing for Life") and Shelly Gordon (see "Preparing Students for Calculus in the Twenty-First Century") should also be noted here. Slightly over one-third of all college students enrolled in two- and four-year college mathematics courses in the fall of 2000 were enrolled in "remedial" courses. Moreover, only a small fraction of students enrolled in even a single term of calculus. Here, then, students register failure of the goal by "voting with their feet" not to complete the goal.

The point is that while the goal of a traditional school mathematics program may have been to prepare students for calculus, the effect has been to miss the mark. Rightly, the *Standards* vision takes a fresh look at what students should know and be able to do as a result of completing a K–12 mathematics program in the twenty-first century. In particular, the aim of the *Standards* is not just to do the same things better. Rather, the *Standards* vision is broader than just preparing students for calculus.

In fact, preparing a subset of students for calculus is only *one* of several important and more explicit goals of the *Standards* vision. In addition to preparing students for further work in mathematics at the collegiate level, the time in school devoted to the study of mathematics in a *Standards*-based program also focuses on preparing mathematically literate citizens, preparing students for the mathematical needs of the workplace,[4] preparing students for the mathematical needs of other disciplines even at the school level, and helping students become independent mathematical learners. Helping students understand the beauty and power of mathematics and its role in our culture and society is also an objective. We want to emphasize that each of these goals applies to *all* students.

A school mathematics program whose major driver is preparation for calculus does not automatically meet these other goals. So, whatever the tradition or motivation was in the past, we conclude that the current answer to the question phrased above should be *no*. For example, college-bound math majors also need to be quantitatively literate and have mathematical tools beyond those usually associated with calculus. Furthermore, rather than filtering students at a young age, *Standards*-based mathematics programs appropriately strive to maintain access to high quality mathematics programs for *all* students throughout their school experience.

It should be mentioned that several of the above goals are reflected in many of the initiatives underway at the introductory college level. These initiatives include rethinking of college algebra, providing quantitative literacy, and determining core mathematics requirements for all students. While it is admirable that some colleges and universities are making the effort to design and implement courses that allow students different course options to meet some of the goals listed above, it is a mistake to view courses whose supporting goals are other than preparing mathematical science majors as less important and terminal. Minimally, if separate undergraduate courses address separate goals listed above, it is a mistake to provide no reasonable pathways from these courses to calculus or to other points of entry into mathematics major. For example, developing courses that precede or parallel calculus courses and stress mathematical modeling, the use of technology, and data analysis are important for today's students—both non-math majors and math majors. Students completing such courses both in high school and in college often have developed many abilities such as geometric understanding, problem-solving ability, ability to recognize mathematical patterns, and so forth—and perseverance that will serve them in successfully completing many of today's calculus courses.

---

[4]According to many sources, the mathematical needs of students in the 21st century workforce go beyond basic computational skills and include being able to recognize the mathematical aspects of situations, being able to apply mathematical concepts in non-routine settings and in complex problems, being able to handle problems that are open-ended or not well-formulated, being able to interpret both qualitative and quantitative data, being able to transform information into multiple representations, and being comfortable using technology. One such recently published source is *Manufacturing in the 21st Century: A View of the Future*; 2001 Conference Proceedings APICS-The Educational Society for Resource Management. See also [20].

In particular, these abilities were sought after in the calculus reform of the early 1990s.

The high school programs mentioned in the appendix address the goals mentioned above in the *same* set of courses rather than in different tracks. We feel that this would be a good tactic for undergraduate mathematics programs as well.

That's all well and good. But, don't we just need to examine the topic differences that there are in *Standards*-based school mathematics programs in order to understand the changes in students' preparation for calculus?

In a word, *no*. Schoolwork in mathematics prior to a calculus course affects preparation for calculus in ways beyond the list of topics studied. In part, as a result of the goals mentioned above, students coming to colleges and universities from *Standards*-based mathematics programs *will view the mathematics they study differently*.

To begin with, students will expect the study of mathematical topics to have intrinsic value as it is encountered—rather than just study topics because they will be useful later. At least in the curricula mentioned in the appendix, the mathematics that students study is usually developed within a context. In this way students see the relevance of the mathematics to a particular problem or situation upon which they are working and where mathematical insight and inquiry are primary goals.

The following example illustrates this point. In the *Mathematics Modeling Our World* (MMOW), secondary school mathematics curriculum students are presented with a problem or situation that serves as a reason and a context for the development of mathematical topics. The example is taken from "Wildlife," which is Unit 5 in the first year of the curriculum.[5]

Example 1   The problem below serves as the basis of Unit 5, "Wildlife," in the curriculum *Mathematics: Modeling OUr World*. It appears in Year 1 of that curriculum [5, Year 1, p. 505] and serves as the context for developing a number of mathematical topics and focuses on several steps in the mathematical modeling process. In particular, the questions below begin the process of establishing assumptions and mathematizing the situation. Year 1 usually corresponds to ninth grade. (*Reprinted with permission.*)

Bullwinkle Returns

Adirondack State Park is a six-million-acre wilderness area in upstate New York. Prior to 1980, the last moose recorded in the park had been shot in 1861. After 1980, however, some moose were again seen there. In 1988, it was estimated that between fifteen and twenty were in the park. In 1993, new estimates put the number at 25 to 30 moose. The New York State Environmental Conservation Department (ECD) conducted a survey at that time to determine what policies the public favored. A majority of the people surveyed favored a "gradual increase in the moose population as the animals migrate from nearby New England states and Canada and an expansion of their numbers through natural reproduction." Conservationists suggested moving 100 moose into the park over a three-year period. The ECD determined that such a plan would cost $1.3 million dollars.

Put yourself in the position of commissioner of the ECD, and suppose that you must make a recommendation to the governor about this situation.

Consider:

1. Pose a specific question concerning these data to which you would like an answer.

2. What additional information do you need in order to answer your question?

The MMOW curriculum emphasizes mathematical modeling. In different units of the curriculum, students address one or more of the steps in the modeling process: *identifying key features of the problem or situation; building a simple mathematical model; exploring the model and deriving mathematical conclusions; testing the conclusions against various criteria or data; revising the model if necessary.*

---

[5]The examples in this paper are excerpted and reformatted versions of material contained in published student materials. We refer the reader to the published texts for more complete information regarding how students encounter these examples.

The initial questions posed in Example 1 begin the process of identifying assumptions, key questions, and useful data. Central questions here include: under certain conditions, can one describe how the moose population will grow? How big will the moose population get?

Here is an excerpt from an activity in the unit "Wildlife." Part of the activity looks at different representations for certain models of population growth. These include recursive, closed form, and parametric representations. In this excerpt from the activity, students encounter simulation as a tool in the modeling process.

Example 2 This is the third student activity in the unit "Wildlife" [5, Year 1, pp. 528–530]. (*Reprinted with permission.*)

In this activity you learn to use simulation in what-if situations. In addition, you examine how modelers can compare models that use different (but related) variables.

1. Consider the pencil-pricing model that you first examined in Lesson 1. Recall that the first pencil cost 25¢, but each additional pencil cost only 17¢. This is an additive model, described recursively as $c_{next} = c_{current} + 17$, $c_{initial} = 25$. Note here that the explanatory variable is the number of pencils and that the initial value corresponds to 1 pencil, not to 0, as is frequently the case. For now, though, pretend that you know neither this equation or the corresponding closed-form equation. Here are some ways you might simulate this situation to gain some understanding if you had never seen it before.

a) Physical. Get several pencils and a lot of change. If these are not readily available, make substitutes. For example, strips of paper could be pencils and smaller squares of paper could be coins. Then act out the assumptions — the cost rules — for various numbers of pencils. Be systematic.

Start with one, then two, and so on. Record the relationship between pencils and price in an organized manner, and then look for patterns. For this situation (pencil and price), what are the strengths and weaknesses of this method? For what kinds of situations might this kind of hands-on simulation be more useful?

b) Calculator. Home screen iteration is a good way to simulate simple models. For the pencil situation, enter the initial price (25). Note that with this method you need to keep track of certain things in your head. Units is one of them, so remember cents. What else must you track mentally? now enter the formula ANS + 17 on the home screen. Repeated pressing of the ENTER key produces a sequence of numbers. What do these numbers represent? Record your results, and compare them to those you obtained in part (a).

c) Computer. One of the best tools for simulating recursive models is a computer spreadsheet. Open your spreadsheet. Set up a table with column headings "Pencils" and "Total Cost" in cells A1 and B1, respectively. In the next cell of the Pencils column (A2), enter the number 1. In the adjacent cell (B2), under the heading total Cost, enter 25 (or 0.25, if you prefer to work in dollars instead of cents). This sets the initial conditions. In the second cell of the Pencils column (A3), enter the formula to add 1 to the previous number; for many spreadsheet programs a formula such as =A2+1 or +A2+1 does the job. In the adjacent Total Costs cell (B3), enter the formula to add 17 to the previous number, such as =B2+17. Then extend these last two formulas down a few lines. Compare the results of your spreadsheet to those from the previous two methods. Comment on strengths and weaknesses of this method.

d) Replying on only a simulation (or two), determine the cost of 25 pencils if the store changed its prices so that the first pencil still cost 25¢, but each additional pencil cost only 13¢. That is, find $c(25)$.

e) Relying again on only a simulation model (or two), determine $c(78)$ if the store changes its prices so that the first pencil costs 27¢ and each additional pencil costs 17.

f) Still relying on only a simulation model (or two), determine the cost of each additional pencil if $c(1) = 23$ (cents) and the store owner wants $c(50$ to be $6.60.

g) The store owner now wants to price pens in a similar fashion. Relying only on a simulation model (or two), determine the cost of the first pen and the (lower) cost of each additional pen if the store owner wants $c(50)$ to be $19.75.

h) Comment on the usefulness of the various types of simulations for investigating variations of the original setting.

During the unit, which takes several weeks to complete, students will develop several mathematical models of population growth. Students will extend their knowledge of recursive functions. They will apply their knowledge of additive models in recursive representations (for example, models of the form $p_{n+1} = p_n + C$) and their linear closed form representations as well as make connections to other topics they have previously encountered (such as parametric equations). They will develop the idea of a multiplicative recursive model (for example, models of the form $p_{n+1} = ap_n$) and develop its closed form representation (exponential function) using functional notation. They will develop the laws of exponents. They will compare additive (linear) models with multiplicative (exponential) models graphically, algebraically (recursively and with closed form) and with tables. They will compare how the "control numbers" a and b affect the graphs of $f(x) = ax + b$ and $f(x) = ab^x$ ($a, b > 0$). They will be able to fit both linear and exponential functions to two data points. They will study exponential growth and decay together with the concepts of constant growth rate, relative growth rate, and growth factor. They will further their understanding of probabilistic modeling and simulation. Technology is used to study long-term behavior of these models.

In Example 1, the contextual situation for development of the mathematics is a real-world problem. Mathematical modeling is a key feature of many of the secondary curricula mentioned in the appendix. Now, a real-world context certainly is one way to effectively eliminate the student question "when will I ever use this?" But, the value of contextual development extends beyond beneficial real-world settings and it is false to characterize *Standards*-based contextual development as merely emphasis on real-world applications of mathematics. Contextual settings may be fanciful or abstract[6] as well as real. (More is said about contexts below.) The point of a context is to set up an environment that is both meaningful for students and where students see (mathematical) inquiry worth pursuing. That is, the "when" in the student question mentioned above in this paragraph can mean "is this *ever* relevant or interesting?"

In the next example, the context for mathematical development was not chosen because of its real-world usefulness. It has both real world and fanciful components. Yet, evaluation information from the field indicates that high school students find the problem interesting and the solution worth pursuing. The mathematical basis of the context was adopted from one of Ross Honsberger's "Mathematical Gems" [6, pp. 43–53]. The problem forms the basis of Unit 2 in the third year of the *Interactive Mathematics Program* (IMP). The unit is titled "Orchard Hideout." A paraphrase of the unit problem is:

> How long after an orchard is planted will the trunks of the trees be so thick that someone standing at the center of a circular orchard cannot see outside the orchard? That is, how long before the center of the orchard becomes an "orchard hideout"?

Example 3   The unit problem paraphrased below is from "Orchard Hideout," Unit 2 in Year 3 (usually eleventh grade) of the *Interactive Mathematics Program* [4, Year 3, pp. 60–66]. The unit problem sets the context for developing a combination of mathematical topics from a variety of areas. (*Reprinted with permission.*)

The particulars that are given include the size and layout of the orchard:

> Madie and Clyde bought a lot in the shape of a circle. They planted their first row of trees along an east-west line through the center of the circle. The trees were equally spaced, except that they left out the tree that would have been located at the exact center of the circle. There were 50 trees to the east of the center and 50 to the west. The trees at the ends of this east-west row were exactly on the boundary of their property.
>
> Then they planted a north-south line of trees through the center, using the same spacing as before and omitting the tree at the center. Again, there were 50 trees to the north of the center and 50 trees to the south. And, again, the trees at the ends of this north-south row were exactly on the boundary of their property.

---

[6]In many cases in the curricula mentioned in the appendix, when an abstract context is used it has been derived from a more concrete situation. This is one way in which students may begin to see the value of abstraction.

They used each of the trees in their north-south row as the center of an east-west row, filling in the orchard with rows of trees. They always used the same distance between trees in every row.

The first activity is for students to make a model of the situation. The second activity is for students to begin to determine what questions need to be asked and to list some assumptions that need to be made. Assumptions made include that the trees in the orchard were all the same size when planted, they all grow at equal rates, the trunks of the trees are perfect right circular cylinders, and it is the trunks that will block the lines of sight (for sure) rather than the branches, etc. In some classes, students have begun to understand the situation by physically modeling the setup for a smaller orchard using the students themselves as trees standing on intersections of grid lines determined by floor tiles. The mathematical model that eventually is built is a two-dimensional representation of the situation with circles to represent the trees centered at each of the lattice points under consideration. The problem then becomes determining a last line of sight from the center of the "orchard" to a point outside the orchard as the circles grow in size. Background knowledge for work on this problem includes an understanding of geometric similarity, right triangle trigonometry in terms of the sine and cosine, and the Pythagorean theorem. In the unit, the mathematics developed encompasses a study of circles (including formulas for the circumference and area of circles as well as the concept of tangent lines), perpendicular lines, coordinate geometry including the distance formula between two points, the distance from a point to a line, the coordinate midpoint formula for a line segment, equations for circles, completing the square, geometric proof (synthetic and coordinate), a statement and its converse, and if and only if statements. This unit requires between two and three weeks to complete.

Although the geometry examples from "Orchard Hideout" may seem unusual to include in a paper on precalculus, our point is that there is more happening mathematically than the development of the content itself. We reiterate our point that the students coming to calculus from a *Standards-based* background will expect to be given more reason to study a mathematical topic than "it will be needed later or for the next course." In this sense, a mathematics course in precalculus that includes primarily bare concepts, technicalities, and computational techniques simply because they will be needed in the next course—that is, calculus—is outmoded.

There is more to point out. In "Wildlife" and "Orchard Hideout," *contextual development* of mathematical ideas is different from the usual *application* of mathematical concepts and techniques. Contexts serve as environments in which mathematics naturally arises and then can be developed. These problems are of a substantial nature. A solution (or several solutions) takes several weeks of study and development. Furthermore, a context is used not only as motivation for studying the mathematics, but also to provide a setting and framework that helps students understand and relate concepts.[7] By contrast, often in traditional school programs, applications usually *follow* the development of mathematical concepts. Moreover, applications are usually comparatively short.

Another result of using the real-world or fanciful contextual development of topics is that students are often presented with a concrete problem or situation *first*—and generalize the mathematics from this particular situation. Thus, such a contextualized approach often results in moving from the concrete to the abstract, or from the specific to the general, rather than the other way around from general theory to specific application.

Perhaps most important, mathematics is fundamentally about inquiry and insight, wonder and explanation, and questioning and making sense. In describing the work of mathematicians, one of the first things the authors of this article would say is that mathematicians answer questions.[8] We believe that when most mathematicians talk about doing mathematics, they are usually referring to a process that is much broader than doing a calculation or providing the technical, logical details in the proof of a result. This

---

[7]We will not delve deeply into the learning theory aspects of contextual development here as it is a digression.

[8]In contrast, the general public would probably say first that mathematicians do computations.

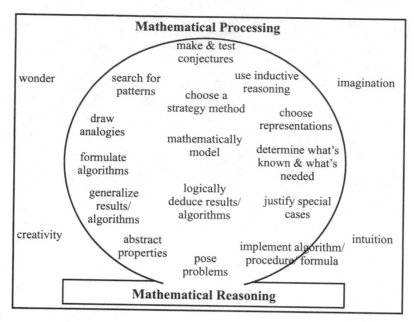

Figure 1. The mathematical reasoning process requires much more than the technical detail in formal proof writing. Mathematical processing includes more than reasoning; it involves wonder, creativity, intuition and imagination among other things. The reason that "implementing an algorithm..." is put on the edge of the circle is that sometimes implementing algorithms requires reasoning, such as when choosing what formulas or algorithms are appropriate or helpful. However, sometimes this is not reasoning, such as in cases where one is given a formula and asked to compute results using certain values of the variables.

mathematical processing can include one or all of the following activities: *exploring examples; searching for patterns; drawing analogies; utilizing intuition; using inductive reasoning; constructing mathematical models; posing problem; abstracting and/or generalizing properties; constructing and testing conjectures; justifying special cases; and finally, providing a deductive, general argument, possibly with additional assumptions or axioms, in order to obtain a mathematical result.* Furthermore, most certainly, wonder, imagination and creativity play a role. (See Figure 1.)

Getting a feel for a mathematical situation and developing insight are extremely important parts of the process. Standard communication of a finished mathematical result in a theorem/proof format is intended to be an efficient method of communication. In some cases, however, if taken by itself, a proof can obscure many of the underlying aspects of the reasoning process or fail to contribute to mathematical understanding.[9]

Here are a couple of additional activities from "Orchard Hideout." By the time students reach this point in the curriculum, they will be expected to justify any conclusions. How many characteristics of reasoning listed above do you think should evidence themselves as students work through the problems? How is "Equally Wet" related to "Garden Path"? How is "Garden Path" related to a solution of the central problem? What familiar geometric theorems should surface in these activities?

Example 4 This activity is an example of a POW (problem of the week). It appears on page 6 in the "Orchard Hideout" unit [4, Year 3, pp. 69–70]. Students are expected to work on this individually for several days, outside of class and write up their results as indicated at the end of the example.

---

[9]As an example, a straight inductive proof that $1 + 2 + \cdots + n = n(n + 1)/2$, where n is a positive integer, verifies that the formula works. But, a classic proof of this formula, often attributed to Gauss, which adds the first n integers twice and then recombines terms to add $n + 1$ a total of n times (thus showing that $2(1 + \cdots + n) = (n + 1)n$), provides more insight into the derivation of the formula than the inductive proof. As a second example: other than for insight, why would Gauss have provided at least five proofs of the quadratic reciprocity law?

Equally Wet

1. Two delicate flowers were planted in a garden. The gardener, Leslie, has a sprinkler that sprays water around in a circle. The closer a flower is to the sprinkler, the more water it gets.

To be sure that her flowers each get the same amount of water, Leslie needs to place the sprinkler where it will be the same distance from each of the flowers.

What are her choices about where to put the sprinkler? Describe all the possibilities. (*Reminder*: The flowers are already in place, and Leslie needs to adjust the position of the sprinkler relative to the flowers.)

2. Now suppose Leslie plants three flowers and wants to know if it will still be possible to place the sprinkler the same distance from all three.

a. Determine which arrangements of the flowers (if any) will make this possible and which (if any) will make it impossible. (As in Question 1, Leslie will be looking for a place to put the sprinkler after the flowers have already been planted.)

b. For those arrangements for which it will be possible, describe how Leslie can find the correct location (or locations) for the sprinkler.

3. What about four flowers? Five flowers? Generalize as much as you can.

You POW is to explain as fully as possible, for various cases, where Leslie can put the sprinkler in order to give the flowers the same amount of water. *Homework 2: Only Two Flowers* gets you started with the first question of the POW.

Write-up

1. *Problem Statement*: State the problem in mathematical language without reference to the context. That is, describe the problem in geometric terms without talking about flowers or sprinklers.

2. *Process*

3. *Solution*

4. *Evaluation*

5. *Self-assessment*

Example 5   This is an in-class activity from "Orchard Hideout" [4, Year 3, p. 84] that follows the POW mentioned in Example 4. It would be expected that students would have completed Example 4 and most likely one or more students would have presented solutions in class. Again, students would be expected to supply reasoned justification for conclusions. (*Reprinted with permission.*)

Down the Garden Path

Leslie (The gardener in *POW 2: Equally Wet*) has decided to plant only two flowers. She has placed them in her two favorite spots in the garden.

Now she has another problem. She wants to make a straight-line path through her garden, with one flower on each side of the path.

Leslie has decided that the two flowers should be the same distance from the path. That way, people walking along the path will see them both equally well (although they may pass by the two flowers at different points along the path).

1. How can Leslie design a path that is equidistant from each flower? Write down simple, step-by-step instructions for her.

2. Is your path the only one possible? Describe all possible straight-line paths that are equidistant from each flower.

A true search for proof usually requires exploration, creativity, cultivated intuition, and good judgement. Students in *Standards*-based mathematics programs will have explicit experience with these *sense-making* aspects of mathematical understanding, including what constitutes mathematical proof and the role of logic in mathematical certainty.[10]

There are several ways the *Standards*-based curricula emphasize the inquiry and sense-making activity of mathematics. An obvious way to promote mathematics as inquiry is text format and lesson structure. Students should be asked many questions—implicitly or explicitly. The types of questions and problems students encounter are extremely important. We quote Howard Eves [3, p. ix].

> There is a distinction between what may be called a *problem* and what may be considered an *exercise*. The latter serves to drill a student in some technique or procedure, and requires little, if any, original thought...In contrast,...a problem, if it is a good one for its level, should require thought on the part of the student. The student must devise strategic attacks, some of which may fail, others of which may partially or completely carry him through...To be suitable, a problem must be such that the student cannot solve it immediately...It is impossible to overstate the importance of problems in mathematics. It is by means of problems that mathematics develops.

Students need to be asked questions and encounter problems that require synthesis of prior knowledge and where the mathematics needed to solve them is not obvious in advance. They need to solve non-routine problems (that is, problems that are not closely related to problems already studied). They need to meet questions and problems that require the refinement of mathematical ideas, extension of concepts, and the development of additional mathematics. The unit problems in Examples 1 and 3 fit in the last category. The types of questions and problems just referred to are usually subsumed under the general heading of "problem-solving." (Problem Solving is one of the standards mentioned in the PSSM.) The point here is that students need to be asked questions and solve problems that get beyond the regular routine of requiring them to simply apply the last mathematical topic or technique studied to solve a problem. They need experience with a wide array of mathematical problems as they actually occur. (Competency with routine applications does not necessarily imply competency with non-routine problems or problems that require synthesis.) In order to promote mathematics as a sense-making activity students repeatedly need to be expected to justify, demonstrate, explain, show, confirm, defend, support, and so on. Such words may be included explicitly in text or implied in the program.

In the *Standards* vision, the level of sophistication and degree of justification required of students increases as students progress through the grades. Moreover, the reasoning activities as outlined in Figure 1 that are required of students as well as the types of problems mentioned above permeate a *Standards*-based mathematics curriculum. Such activities are not confined to particular topics (like geometry, for instance). The majority of student reasoning will be done in informal ways.[11] The use of the term informal, here, means that the reasoning need not conform to a specific format, such as a two-column proof. It does not imply a lack of soundness of argument or explanation. Upon completion of a school mathematics program, it would be expected that students would know the difference between an assertion (conjecture), statistical support (or support offered by a few examples) and deductive certainty. In addition, they should be able to develop and evaluate some mathematical arguments and proofs as well as select and use various types

---

[10]Logical deduction associated with mathematical proof can be used as a tool for exploration as well as verification. For instance, suppose I seek to determine if a given conclusion follows from a certain hypothesis. One strategy may be to see what I can (easily) prove with the given hypothesis and then look for how these results are related, if at all, to what I want to prove. This is often a good strategy when attempting to prove geometric theorems. However, students may not explicitly identify and experience this aspect of logical deduction at the K–12 level.

[11]In the era preceding the Standards, many schools emphasized techniques such as two-column proof formats and propositional logic proofs using truth tables. The intent was to make the logic more transparent and the thinking easier. But, the effect of this approach appears to have been different. Too much concentration on the form early on may lead students away from a desire for real insight into a mathematical problem—a desire that helps provide direction in thinking. Instead of mathematical proof becoming a useful tool to solve a meaningful question or add to mathematical understanding, for many students, it appears to have become, at best, a detached, confusing, and haphazard activity.

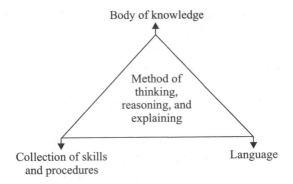

Figure 2. In the outside world, mathematics is often described as a body of knowledge, a collection of skills and procedures, a language, etc. While all of these are aspects of mathematics, each derives from the fact that at its core, mathematics is fundamentally a way of thinking about and explaining certain phenomena.

of reasoning and methods of proof. For further detail, consult the PSSM, particularly the standard on Reasoning and Proof.

What is important to note here, both in terms of student background and student expectations is that, as a result of well-implemented *Standards-based* curriculum, students should see mathematics (and its usefulness) as something *beyond* a collection of skills, procedures, definitions, and facts. They should see mathematics as both content and *process*—activity requiring both the utilization of content and reasoning skills. To give a bit more detail in the term "process" as used here, both the 1989 *Curriculum and Evaluation Standards* and the 2000 *Principles and Standards for School Mathematics* have standards relating to content and process. In particular, in the PSSM, there are five process standards that should permeate the K–12 mathematics curriculum: Problem Solving, Reasoning and Proof, Communication, Connections (both between mathematical ideas and between mathematics and other areas), and Representation.

The authors see mathematical thinking, explaining, and communicating as central to the discipline. In current efforts to improve mathematics education, we maintain that having a major emphasis on *student* activity that involves the process of doing mathematics is a key shift from traditional approaches. (See Figure 2.)

We already have referred to the first two "process" standards in PSSM (Problem Solving and Reasoning and Proof) in our discussion. The Communication Standard refers to the ability of students to organize and consolidate their mathematical thinking as well as coherently and precisely express it. They also need to be able to analyze the mathematical thinking and strategies of others. The Connections Standard emphasizes the need for students to make connections between mathematical ideas as well as to utilize mathematical ideas and thinking in contexts and applications outside mathematics itself. The Representation Standard emphasizes the need for students to be able to create (model), select, apply, and translate among mathematical representations to solve problems. This refers to visual (graphical, geometric, diagrammatic, and so forth), symbolic (including, among other things, both recursive and closed form algebraic representations), tabular, and verbal (written) representations. Evidence of the process standards should be apparent even in the few activities that we have chosen to include in this paper.

To return to the question: should we just consider the variation in topics between new and old pre-calculus courses? No. Mathematics is more than just topics. We have argued here that mathematical processing—as broadly defined above—should be considered with regard to student preparation as they enter calculus.[12] We also assert that this type of inquiry based "mathematical processing" should be a central part of any further study of mathematics.

---

[12]The process standards not only support the inquiry basis of mathematics, they also provide means to address learning deficiencies that many U.S. students have. These deficiencies include the lack of long-term memory of the mathematical topics studied and the inability to transfer knowledge to new settings.

If students believe that mathematics is fundamentally about inquiry, students should be ready to learn new concepts and techniques as they need them—and learn them relatively quickly. The downside, if there is a downside, is that students will expect some setting in which they see intrinsic value in learning new material. In addition, if students believe and have encountered work emphasizing that mathematics is about understanding, if students have had experience discussing and communicating about and with mathematical ideas, then they are better positioned to learn some mathematics independently. As mentioned above, this is one of the goals suggested in the *Standards*.

Furthermore, an approach that focuses attention on mathematics as a sense-making activity is consistent with the reason that calculus is one of the greatest inventions of the human mind. The theory and techniques of calculus epitomize the essence of Figures 1 and 2. From a mathematical perspective, the marvel is not just the techniques in themselves, but that the conceptualizations of limit, Riemann sums, the derivative, the integral, etc. can be connected in such a way that a whole collection of computational techniques can be derived. Wouldn't it be wonderful if students felt this incredible accomplishment instead of feeling the purpose of the course was *only* to equip them with the collection of computational techniques that they could utilize when asked to do so?

In summary, students who have completed a *Standards*-based program such as those mentioned in the appendix should come to college with a new set of abilities and expectations as a result of attention to the process standards. At a minimum, the students should:

- Have a view of mathematics that contains an approach to inquiry methods and strategies
- Be competent and confident problem-solvers
- Have a willingness to tackle and persist with complex problems
- Have experience discussing and expressing mathematical ideas (in words and symbols)

Content also is important and there is a sequential nature to the development of mathematical content. How does the content in *Standards-based* school mathematics education lay a foundation for calculus?

In the United States we do not have a national school mathematics curriculum. What is taught in each grade differs from state to state, sometimes from school district to school district within a state, sometimes from school to school within a district and sometimes from classroom to classroom within a school.[13] The situation is the same at the college level! Focusing just on introductory calculus, at present there is not just one calculus course or sequence that students are likely to take in college; there are many different possibilities nationally and often within the same institution. As a first calculus course there is traditional college calculus, several types of reform calculus, business calculus, honors calculus, etc. Even the same course, such as traditional calculus, varies in course content institution by institution. We will not digress here to debate the issue of a national curriculum at any level. We simply mention that the right to such local decision-making with regard to curriculum, on both sides of this coin, is often fiercely maintained. This state of affairs defines a portion of the terrain to be navigated by any curriculum reform and makes perfect articulation between pre-calculus courses at the school level and calculus in college extraordinarily difficult if not impossible to achieve on a large scale.

The *Standards* vision does not detail what specific topics should be taught at each grade level. It does not propose to be a national curriculum. This vision, which sets professional guidelines and goals, lies in-between isolated local decision-making about curriculum and a national curriculum. However, there is guidance on content within the grade bands. In addition to the five Process Standards mentioned above, the PSSM contains five Content Standards for the grade bands Pre-K–2; 3–5; 6–8; and 9–12. The Content

---

[13]There are some groups that are working on the problem. For example, at this writing, the organization Achieve has a draft document outlining what it suggests (all) mathematics students should have by the end of 8th grade. This does not imply national consensus, however.

Standards detail five content areas to be addressed at each of these grade bands. They are: Number and Operations, Algebra, Geometry, Measurement, and Data Analysis.[14]

We do not intend to summarize the PSSM content areas in this paper. We strongly refer the reader to that document for a detailed discussion of specific content. However, at this point we will offer a brief look at a small part of one content standard as an example. Under the 9–12 Algebra Content Standard it is stated "all students should... understand and compare the properties of classes of functions, including exponential, polynomial, rational, logarithmic, and periodic functions" [12, p. 296]. Example 6 (following pages) is one of the activities from year three of *Contemporary Mathematics in Context* (CMiC) that illustrates the type of activity students might do in the process of addressing this expectation.

The Content Standards suggest a reorganization and alteration of the content of school mathematics. Traditionally, in most states, data analysis received little or no attention for calculus-bound students.[15] Now, there is significant inclusion of content in this area throughout the pre-K–12 curriculum. Secondly, since each of the five aforementioned content areas are addressed at each grade band, it is clear that a *Standards*-based curriculum is sequenced very differently from the very familiar algebra, geometry, algebra II, precalculus curriculum that still predominates the national landscape in high school. An intentional effect of the *Standards*-based approach is to move away from the tendency to isolate topics. Rather, the approach is an attempt to unify content. This harmonizes with the Connections Standard. It also serves as a method to enable students to utilize prior knowledge in new contexts—as topics are revisited at deeper levels in later courses.

Curriculum programs that include content from more than one content area within the same year are often referred to as "integrated" programs. Several curriculum programs, including those mentioned in the appendix, actually mix content areas within the same unit of study. The "Wildlife" unit is usually encountered in ninth grade. The content includes material from algebra (laws of exponents), precalculus (functions, particularly linear and exponential), discrete mathematics (recursive relations) and probability. "Orchard Hideout," typically encountered in the third year of high school, includes content from geometry, trigonometry, algebra, and coordinate geometry (precalculus).

Because a vast number of mathematical problems involve more than one content area naturally— including many of those that require mathematical modeling—an integrated approach to curriculum extends the variety of problems students can work on and widens the possibilities for sequencing course material.[16] Actually, several aspects of the *Standards* vision suggest consideration of different sequencing and different choices of mathematical topics. In the curricula we have listed in the appendix a content-organizing principle might be large problems, (this is one of the themes in IMP), the process of mathematical modeling (MMOW), the incorporation of technology (SIMMS), and/or other principles. (For example, CMiC organized the material, in part, based on consideration of the question: "what if this course were the last mathematics course a student would take?" In this respect, CMiC was partially responding to the statement mentioned in *Everybody Counts* that indicated that students were leaving the study of mathematics at exponential rates as soon as the study was not required [14, p. 6].

We will give brief mention of a few of the topics that are probably in the background of students with a *Standards*-based education that are relevant to the preparation for calculus and that may not be in the background of more traditionally educated students. We use the word probably because, as we mentioned above, we do not have a national curriculum in this country. Even looking at textbook topics does not necessarily provide a perfect match to what students study. However, students most certainly will have a significant understanding of the concept of function. They will be able to move among representations of

---

[14]The 1989 Curriculum and Evaluation Standards from NCTM included content standards on Discrete Mathematics for grades 9–12. In the updated PSSM, discrete mathematics is distributed across the other content standards and across grades pre-K–12.

[15]Although students may elect to take an AP Statistics course—an option that is offered in an increasing number of schools nationwide.

[16]The curricula mentioned in the appendix each offer a different coherent approach to the order, and sometimes year, in which material is presented.

Example 6   This activity appears in Unit 6 "Families of Functions" in Year 3 (usually eleventh grade) of the *Contemporary Mathematics in Context* secondary school curriculum [2, Year 3, pp. 431–433]. In this activity students organize information about function types they have studied. The term "NOW-NEXT equation" refers to a recursive relationship. (*Reprinted with permission.*)

Investigation 3: It's All in the Family

Studying the causes and effects of change in Earth's climate requires modeling and forecasting change in many variables. The functions that model relations among those variables and their patterns of change over time are often drawn from several important families that you've studied in earlier work: linear, exponential, power, and periodic. You've probably discovered that, to use those functions as descriptive and predictive models of change, it's important to know well the numeric, graphic, and symbolic patterns that are typical of each type.

The activities in this investigation ask you to construct and then extend an outline of the functions toolkit that you've been building throughout your study of algebra and trigonometry. As you complete this unit and other units in the *Contemporary Mathematics in Context* curriculum, you should add other function families to this toolkit outline.

1. On a copy of the functions toolkit outline on the next page, create your own toolkit by doing the following for each basic type of function.

a. Give an example of a typical symbolic rule to calculate outputs from given inputs. Describe the domain and range of your sample function.

b. Provide a table of sample *(input, output)* number pairs.
  • Explain how the table pattern can be predicted from the symbolic rule and how the rule can be predicted from the table pattern.
  • If possible, describe the pattern in the output values using a *NOW-NEXT* equation.

c. Sketch a graph of the function. Explain how the graph pattern can be predicted from the symbolic rule and table pattern and how the rule and table pattern can be predicted from the graph.

2. Looking back at your toolkit of functions, you can see that the general symbolic rule for each function family involves one or more constants *a* and *b*, called **parameters**. For each function family, describe how different values of the parameters match different patterns in tables and graphs.

3. Examine this graph of the inverse power function $f(x) = \dfrac{1}{x}$.

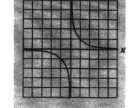

a. Note that for large values of x, either positive or negative, the graph gets closer and closer to the x-axis. Explain why the distance between the graph and the x-axis gets closer and closer to zero.

b. The graph of $f(x) = \dfrac{1}{x}$ is said to have the x-axis as a **horizontal asymptote**. This graph also has a **vertical asymptote**, which is the y-axis. Explain why this is the case.

c. Describe the asymptotes of the family of power functions.

d. What other function families have graphs with *asymptotic behavior*? What are the asymptotes?

e. What conditions in problem situations suggest function models with one or more asymptotes?

Checkpoint

In this investigation, you organized, in a systematic way, characteristics of several important families of the functions.

a. Compare your examples of linear, exponential, power, and periodic functions to those of other groups. Explain how the specific parameters in your function rules relate to the patterns in your sample tables and graphs. Then, as a class, develop general statements for each function family summarizing the effects of the parameters in symbolic rules on corresponding tables and graphs.

b. Write a general statement summarizing how the integer exponent in direct power models relates to the patterns in corresponding tables and graphs. Write a corresponding statement about inverse power models.

c. Where do quadratic functions fit in your functions toolkit?

Be prepared to share your analysis and provide arguments supporting your generalizations.

On Your Own

In Course 2 of *Contemporary Mathematics in Context*, you explored the square root function $f(x) = \sqrt{x}$, where $x \geq 0$.

a. Include this function model in your toolkit. Supply a sample table of (input, output) pairs, a graph, and information on the domain and range.

b. Describe at least two specific problem contexts that you think are good illustrative examples of this function model.

Functions Toolkit

Function Family/Example	Table	Graph
Linear Functions $f(x) = a + bx$ $f(x) = 0.5x + 3$ domain: all real numbers range:	x $f(x)$ 0 3	
Exponential Functions $f(x) = a(b^x)$, $a \neq 0$, $b > 1$ or $0 < b < 1$ $f(x) =$ domain: range:	x $f(x)$	
Power Functions Direct: $f(x) = ax^n$, $a \neq 0$, n a positive integer $f(x) =$ domain: range:	x $f(x)$	
Inverse: $g(x) = ax^n$, $a \neq 0$, n a negative integer $g(x) =$ domain: range:	x $g(x)$	
Periodic Functions $s(x) = a \sin x$, $a \neq 0$ $s(x) =$ domain: range:	x $s(x)$	
$c(x) = a \cos x$, $a \neq 0$ $c(x) =$ domain: range:	x $c(x)$	

functions in terms of graphs, tables and symbolic representations. This understanding extends to recursively-defined sequences. (Students may or may not have seen a formal definition of a function as, for example, a set of ordered pairs with certain properties. Similarly, they may or may not have seen a formal definition of domain and range.) They will have studied several families of functions including, but not necessarily limited to, linear functions, quadratic functions, power functions, exponential functions, logarithmic functions, and periodic functions (especially the sine, cosine, and tangent). They will have some experience with more general polynomials and rational functions. They will have made comparisons between families and within families. For many of these families they will be able to describe what certain parameters do. See Example 6 above. Moreover, given a function $y = f(x)$, students will understand how to symbolize vertical and horizontal translations of this function. They will have had experience with curve fitting (for example, linear and exponential regression and possibly other fits involving quadratic functions, power functions, or general polynomials). They will have had experience arithmetically combining and composing functions. They will understand the concept of the inverse of a function. They will have had experience with functions of more than one variable. They will have had extensive experience creating and interpreting algebraic expressions including relations with several variables. They will have used matrices in a variety of settings including their use in solving systems of linear equations and geometric transformations of the plane (and possibly 3-space). They will have experience with the usual Cartesian coordinatization of 2- and 3-space (and possibly polar and spherical coordinates as well). They will have studied the distance formula (at least in 2-space). They will probably have had some study of position vectors and/or parametric equations. In statistics, they will have studied normal (and possibly other) distributions and will understand how the area under the curve is related to probability.

There are other differences that bear mention. Prior to a course that might be labeled precalculus, students will have studied trigonometry in terms of right triangles and in relation to points on the unit circle in the plane. Students will probably have concentrated on the sine, cosine, and tangent functions. They will have derived some of the more elementary identities such as $\sin(-x) = -\sin(x)$; $\cos(-x) = \cos(x)$; $\sin^2(x) + \cos^2(x) = 1$; $\cos(\pi/2 - x) = \sin(x)$; etc. They will probably have studied the Law of Sines and the Law of Cosines. They will understand radian measure and probably have had some exposure to the concept of inverse functions for the (restricted) sine, cosine, and tangent. They may not have spent a significant amount of time deriving other trigonometric identities.[17] For example, they may or may not have studied the trigonometric identities relating to the sum or difference of angles. Students will probably have studied the equation for a circle. They may or may not have studied the other conic sections. However, they will have studied a parabola in the sense that it is the name given to the graph of a quadratic function.

Symbolic fluency is a goal of the PSSM vision. (See [12, pp. 301–302], for example.) However, there may be less experience in students' background in simplifying intricate algebraic expressions. So, for example, while students should be able to convert an expanded quadratic expression to vertex form by completing the square, expand the factored product of two binomials, or apply the laws of exponents to simplify $x^6 y^{10}/(x^2 y)$, they may or may not have had significant experience with simplifying expressions such as $(2x^2 + 5x + 2)^2(6y + 3)/(x + 2)^3(2y + 1)$ by hand.

The description of content changes above is not intended to be complete, but rather an indication of some shifts in content described in detail in PSSM and realized in the curricula listed in the appendix. For more information on these curricula, we suggest you look at the descriptions of content or sample lessons at the following web sites:

- high school: http://www.ithaca.edu/compass:

- middle school: http://www.showmecenter.missouri.edu:

- elementary school: http://www.arccenter.comap.com:

---

[17]The breadth of coverage of trigonometric identities varies among the secondary school curricula mentioned in the appendix.

It is important to realize that traditional content that receives less emphasis (or that is not included) in *Standards*-based curricula is not necessarily considered to have significantly lesser value. There is a great deal of accessible mathematics worth inclusion into the school mathematics curriculum. Furthermore, the U.S. penchant for including so many topics in curricula has been roundly criticized by some of the TIMSS researchers, [18] and [7] for example. Clearly, one must make choices. The *Standards'* working groups made such choices in light of the multiple goals mentioned at the beginning of this paper, among other things. Choices lead to both the content strands (Content Standards) and general topics within the strands. The richness of mathematics makes the goal of helping students become independent mathematical learners even more important.

It is suitable at this point to think about another organizing principle often used, tacitly if not explicitly, in more traditional content organizing schemes. This is the principle that a topic should be included in the curriculum just because it becomes "developmentally appropriate" to include it. The principle often appears in terms of a criticism. (For example, "that curriculum is a bad one. They don't do the quadratic formula until eleventh grade.") While the authors of this article agree that no mathematical topic should be introduced before students can comprehend its meaning and significance, we also feel it does no *a priori* harm to encounter some topics at a time beyond the instant the topic can be grasped by the learner. Faced with the evidence of poor retention of skills and knowledge that exists, it seems to make sense that we move the study of some topics closer to the point at which students will use them.

In fact, we suggest that it might be acceptable to move some topics out of the curriculum preceding calculus and defer them until the point that students will actually use them in a significant way. This parallels the way much mathematics enters the curriculum in some of our client disciplines. Discussions with engineering faculty indicate that they do not object to, nor shy away from, teaching relevant mathematics in engineering classes. For instance, one faculty member teaching nano-technology to juniors indicated that he often taught topics from data analysis and partial differential equations when they were needed. As a second example, when students are studying circuits, they learn something about complex numbers in their engineering class. In these examples, students may or may not have encountered the mathematical topics in prior mathematics courses. Regardless of that fact, a perusal of engineering texts suggests that mathematics topics are taught as new rather than review, that is, without assuming prior knowledge. Let us ignore the particular mathematical topics in these examples, but refer to the practice of studying some mathematical topics at a point where they will be of significant and immediate use as the engineering model. Let's give an example of how this model might be employed to alter precalculus content. The identity $\sin(x + y) = \sin(x)\cos(y) + \sin(y)\cos(x)$ is useful in calculus. It is usually studied in some course prior to calculus in a section labeled "trigonometric identities." Typically, however, one of the first times this identity is used in a serious way is during a derivation of the derivative for the sine function from the definition of the derivative. Suppose we move the study of this trigonometric identity into the calculus course at the point in the curriculum where the derivative of the sine function is examined. Furthermore, let us suppose that students have some ability to read and understand mathematics by the time they reach this point in calculus. That is, let us assume that the "independent learning" goal of the *Standards* had been realized to some degree. Under such circumstances, we suggest that it is possible that students would have a better understanding of this identity by reading about it themselves at a point where it is relevant. We suggest further that this alternative approach not only reinforces student responsibility for learning, it directly addresses the issue of poor retention of knowledge as well.[18] Furthermore, if we can be freed from the assumption that teachers need to explain all new material to their students, this new sequencing need not sacrifice significant class time.

Here, we are suggesting that precalculus need not be a gatekeeper course in the sense that the course ensures that students have *all* the non-calculus mathematical content knowledge needed and useful in the

---

[18]The identity $\sin(x + y) = \sin(x)\cos(y) + \sin(y)\cos(x)$ has other uses, particularly in physics. However, our placement in first semester calculus should precede its typical usage in the physics curriculum.

study of calculus. Giving up old assumptions and embracing new curriculum design principles may be hard—partially because it requires us to give up some cultural traditions. However, as mentioned before, current data on knowledge retention of students begs us to rethink the sequencing of material!

Let's go a step further in re-conceptualization. As an exchange it makes sense to move some of the ideas of calculus into the pre-calculus curriculum. Indeed, curricula designed for *all* students with attention to the *Standards* vision include some of the big themes in calculus. Students in these curricula will have encountered the idea of a *rate of change* in several contexts. They will have seen linear models where the coefficient of x can be interpreted as a rate of change. They will have studied exponential growth in terms of rate of change. They will have had experience approximating rates of change from graphical and numerical data. School students will probably have had experience with recursive relations such as $p_{n+1} = p_n + C$ and $p_{n+1} = rp_n$ and discussed them from the point of view of incremental change. They should understand what it means to say that the "rate of inflation is increasing." They will have considered the idea of optimization from a non-calculus point of view. They may have had some exposure to approximating irregular planar area. They will probably have had experience deducing some information about velocity from a graph of distance as a function of time.

In some cases, students may have studied other concepts or topics found in calculus. For example, in IMP and possibly SIMMS[19] students are informally introduced to the concepts of instantaneous rate of change and limit. In MATH Connections students study the volume of solids of revolution using Cavalieri's Principle, among other things. In MMOW, students study ideas relating to numerical integration. We note that many of these ideas presented in these paragraphs on content are consonant with ideas that have emerged with reform efforts focused on courses in the first two years of undergraduate mathematics education including calculus, college algebra, and precalculus. (See other papers in this volume, for example.)

What other things influence students' preparation for calculus?

Certainly the way students interact with the discipline matters. That is how students acquire mathematical processing abilities as well as how the skills and knowledge inherent in the discipline sets up expectations and, in that way, affects preparation for calculus.

This leads us to talk briefly about pedagogy. If we want students to have some proficiency in mathematics as a method of inquiry as well as a collection of skills and a body of knowledge, it stands to reason that students should be involved in inquiry activities. If we want to facilitate the ability of students to think mathematically, communicate their mathematical ideas coherently and with clarity, we are defining mathematics, in part, as an activity. Students in *Standards*-based programs are very likely to have experienced pedagogical strategies that involve collaborative or cooperative groups (from pairs to the entire class) as well as lecture and individual learning formats.

But, it is not the particular pedagogical strategies that we want to discuss here. Rather, it is the set of student expectations that accrue. In future courses, students with *Standards*-based backgrounds will expect to be able to spend time sharing *their* ideas, having an opportunity to ask and discuss questions and having an opportunity to struggle with ideas in a way that initial mistakes or false starts are expected, rather than a sign of incompetence. It seems reasonable to expect, and there is some evidence to bear this out, that students who transition from a *Standards*-based mathematics program to mathematics courses where the major focus is learning definitions and practicing procedures soon become bored and leave mathematics. We don't claim that there is enough evidence to verify this assertion. But, the authors have witnessed cases where some of the (potentially) best undergraduate mathematics students leave (traditionally taught) mathematics for more "hands-on" majors where they could get more "involved" or "creative." This is a sad reflection on our discipline.

---

[19]The SIMMS curriculum allows for different choices of content as several places in the curriculum. So the above statement assumes units on limits and the derivative were selected as part of the curriculum.

We have saved one of the most obvious characteristics of current change at every level of mathematics education for last. One of the six principles in the PSSM is a statement about technology. Specifically, the Technology Principle states that "Technology is essential in teaching and learning mathematics; it influences the mathematics that is taught and enhances student's learning" [12, p. 11]. Students about to enter calculus who have been educated with at least secondary school *Standards*-based programs will have used technology frequently in their study of mathematics. The most common technology in use in school mathematics classrooms is the graphing calculator, although other technology (spreadsheets, geometric drawing utility, and so forth) also appear. Students may be used to having this technology available as needed (or wanted). Students will have used technology in all the content areas. Among its uses in *Standards*-based programs, technology provides ways to do tedious calculations, so that students can concentrate on higher level aspects of problems (for example, actually multiplying matrices or computing the standard deviation from data). It also assists in problem solving (for example, by providing a medium for exploration and collection of data for conjectures as in the exploration of long-term behavior of a recurrence relation or the effect of a parameter change in a quadratic function). Technology also provides ways to do simulations.

But, our objective is not to debate the use of technology or even discuss all of its uses. Rather, our objective is to point out that being able to use technology as a tool in the study of mathematics is an expectation that students bring to the study of calculus and to other college-level courses. We believe this to be true of a large segment of the school student population whether or not they are enrolled in a fully implemented *Standards*-based school mathematics program.

In conclusion, we have attempted to describe and give a perspective on the *Standards* vision as it relates to the preparation of school students, particularly secondary students, for calculus. Furthermore, we feel some of the ideas we have mentioned might help us improve mathematics education at the undergraduate level. We realize that making transitions in education is not easy and certainly not a one-step process. Many issues are involved. However, from our experience, the largest issue in transitioning from more traditional school mathematics to college calculus is not content knowledge. It is the inability or unwillingness of students to think and their lack of understanding of mathematics as a sense-making activity. We feel that the *Standards* vision has set us on a course to correct this.

Acknowledgements: The work reported in this paper was supported, in part, by grants from the National Science Foundation (ESI-0001377 and ESI-0137772). Views and conclusions expressed are those of the authors and not necessarily those of the Foundation.

References

1. Conference Board of Mathematical Sciences, "The Mathematical Sciences Curriculum K–12: What Is Still Fundamental and What Is Not." In *Educating Americans for the 21st Century: Source Materials*. National Science Board Commission on Precollege Education in Mathematics, Science and Technology, CBMS, Washington, DC, 1983.

2. Coxford, A.F., J. T. Fey, C.R. Hirsch, H.L. Schoen, G. Burrill, E.W. Hart, B.A. Keller, and A.E. Watkins, *Contemporary Mathematics in Context: A Unified Approach.* Glencoe/McGraw-Hill, Columbus, OH. 1997–2001.

3. Eves, Howard. *A Survey of Geometry*. Vol. 1. Allyn and Bacon, Boston, MA. 1963.

4. Fendel, Dan, Diane Resek, Lynne Alper, and Sherry Fraser, *Interactive Mathematics Program*. Key Curriculum Press, Emeryville, CA. 1997-2000. Year 3, pg. 60–61.

5. Garfunkel, Solomon, Landy Godbold, Henry Pollak, *Mathematics: Modeling Our World*. W.H. Freeman and Company, New York, New York. 1998–2000. Year 1, pg. 505.

6. Honsberger, Ross, *Mathematical Gems I.* "The Orchard Problem," Mathematical Association of America, Washington, DC, 1973, pp. 43–53.

7. Mid-Atlantic Eisenhower Consortium for Mathematics and Science Education, Research for Better Schools, *The Third International Mathematics and Science Study: TIMMS: A Sourcebook of 8th Grade Findings*, Mid-Atlantic Eisenhower Consortium for Mathematics and Science Education, Philadelphia, PA, 1997.

8. National Commission on Excellence in Education, *A Nation at Risk: The Imperative for Educational Reform,* Government Printing Office, Washington, DC, 1983.

9. National Council of Teachers of Mathematics, *Assessment Standards for School Mathematics*, National Council of Teachers of Mathematics, Reston, VA, 1995.

10. ——, *Curriculum and Evaluation Standards for School Mathematics*, National Council of Teachers of Mathematics, Reston, VA. 1989.

11. ——, *Professional Standards for Teaching Mathematics*, National Council of Teachers of Mathematics, Reston, VA, 1991.

12. ——, *Principles and Standards for School Mathematics*, National Council of Teachers of Mathematics, Reston, VA, 2000.

13. National Research Council, *Adding it Up: Helping Children Learn Mathematics*, National Academy Press, Washington, DC, 2000.

14. National Research Council, Mathematical Sciences Education Board, *Everybody Counts: A Report to the Nation on the Future of Mathematics Education,* National Academy Press, Washington, DC, 1989.

15. ——, *Reshaping School Mathematics*, National Academy Press, Washington, DC, 1991.

16. Ruddock, Maryann Steele, "Are All 'Bs' Alike? The Effect of a Prerequisite Math Course on Calculus II Grades," presented at Southern Association for Institutional Research/Society for College and University Planning, San Antonio, Texas, 1994.

17. ——, "Are Prerequsite Courses in Mathematics Effective Routes to SEM Degrees?" presented at Southern Association for Institutional Research, Hot Springs, Arkansas. 1997.

18. Schmidt, William H., Curtis C. McKnight, and S.A. Raisen, *A Splintered Vision: An Investigation of U.S. Science and Mathematics Education*, Kluwer Academic Publishers, Boston, MA, 1996.

19. Schoen, Harold, Christian Hirsch, "Responding to the Call for Change in High School Mathematics: Implications for Collegiate Mathematics," *The American Mathematical Monthly*, Vol. 110, Number 2, Feb. 2003, pp. 109–123.

20. U.S. Department of Labor, *What Work Requires Schools: A SCANS Report for America*, Government Printing Office, Washington, DC, 1991.

Appendix

NSF-funded Multi-year *Standards*-based Instructional Development Materials

The following list includes the published name of each curriculum, a project name in parentheses that may be familiar to some, development location, beginning list of developers, and publisher.

1. Elementary instructional development projects

   - *Everyday Mathematics*, (EM or UCSMP elementary), University of Chicago School Mathematics Project, Max Bell, et al., SRA/McGraw-Hill.
   - *Investigations in Data, Number and Space*, TERC, Susan Jo Russell, et al., Scott Foresman.
   - *Math Trailblazers; A Mathematical Journey Using Science and Language Arts*, (TIMS), IMS Project University of Illinois at Chicago, Phillip Wagreich, et al., Kendall/Hunt.

2. Middle School instructional materials projects

   - *Connected Mathematics*, (CMP), Michigan State University, Glenda Lappan, et al., Prentice Hall.
   - *Mathematics in Context*, (MiC), National Center for Research in Mathematical Sciences, University of Wisconsin-Madison, Thomas A. Romberg, et al., Holt, Reinhart & Winston.
   - *MathScape*, (STM), EDC, Glenn Kleiman, et al., Glencoe/McGraw-Hill.
   - *Math Thematics*, (STEM), Montana State University, Rick Billstein and Jim Williamson, McDougal Littell.

3. Secondary School instructional materials development projects

   - *Mathematics: Modeling Our World*, (ARISE), COMAP, Inc., Sol Garfunkel, et al., W.H. Freeman.
   - *Contemporary Mathematics in Context: A Unified Approach*, (CPMP or Core-Plus), Arthur F. Coxford, et al., Glencoe/McGraw-Hill.
   - *Interactive Mathematics Program*, (IMP), San Francisco State University, Dan Fendel, et al., Key Curriculum Press.
   - MATH *Connections, A Secondary Mathematics Core Curriculum*, (MATH Connections), CBIA, William P. Berlinghoff, et. al., IT's ABOUT TIME, Inc.
   - *Integrated Mathematics: A Modeling Approach Using Technology*, (SIMMS), Montana Council of Teachers of Mathematics and the Systemic Initiative for Montana Mathematics and Science; Montana State University, Johnny Lott, et al., Kendall Hunt.

Theme 3. The Needs of Other Disciplines

As Deborah Hughes Hallett observes in her paper, "What Have We Learned from the Calculus Reform Movement?" (which appears earlier in this volume): "In the long run, the largest impact of calculus reform is likely to be the creation of a community of mathematicians who innovate and reflect on their teaching—and who do so in collaboration with faculty in other disciplines and across institutional boundaries." One of the challenges confronting refocusing the courses below calculus is for mathematicians to understand and to respond to the needs of partner disciplines. The three papers in this section address this challenge. Bill Barker and Susan Ganter summarize the outcomes of the *Curriculum Foundations Project*, which held a series of eleven disciplinary workshops that focused on the needs of the partner disciplines during the first two years of undergraduate instruction. Rich West notes that as a result of a new curriculum, which was introduced in 1990, the senior leadership at West Point strived to find a balance between teaching students concepts and helping them master skills. He describes the compromise that resulted after many discussions with their partner disciplines. Allan Rossman observes that many calculus reform projects emphasize applications, which in turn often involve genuine data. He argues that the use of real data is an important aspect of refocusing precalculus.

Bill Barker and Susan Ganter describe the portions of the Curriculum Foundations Project that are most relevant to college algebra and precalculus. They note, "The recommendations of the Curriculum Foundation Conference are very much in keeping with the philosophy and principles of the conference *Rethinking the Preparation for Calculus*." In particular, the participants were aware of the obvious mismatch between a curriculum designed to prepare students for calculus and the reality that very few of these students subsequently enroll in calculus. Consequently the participants made the following recommendation (which is echoed in many of the papers in this volume): "Replace traditional college algebra with a course stressing problem solving, mathematical modeling, descriptive statistics, and applications in the appropriate technical areas. De-emphasize intricate algebraic manipulation."

It is clear that major changes are needed in the teaching of precalculus courses. It is also clear—and fortunate—that there is an able and talented group of mathematicians who are committed to the huge task of designing appropriate new courses, and convincing the mathematics community of the need for such changes. Finally, it is clear from the Curriculum Foundations Project that the partner disciplines will enthusiastically support these changes. Mathematicians need to continue to involve colleagues from these disciplines as we take action. They can provide much information to enrich the courses, they have a major stake in the success of our ventures, and they can provide a great deal of political muscle. And a lot of political muscle will be needed to move the mountains before us.

Fundamental Mathematics: Voices of the Partner Disciplines
William Barker and Susan L. Ganter

In 1990 West Point adopted a new curriculum that changed the mathematics courses that all students must take during the first two years. Rich West describes how, as a result of the new curriculum, the senior leadership at West Point were "torn between teaching students to understand and use the tools of mathematics in solving real-world problems as opposed to mastering the skills (usually precalculus) that they had already learned." He describes the compromise that resulted after many discussions with their partner disciplines: the partners would concur on the needed skills, and the mathematics faculty would ensure that these essential skills were either internalized or memorized. The resulting accountability mechanism was a series of fundamental skills tests given throughout the first two years.

> Changes in student mathematical capabilities can be reliably and relatively uniformly achieved, if these changes become institutional requirements that are monitored while providing strong support systems for students. Further, changes in mathematical curriculum or requirements for student preparation must be made in consultation with client disciplines.

> *Skills Vs. Concepts at West Point*
> Rich West

Allan Rossman notes that many calculus reform projects emphasize applications, which in turn often involve genuine data, and he observes that the use of real data is an important aspect of refocusing precalculus. He demonstrates that statistics can provide a rich application area in which students can study interesting examples of functional behavior and justifies why data analysis should be an integral part of precalculus.

> Data analysis can play an important role in enhancing students' learning experiences in precalculus. Genuine data often provide motivation and interest for students, and they reveal that concepts of precalculus do have application to analyzing data from a variety of disciplines as well as from everyday life. Beyond its usefulness for teaching precalculus ideas, data analysis also provides the opportunity for students to discover and explore some important statistical principles with which all educated citizens should be comfortable.

> *Integrating Data Analysis into Precalculus Courses*
> Allan J. Rossman

16

Fundamental Mathematics:
Voices of the Partner Disciplines

William Barker and **Susan L. Ganter**
Bowdoin College *Clemson University*

The Curriculum Foundations Project

Given the impact of mathematics instruction on the sciences and quantitative social sciences—especially instruction during the first two years—there is a need for significant input from these partner disciplines when revising the undergraduate mathematics curriculum. The committee Curriculum Renewal Across the First Two Years (CRAFTY), a subcommittee of the MAA Committee on the Undergraduate Program in Mathematics (CUPM), has gathered such input through the *Curriculum Foundations Project*. The primary component of the Curriculum Foundations Project has been a series of eleven disciplinary workshops held across the country from November 1999 to February 2001 (see Figure 1).

Each Curriculum Foundations workshop consisted of 20–35 participants, the majority chosen from the discipline under consideration, the remainder chosen from mathematics. Each workshop produced a report

---

Physics and Computer Science
    Bowdoin College, Maine, October, 1999
Interdisciplinary (Mathematics, Physics, Engineering)
    United States Military Academy, West Point, November, 1999
Engineering
    Clemson University, May, 2000
Health-related Life Sciences
    Virginia Commonwealth University, May, 2000
Technical Mathematics (at two sites)
    Los Angeles Pierce College, California, October, 2000
    J. Sargeant Reynolds Community College, Virginia, October, 2000
Statistics
    Grinnell College, October, 2000
Business, Finance and Economics
    University of Arizona, October, 2000
Mathematics Education
    Michigan State University, November, 2000
Biology and Chemistry
    Macalester College, November, 2000
Mathematics Preparation for the Major
    Mathematical Sciences Research Institute, February, 2001

Figure 1. MAA Curriculum Foundations Workshops

addressing a series of questions formulated by CRAFTY.[1] The same core set of questions was provided at each workshop, the purpose of which was to learn the needs of the featured discipline relative to the initial two years of undergraduate study in mathematics.

In November 2001, invited representatives from each disciplinary workshop gathered at the U.S. Military Academy in West Point, NY for the final Curriculum Foundations Conference. The discussions resulted in *A Collective Vision*, a set of commonly shared recommendations for the first two years of undergraduate mathematics instruction [1].

The individual reports and the *Collective Vision* recommendations have been published and will be widely circulated within both the specific disciplines and the mathematics community. The disciplinary reports and the *Collective Vision* are currently being used by CUPM to inform the preparation of the *MAA Curriculum Guide* 2004 [2], a set of recommendations for the mathematics community on the undergraduate curriculum that will focus on desired student outcomes. However, the workshop reports and the *Collective Vision* have value independent of the *MAA Curriculum Guide* 2004, since they serve as resources for multi-disciplinary discussions at individual institutions.

Workshop participants from the partner disciplines were extremely grateful—and surprised—to be invited by mathematicians to state their views about the mathematics curriculum. That their opinions are considered important and will be taken seriously in the development of the *MAA Curriculum Guide* 2004 only added to their enthusiasm for the project, as well as their interest in continuing and expanding conversations with the mathematics community.

The Curriculum Foundations Project and precalculus instruction

The Curriculum Foundations workshops focused on the needs of the partner disciplines for all mathematics experiences in the first two years of undergraduate instruction. Though many of the workshop participants focused primarily on calculus, and many others centered on college algebra and basic data analysis, few singled out precalculus by name. As others have observed, precalculus is an invisible course acknowledged only indirectly through its supporting role for calculus.

However, most of the recommendations from *A Collective Vision* have a universality that applies to all undergraduate mathematics instruction, including precalculus. Moreover, the recommendations are very much in keeping with the philosophy and principles of the conference *Rethinking the Preparation for Calculus*, held October 2001 in Arlington, VA. This is very encouraging and provides evidence that the partner disciplines are allies in the struggle to reformulate the content and pedagogy of the foundational mathematics courses such as college algebra and precalculus.

The following sections discuss the portions of *A Collective Vision* that are most relevant to instruction in college algebra and precalculus.

Conceptual understanding and problem solving

Rethinking Precalculus assumes a collection of basic principles about precalculus courses. One is that such a course needs "...to prepare students for calculus both *conceptually* and *algebraically*. It is not enough just to emphasize developing manipulative skills; students need help to develop the conceptual skills needed to understand and apply the basic calculus concepts" [3]. Colleagues in the partner disciplines most strongly support this principle, as illustrated in *A Collective Vision*:

---

[1]Individual workshop reports also can be obtained electronically at
http://academic.bowdoin.edu/faculty/B/barker/dissemination/Curriculum_Foundations/

Emphasize conceptual understanding

- Focus on understanding broad concepts and ideas in all mathematics courses during the first two years.
- Emphasize development of precise, logical thinking. Require students to reason deductively from a set of assumptions to a valid conclusion.
- Present proofs when they enhance understanding. The fundamental connection between proof and understanding must be highlighted.

Strive for depth over breadth

- Explore locally what topics can be omitted and teach the remaining topics in more depth.

There is a common belief among mathematicians that the users of mathematics (engineers, economists, etc.) care primarily about computational and manipulative skills, that we mathematicians are forced to cram our courses full of algorithms and calculations to keep them happy. Perhaps the most encouraging discovery from the Curriculum Foundations Project is that this stereotype is largely false. Though there are certainly individuals from the partner disciplines who hold the more strict algorithmic view of mathematics, the disciplinary representatives at the Curriculum Foundations workshops were unanimous in their emphasis on the overriding need to develop in students a conceptual understanding of the basic mathematical ideas and methods. They felt that topics can and should be eliminated to achieve this depth of understanding. For example, most workshop reports emphasize that the conceptual ideas of derivatives and integrals are critically important, while specific techniques of calculation are secondary. Time can be gained for teaching such conceptual underpinnings by focusing only on calculations using the most basic functions (e.g., exponential, polynomial, and trigonometric functions) and eliminating more tedious calculations that students will not likely remember anyway.

The partner disciplines also value the precise, logical thinking that is an integral part of mathematics. In general, they do not share the mathematician's unconditional love of formal proof, but logical arguments and deduction are acceptable if they enhance understanding of the underlying concepts. However, colleagues in the partner disciplines are skeptical about early collegiate mathematics instruction that is inappropriately tilted in the direction of formal proof. They believe that this emphasis is confusing to students at this level and negatively affects their ability to understand and apply mathematics. Mathematicians also should be skeptical about such curricular design.

Emphasize problem-solving skills

- Develop fundamental computational skills, but emphasize *integrative skills*: the ability to apply a variety of approaches to a single problem, to apply familiar techniques in novel settings, and to devise multi-stage approaches in complex situations.

Fundamental computational skills are important and must be developed in students. However, colleagues in the partner disciplines are well aware that applying mathematics to unfamiliar problems requires far more than computational skills. Mathematics courses must include more sophisticated problem-solving experiences than ones in which students simply look in the book for a problem of the same type and change the numbers.

Modeling and Applications

Another basic principle for precalculus mathematics as outlined in *Rethinking Precalculus* is the need for students "...to see an emphasis on mathematical modeling to learn how mathematics is connected to the real world. The basic mathematical concepts and methods should be developed in contexts to help the students transfer their learning outside the mathematics classroom" [3]. Support for this principle is illustrated in *A Collective Vision* as follows:

Emphasizing Mathematical Modeling

- Expect students to create, solve, and interpret mathematical models.

- Provide opportunities for students to describe their results in several ways: analytically, graphically, numerically, and verbally.

- Use models from the partner disciplines so that students can see mathematics in context.

The importance ascribed to mathematical modeling by *every* disciplinary group in *every* workshop was quite striking—and is one of the strong recommendations of the Curriculum Foundations Project. The addition of mathematical modeling to precalculus courses will be enthusiastically supported by the partner disciplines.

In fact, workshop participants were so excited by the use of real models in mathematics courses that they volunteered to help with the development of such models. The support for interdisciplinary cooperation with mathematics departments was so strong in all the workshops that mathematicians at individual institutions should be encouraged to seek out such support, as it is likely to exist on their own campuses. Here are the recommendations from *A Collective Vision*:

Improve Interdisciplinary Cooperation

- Seek projects from the partner disciplines.

- Encourage interdisciplinary guest lecturing.

- Increase team teaching opportunities.

- Seek ways to overcome the transfer problem from mathematics courses to courses in other disciplines.

The transfer problem refers to the difficulty students have recognizing when a problem in a non-mathematics discipline is related to material studied in mathematics courses—let alone remembering the techniques that can be used to solve the problem. Colleagues in the partner disciplines believe that exposing students in mathematics courses to some of the discipline-specific contexts for various mathematical topics will have a positive effect on the transfer problem.

Communication

Another theme that ran through nearly every disciplinary workshop was the importance of students being able to communicate mathematical and quantitative ideas. Here is the relevant portion of *A Collective Vision*:

Emphasize communication skills

- Incorporate development of reading, writing, speaking, and listening skills into courses.

- Require students to explain mathematical concepts and logical arguments in words. Require them to explain the meaning—the hows and whys—of their results.

The importance of communication skills was emphasized time and again in the workshops. Though there are many successful examples of writing and speaking in the mathematics classroom, there is still a need for more universal implementation of these activities. It is viewed by many as hard, time-consuming, and foreign to our training as mathematicians—and rightly so. However, these skills are critical to students and, therefore, all faculty have a responsibility for incorporating these skills in the classroom—even if it is hard, time-consuming, or foreign. Such activities can take the form of written lab assignments, technical reports, group projects, professional presentations in class, short essays on exams, and the like.

Technology

Rethinking Precalculus states that, "Precalculus courses should help students learn to use modern technology wisely and appropriately" [3]. Workshop participants expressed the same sentiment in more detail:

Emphasize the use of appropriate technology

- Stress choosing tools appropriate to the desired task: mental calculation, paper and pencil calculation, and technology-based calculation.

- Stress intelligent and careful interpretation of results obtained from the use of technology.

- Be aware of the preferred tools of the partner disciplines. Spreadsheets are often preferred.

Colleagues in other disciplines do not need to be convinced of the importance of technology in mathematics instruction: they know how critical technology is to their own fields, and hence how important it is for students to master appropriate technological tools. And, they are well aware of the importance of choosing the right tool for each problem-solving activity and learning how to properly interpret the results obtained using technology. Blind, unquestioning belief in the results obtained from a calculator or computer can be disastrous. None of this is surprising.

However, a more surprising discovery was that spreadsheets are the technology of choice for a large number of the partner disciplines. Although the workshop reports purposely stopped short of recommending spreadsheets as the primary technology in mathematics instruction, their widespread use is relevant to the technology choices made in precalculus. A related observation was the unimportance of graphing calculators—very few workshop participants reported their use in disciplinary courses. If graphing calculators are chosen as the technology for a course, it must be understood that this is done for pedagogical reasons, not to support uses in other disciplines.

Instructional techniques

As stated in *Rethinking Precalculus*, there is a need to base effective pedagogy on educational research: "Current research into the learning process has much to tell us about how students acquire fundamental precalculus (mathematical) concepts. Only a small minority of students learn mathematics the way we did" [3]. While educational research was not cited directly, colleagues in the partner disciplines recognize the need for alternatives to traditional, non-interactive lecture courses. In their words,

Use a variety of teaching methods

- Different students have different learning styles. In particular, encourage the use of *active learning*, such as

  - in-class problem-solving,

  - discussion,

  - collaborative group work, and

  - out-of-class projects.

Given that many students who enroll in precalculus courses have had difficulty with mathematics in the past, it would seem that they may be the students most in need of innovative, active learning and teaching techniques.

Student assessment

The important relationship between assessment and student learning was discussed extensively in the workshops; i.e., how and what you assess directly affects how and what students learn. Because assessment can be difficult, time-consuming, and tedious, instructors often put less thought and effort into this aspect of course design. This is unfortunate, since effective assessment is necessary for effective learning. Workshop participants focused on this issue as follows:

Emphasize the use of appropriate assessment

- Employ a variety of assessment strategies to measure achievement of course objectives (problem-solving, conceptual understanding, algorithmic skills).
- Use conceptual questions on examinations, not just questions that a calculator could solve.
- WYTIWYG: "What you test is what you get."
- Encourage *institutional assessment* of the effectiveness of programmatic changes.

WYTIWYG became almost a mantra at the final Curriculum Foundations Conference because of the importance ascribed to it. It was also instructive to see the emphasis placed on the use of conceptual questions on examinations as opposed to algorithmic computations. It underscores the importance assigned to the development of conceptual understanding by colleagues in the partner disciplines.

Whither college algebra?

Rethinking Precalculus states as an obvious principle the need for precalculus courses (and, by extension, college algebra courses) to be changed to meet the needs of the current student population. "Precalculus courses serve two distinct student populations: The overwhelming majority for whom precalculus is a terminal course and the relatively small minority for whom it is a gateway to higher mathematics. We need to identify and meet the needs of both populations" [3]. It is expected that the *MAA Curriculum Guide* 2004 will recommend the reorganization of college algebra/precalculus courses into new configurations better designed to meet the needs of various student populations. Similar ideas were put forth in many of the disciplinary workshop reports, although the group assembled for the final Curriculum Foundations Conference did not choose to make such a recommendation. However, aware of the obvious mismatch between a curriculum designed to prepare students for calculus and the reality that very few of these students subsequently enroll in calculus, the Curriculum Foundations Conference participants recommended the following:

Replace traditional College Algebra with a course stressing problem-solving, mathematical modeling, descriptive statistics, and applications in the appropriate technical areas. De-emphasize intricate algebraic manipulation.

Conclusion

It is clear that major changes are needed in the teaching of precalculus courses. It is also clear—and fortunate—that there is an able and talented group of mathematicians who are committed to the huge task of designing appropriate new courses, and convincing the mathematics community of the need for such changes. Finally, it is clear from the Curriculum Foundations Project that the partner disciplines will enthusiastically support these changes. Mathematicians need to continue to involve colleagues from these disciplines as we take action. They can provide much information to enrich the courses, they have a major stake in the success of our ventures, and they can provide a great deal of political muscle. And a lot of political muscle will be needed to move the mountains before us.

References

1. Ganter, S.L. and Barker, W., "A Collective Vision," In S. Ganter and W. Barker (Eds.) *A Collective Vision: Voices of the Partner Disciplines*, MAA Reports Series, Mathematical Association of America, Washington, DC, 2003.

2. Report by the Committee for the Undergraduate Program in Mathematics (CUPM), *CUPM Curriculum Guide 2004*, MAA Reports Series, Mathematical Association of America, Washington, DC, 2004.

3. Narayan, J. and Narayan, D. "The Conference: Rethinking the Preparation for Calculus," in this volume.

17

Skills versus Concepts at West Point

Rich West
Francis Marion University

Introduction

In 1990 West Point adopted a bold, new curriculum that changed the mathematics courses that all students must take during the first two years. The senior leadership at West Point has always struggled with the issue of emphasis: skills or concepts. With the new curriculum, we were torn between teaching students to understand and use the tools of mathematics in solving real-world problems as opposed to mastering the skills (usually precalculus) that they had already learned. After many discussions with our partner disciplines, we came up with a compromise: the partners would concur on the needed skills, and the mathematics faculty would ensure that these essential skills were either internalized or memorized. The resulting accountability mechanism was a series of fundamental skills tests given throughout the first two years. Based on placement tests and observations of students' work done in class, the precalculus skills on entry to West Point are usually disappointing and at different levels. So, in consultation with our partner disciplines, the department of mathematics established and published for students and faculty the minimal math skills standards. To hold the students accountable, they must achieve an 80% or better on a gateway exam before being allowed to move on to the second course in the mathematics program, a prerequisite for physics and all of the engineering courses. Because of the involvement of many different faculty, this program has now been in place for ten years. It is not perfect, but it allows the faculty to focus on mathematical modeling and concepts that our students need, places responsibility for review on the students, and informs our partner disciplines while they embrace our move to reform. In short, this West Point case study makes two broad conclusions. Changes in student mathematical capabilities can be reliably and relatively uniformly achieved, if these changes become institutional requirements that are monitored while providing strong support systems for students. Further, changes in mathematical curriculum or requirements for student preparation must be made in consultation with client disciplines.

Perspective

At West Point, all students are required to take four sequential semester-long mathematics courses during the first two years. This mathematics program, in conjunction with a semester of computer science, a year of chemistry during freshman year, and a year of physics during sophomore year, forms the basis for a required five-course engineering minor taken during junior and senior years. All students must take, or validate from previous success, each of these courses. These requirements have been in effect since 1988.

160

In 1990, the mathematics department, with the concurrence of all departments at West Point, implemented a bold, new curriculum for their four courses during the first two years. Briefly, the four courses are: discrete dynamical systems and introduction to calculus, calculus I (which finishes differential calculus, covers integral calculus, and introduces differential equations through systems of differential equations), calculus II (which covers most of multivariate calculus and vector calculus), and probability and statistics. A good college preparation background is required for entrance into West Point. Math SATs average about 650 each year. High school courses must include intermediate algebra and trigonometry and/or precalculus. Typically, about fifty percent of an entry class has taken some form of calculus in high school or college. Although precalculus topics are covered in the first course, all students must take this first course because of the heavy emphasis on mathematical modeling. In addition, there is a small precalculus program that involves about sixty students each year. This will be discussed more later.

Need

The mathematics and chemistry faculty at West Point that teach freshman courses were always disappointed with students' entry-level skills. If not corrected early in the four-year sequence, these deficiencies could have negative impacts later in a student's college education. In fact, the reform courses started in 1990 were put in place essentially because most of the mathematics, science and engineering faculty were disappointed in the mathematics preparation of the students in junior and senior level courses. In this new curriculum, calculator and computer technology were used to compress the desired mathematics coverage and in many cases compensated, or covered up, for these poor skills. Further, we sought to gain more efficency and enhancement in coverage by designing a four-course integrated program rather than four sequential courses. So, while we were expanding coverage to get at concepts, poor skills may have slipped by. In support of our reform ideas, mathematics reform across the nation provided better teaching ideas and legitimacy in dealing with other disciplines. But still the lack of skills, which our colleagues from other disciplines expected, remained a problem.

In the past at West Point, several programs to help fix this need for "remediation" had been implemented and discarded. Usually we felt that time spent specifically on skills was time taken away from problem-solving, a more fruitful goal. Almost all mathematics taught at West Point is presented in some form of application from science, engineering, or future mathematical fields. Our approach has always been to show students that skills are necessary to solve the contextual problems of the courses and disciplines they will see in their futures. This did not change in the reform curriculum, and we felt initially that the precalculus content of the first course would address this need for review. It did somewhat; however, the modeling was so new to all the students that the weaknesses seemed to become worse rather than better. As a result, we decided that while skills were necessary, remediation would be a student responsibility, and in 1992 we implemented a new skills program.

Solution

Our idea was to define and publish clear, achievable standards and then hold the students accountable until they meet or surpass these standards. To help prepare students to meet these standards, we used a programmed text that we had developed long ago and the "Are You Ready for Calculus?" software developed at the University of Arizona. To determine if students met the established standards, we used ideas from the gateway exams implemented at Duke University and at the University of Arizona. To better relate these ideas to West Point and its sequential curriculum, I must first describe the precalculus program.

As I said before, for years the mathematics faculty had been disappointed in the retained skills of the incoming freshmen. Each year during the summer (freshmen come to stay at West Point in early July), the department of mathematics administered placement tests. The purpose of one of the tests was to identify the group of students who had the weakest algebra and trigonometric skills. These students were placed into a special precalculus course to help them obtain the skills they needed for calculus. This made sense before the reform, because the first course was calculus. As a result, these students were one semester behind in their mathematics, physics, and engineering courses, but they usually made up this deficit during their junior or senior years. For years the program was considered successful, because it graduated students at the same rate as the regular students.

Before implementing a program of gateway exams, the sequence of tests and decision points was presented to and approved by a committee of all of the mathematics, science and engineering heads (called the MSE Committee). They in turn set up a subcommittee with representatives from mathematics, physics, chemistry, and two of the engineering departments. This committee met for over a semester to hammer out the standards for fundamental skills on entry. Once approved by the MSE Committee, the standards were published for the mathematics, science, and engineering faculty and the students. These written standards are now mailed to all students once they are accepted at West Point (usually by March or April). At the same time, students are asked if they would like a floppy disk with the "Are you Ready for Calculus?" software mailed to them. Notice that the responsibility for review is put on the student.

The gateway exam

We decided to use this placement test as an opportunity to administer our first gateway exam. Since each student is required to pass the gateway exam at the 80% level at least once during the first semester, this placement exam could be used as an opportunity to achieve this 80%. There were 25 multiple-choice questions and students were given 45 minutes to complete the test. Those that did the poorest became candidates for our precalculus course. A member of the mathematics faculty who sits on the Admissions Committee subsequently reviewed their admissions records. A list of candidates was prepared and the department head, who also sits on the Admissions Committee, would decide on those who would be placed in this precalculus course.

All other students started in discrete dynamical systems. Their corrected exams were returned to them on the first day of classes. General questions on this exam were covered in class, but specific questions were gone over one-on-one with their instructor. The students had a week to prepare for their first college exam with content similar to their corrected summer tests.

Subsequently, throughout the semester, the gateway exam was administered at least four times. So the students had a total of five opportunities to achieve the required 80%. Originally, in implementing this program of tests, we allowed students who passed to be exempt from the other exams. However, for motivation and because of the sequential design of curriculum, we found it necessary to include these exam grades in their course grade. Therefore, all students were required to take all five tests. Good students who do well are rewarded. If a student passed discrete dynamical systems and still failed to pass the gateway, then, subject to the assessment of his or her instructor and the department head, the student became a candidate for the precalculus course in the second semester.

The standards

The mathematics department each year reviews the standards before they are published. In addition, a committee, with representatives from mathematics, physics, chemistry, and all the engineering departments, reviewed these standards in detail in 1996 and again in 1998. The program has evolved some as I will relate later in Lessons Learned, but in spirit it remains intact and is still being executed. Appended to this article are the published "Required Mathematical Skills for Entering Cadets." These are excerpted from the "Core Mathematics at USMA" pamphlet published in 1998. In general terms, they consist of basic algebra, basic trigonometry, graphs of lines and basic nonlinear functions, basic functions, and perimeter, area and volume formulas. These are the skills that we and other disciplines see as important from a precalculus curriculum. The pamphlet is published each year and a copy is presented, orally and in writing, to each new faculty member in the mathematics, science and engineering departments.

Fundamental skills exams

At the same time that the mathematics department was developing the precalculus skills standards and working through the gateway exam procedures, we saw an opportunity to set the skills standards for the whole four-course program. As a result, the same mathematics, science, and engineering subcommittee that established the gateway standards articulated skills for the entire mathematics program. Since these standards are not a focus of this article or the conference it supports, they are not included. However, the process of establishing the standards with our partner disciplines, and establishing a timeline by when students should have internalized these skills, have raised awareness of students and faculty, especially those in other disciplines, about responsibilities and expectations. Each of the three subsequent core mathematics courses has at least one fundamental skills exam as their form of accountability. In addition, at the beginning of physics and some of the engineering science courses (usually junior year), instructors use these math skills tests to assess their students' retention and communicate the minimum math skills needed throughout their course(s).

Lessons learned

As I mentioned above, initially, we scheduled makeup gateway exams for the students who had not achieved 80%. We learned that freshman were not likely to take on a protracted study program in preparation for these exams. As a result, since that first year we have included these exams and grades as a part of the discrete dynamical systems course and required all students to take them. This rewards the good students and identifies those who are weak in skills, achieving the 80% only once in the semester (usually after five attempts).

Further, after ten years of executing this exam, our numbers have been fairly consistent. Out of entering classes of approximately 1,000 to 1,200, forty to fifty students have been selected for the first semester precalculus course and ten to eighteen have been selected for the second semester precalculus course. The graduation rates of these students have actually been better than those who follow the regular course sequence. We use the gateway exam as part of the final exam for both precalculus courses. As a result, students who fail the final exam and the course are usually separated from the Academy.

As the client disciplines embrace and take advantage of our reformed program, the skill standards have changed somewhat over ten years due mostly to input from these MSE disciplines. Thus, the exams are adjusted to the standards. Likewise, the precalculus course is no longer just preparing students for calculus I, but for an intergrated four-course program with mathematical modeling and conceptual understanding required throughout. Thus, the precalculus course has changed to better integrate it into the core mathematics program.

Implications

The reform of the core mathematics curriculum that was initiated in 1990 took a lot of salesmanship to our partner disciplines. The mathematics department had to make sure that it met the end objectives of the core mathematics program for all of our partner disciplines. The keys to success of this major change, besides inspired leadership, were the involvement of the entire faculty in developing the reform and communication with our partner disciplines. The biggest obstacle to change, besides change itself, was fixing the disappointment that got us into the reform in the first place. As it turned out for the partner disciplines, this disappointment was in the area of what we in the mathematics department call skills. So as we involved the partner disciplines in developing the standards for skills and articulated our programs for accountability through gateway exams and the series of fundamental skills exams, we were minimizing their greatest fear. They were afraid that a concept-based curriculum would reduce the algebra and trigonometry skills. Once involved in the process of development, they supported the implementation of accountability and were pleased with the other benefits of the concept-based curriculum, such as mathematical modeling and aggressive, flexible problem solving.

Further, as technology improves and evolves toward a handheld computer algebra system such as the TI-89, which West Point students have been issued for two years now, the standards for skills will evolve and may become less important to our partner disciplines. In the meantime, our communal review and setting of standards has become a great success and a good compromise of skills versus concepts.

Conclusions

Discussions continue at West Point about the give and take in a compressed curriculum. As long as these discussions involve the desires of our partner disciplines that we service with the core mathematics program, a happy medium between skills and concepts can be achieved. The gateway exam used to account for precalculus skills has been a good compromise and an effective catalyst for change. One overflow benefit has been that most mathematics courses at West Point and all mathematics courses that I teach utilize a gateway exam early in the course to establish the skills needed for success in the course being taught.

Note: The author was a member of the faculty at West Point from 1989 to 1999.

References

Department of Mathematics, United States Military Academy, West Point, NY, "Core Mathematics at USMA," US Government Printing Office, West Point NY, 1998.

Appendix

Required mathematical skills for entering cadets

Cadets enter the US Military Academy with different backgrounds of study in mathematics. To successfully begin work in their math/science/engineering (MSE) courses at USMA, incoming cadets must arrive with the knowledge of certain mathematical skills and concepts. The MSE Committee identified the items below as those fundamental skills and concepts that all entering cadets must possess. These skills are tested soon after each cadet arrives at USMA; therefore, all entering cadets are encouraged to evaluate their mathematics skills and to work at remediating any deficiencies before arriving at USMA.

During Cadet Basic Training, the Department of Mathematical Sciences administers a gateway test to assess these fundamental skills. The math dept administers other gateway tests to all fourth class cadets during the first semester in MA103 (Discrete Dynamical Systems and Introduction to Calculus). Cadets are given opportunities for self-remediation during the first semester through self-paced text and diskettes. Failure to pass at least one gateway test at mastery level (i.e., greater than 80%) may require remediation in MA101 (Precalculus) during the second semester instead of moving directly into MA104 (calculus). Those cadets in MA101 who do not demonstrate proficency in the fundamental mathematics skills by the end of the fourth class year may be separated from USMA.

Almost any high school algebra book is a suitable reference for these fundamental mathematics skills; a DOS-based disk tutorial is also available (3.5″ format) by writing to the Department of Mathematical Science, US Military Academy, West Point, NY 10996, or by calling the Department at (845)938-4603/5673.

Note: All calculations must be done **without** the use of technology (i.e., calculator) unless otherwise stated.

1. Algebra and Real Numbers

a. Manipulate and reason with symbols in algebra.

b. Understand the relationship between velocity, distance, and time. (On a 40-mile car trip to Middletown, NY, you drive the first twenty miles at 40 mph and the last twenty miles at 60 mph. What is your average speed during the trip to Middletown?)

c. Be familiar with the following important algebraic properties of the real number system: identity, associativity, commutivity, inverse, and distributivity.

d. Express numbers using scientific notation. (Express 0.004312 in scientific notation.)

2. Integer and Rational Exponents

Manipulate algebraic expressions containing integer and rational exponents. (Simplify $4^{-\frac{3}{2}} \times 27^{-\frac{2}{3}}$.)

3. Radicals

a. Express an algebraic expression in rational exponent form. (e.g., Transform $1\sqrt{x+2}$.)

b. Convert between radical and rational exponent form. (Simplify $a^2 b \sqrt{90 b^7 a^6}$.)

4. Algebraic Expressions

a. Be able to add, subtract, multiply, and divide algebraic expressions. (Find the remainder when $x^3 - 7x^2 + 9x$ is divided by $x - 2$.)

b. Simplify algebraic expressions. (Expand and simplify $(x-3)(x-2)(x-1)$.)

5. Factoring / Prime Numbers

a. Write a number as the product of factors. (Write 42 as the product of prime factors.)

b. Solve for the roots of a polynomial by factoring. (Find the roots of $x^2 - 5x + 6 = 0$ by factoring.)

6. Linear Equations, with Inequalities and Absolute Values

a. Solve two simultaneous linear equations by graphing and by backsubstitution. (By graphing estimate the point of intersection of the lines $2x + 3y = 7$ and $-x + y = 4$. Verify your result using backsubstitution.)

b. Solve linear inequalities graphically and algebraically. (Given $5(3 - x) > 2(3 - 2x)$, solve for x.)

c. Solve linear equations with absolute values. (Solve $|x - 4| \geq 3$ for x.)

7. Polynomials and Rational Inequalities

a. Solve simple polynomial inequalities. (Solve $x^2 + 3x + 6 > x - 4$.)

b. Solve simple rational inequalities. (Solve $\frac{x-3}{x+1} < 2$.)

8. Quadratic Equations

Memorize and apply the quadratic formula to find real and complex roots. (Find the roots of $3x^2 + 2x = -1$.)

9. Graphs: Basic Tools

a. Graph equations and inequalities. (Provide a sketch of the function $3x^2 - 2x + 7$ for $1 < x < 5$.)

b. Properly label a graph (axes, intercepts, asymptotes, and roots).

10. Straight Lines

a. Find the slope of a line given a line or an equation of a line. (Find the slope of the line $4x - 3y = 8$.)

b. Determine the equation of a line. (Find the equation of a straight line with slope $-1/3$ passing through the point $(2, 1)$.)

c. Graph a line. (Graph the line $5y - x = 3$.)

d. Find the equation of a line that is parallel or perpendicular to a given line. (What is the slope of the line parallel to the line $2y - 3x = 7$?)

11. Functions

a. Identify the independent and the dependent variable of a function.

b. Determine the domain and range of a function. (Find the domain of the real-valued function $g(x) = 1/(x^2 - 2)$.

c. Evaluate a function at a point. (Given $f(x) = 1/(x^2 + x - 6)$, find $f(3.14)$.)

d. Be able to evaluate composite functions. (Given $h(r) = 3r^2$ and $g(s) = 2s$, find $h(a+2) - g(2a)$.)

12. Systems involving Quadratic Equations

Solve a system in two variables by backsubstitition. (Given $y = 3 - x^2$ and $y = 3 - 2x$, solve for x.)

13. Trigonometric Functions

a. Define each of the six trigonometric functions ($\sin(\theta)$, $\cos(\theta)$, $\tan(\theta)$, $\cot(\theta)$, $\sec(\theta)$, $\csc(\theta)$) in terms of the sides of a right triangle. (For example, $\cos(\theta) = x/r$ where x is the adjacent side and r is the hypotenuse.)

b. Define each of the six trigonometric functions in terms of sine and cosine. (For example, $\tan(\theta) = \sin(\theta)/\cos(\theta)$.)

c. Memorize and use the 30/60/90 and 45/45/90 degree reference triangles. (Given $\theta = 2\pi/3$ radians $=$ 120 degrees, determine $\cot(\theta)$, $\sec(\theta)$, and $\sin(\theta)$.)

d. Understand the relation between degrees and radians. (For example, 2π radians $= 360$ degrees)

e. Know and apply the trigonometric identity $\sin^2(\theta) + \cos^2(\theta) = 1$. (Express $\left(1 - \frac{1}{\csc(\theta)}\right)^2 + \cos^2(\theta)$ only in terms of the $\sin(\theta)$).

14. Graphs of the Trigonometric Functions

a. Know the domain and ranges for the sine, cosine, and tangent functions.

b. Graph the sine, cosine, and tangent functions.

15. Logarithmic and Exponential Functions

a. Memorize and apply the fact that the logarithmic function $y = \log_a x$, $a > 0$, $a \neq 1$, is the inverse of the function $y = a^x$—that is, $\log_a x = y \Leftrightarrow a^y = x$. (Evaluate $\log_3 27$.)

b. Memorize the properties of the logarithmic and exponential functions; use them to simplify logarithmic expressions. (Express as a single logarithm: $0.5 \log_{10} x - \log_{10} y$.)

c. Solve simple logarithmic and exponential equations. (Solve the equation $3^{x+4} = 4$.)

16. Analytic Geometry

a. Memorize and apply the distance formula between two points. (Find the distance between the two points $A(1, 2)$ and $B(-5, -3)$.)

b. Memorize and apply the circumference and area formulas for circles, triangles, and rectangles. (For example, if you double the radius of a circle, what happens to its circumference?)

c. Memorize and apply the surface and volume formulas for cylinders and rectangular solids, and be able to apply the surface and volume formula for a sphere.

d. Apply the fact that corresponding sides of similar triangles are proportional. (A rectangle with base x and height 5 is inscribed in an isosceles triangle with base 10 and height 20. Determine x.)

e. Memorize the Pythagorean Theorem and be able to apply it to simple geometric problems. (Given a rectangle that is 4 ft by 7 ft determine the length of the diagonal.)

Algebra

1. $x^2 + bx + c = 0 \quad x = \frac{-b \pm \sqrt{b^2 - 4ac}}{2a}$

2. $a^b \cdot a^c = a^{b+c}$

3. $(a^b)^c = a^{bc}$

4. $\frac{a^b}{a^c} = a^{b-c}$

5. $y = \log_b x \Rightarrow x = b^y$

6. $\log_b bx = b^{\log_b x} = x$

7. $\log_b x^a = a \log_b x$

8. $\log_b ac = \log_b a + \log_b c$

9. $\log_b \frac{a}{c} = \log_b a - \log_b c$

10. $\log_b a = \frac{\log_c a}{\log_c b}$

Analytic Geometry

Rectangle: Area $= lw$ Perimeter $= 2l + 2w$

Circle: Area $= \pi r^2$ Circumference $= 2\pi r$

Rectangular Solid: Volume $= lwh$ Surface Area $= 2lw + 2lh + 2hw$

Cylinder: Volume $= \pi r^2 l$ Surface Area $= 2\pi r^2 + 2\pi r l$

Sphere: Volume $= \frac{4}{3}\pi r^3$ Surface Area $= 4\pi r^2$

Distance between two points (x_1, y_1) and (x_2, y_2) is $d = \sqrt{(x_1 - x_2)^2 + (y_1 - y_2)^2}$

Trigonometry

With reference to the right triangle:

2π radians $= 360$ degrees

$\sin \theta = \frac{y}{r}$ $\cos \theta = \frac{x}{r}$ $\tan \theta = \frac{y}{x}$

$\tan \theta = \frac{\sin \theta}{\cos \theta}$ $x^2 + y^2 = r^2$ $\sin^2 \theta + \cos^2 \theta = 1$

$\cot \theta = \frac{1}{\tan \theta}$ $\sec \theta = \frac{1}{\cos \theta}$ $\csc \theta = \frac{1}{\sin \theta}$

Relationships

Distance = average rate × time
Corresponding sides of similar triangles are proportional

Properties of Functions

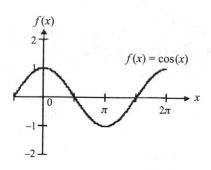

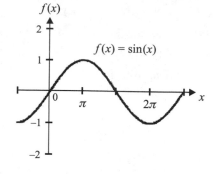

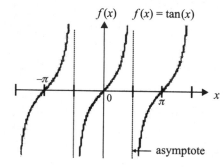

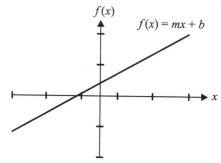

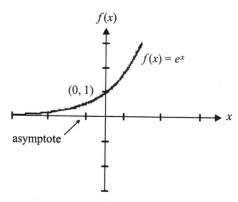

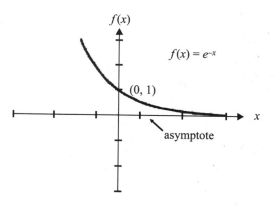

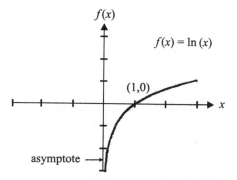

18

Integrating Data Analysis into Precalculus Courses

Allan J. Rossman
California Polytechnic State University

Introduction

The statistics education reform movement of the past fifteen years has emphasized genuine data, conceptual understanding, and active learning (see [3] and [9] for overviews). These features have also been hallmarks of calculus reform efforts, although the use of genuine data is naturally more prevalent in statistics courses than in calculus, for statistics has data analysis at its very core. Yet many calculus reform projects emphasize applications, which in turn often involve genuine data.

The articles in this volume reveal that these features, including the use of real data, are prominent in the call for rethinking precalculus courses as well. Indeed, Gordon uses college enrollment data to support his argument that students should analyze real data in preparation for calculus [6]. Examples of precalculus books that feature genuine data prominently include Gordon et al.'s *Functioning in the Real World: A Precalculus Experience* [7], Connally et al.'s *Functions Modeling Change: A Preparation for Calculus* [4], and Swanson et al.'s *Precalculus: A Study of Functions and Their Applications* [14]. All three of these texts present many examples of real data for students to analyze.

Data analysis plays an even larger role in an earlier book: Barrett et al's *Contemporary Precalculus: Functions, Data Analysis, and Matrices*, written by faculty of the North Carolina School for Science and Mathematics and aimed for a high school audience [2]. This book introduces data analysis in its very first chapter, while the central concept of function waits until the second chapter to make its appearance. A later chapter is fully devoted to data analysis, and real data pervade most of the book. The second edition of the book goes so far as to include interval estimates for predictions made from data-based models.

The predominant use of data in these books involves fitting functions to model data. This skill is especially important to the many client disciplines in the social and natural sciences where data abound. For these disciplines, one of the primary goals of a precalculus course is to help their students to understand principles and acquire skills of data analysis. This interest in data analysis is a consistent theme in the Curriculum Foundations reports from partner disciplines [5].

In the first section below I argue that data can play an even more central role by helping students learn fundamental ideas of precalculus, particularly the crucial concept of function. Moreover, I suggest that statistics provides a rich application area in which students can study interesting examples of functional behavior. Then in the following section, I return to the issue of modeling data with functions by proposing a series of data analysis principles to be taught in precalculus. Finally, I conclude the article by considering the role of precalculus in general education and the importance of data analysis for educated citizenship.

Using data to study functions

The concept of function is the most important and foundational idea in precalculus. Real data can not only help to motivate students to study functions, but can also help them to develop their understanding of what a function is and how functions behave.

Example 1: Scrabble Names

In this activity, students are asked to count how many letters are in their last name and then to determine how many Scrabble points are in their last name (by summing the point values of the letters according to the board game *Scrabble*). Students record this data for themselves and for their classmates. Then they consider whether the data they have just gathered constitute a function. They should realize that the number of letters is a function of the name, and the number of *Scrabble* points is also a function of the name. Students also gain practice applying the notions of domain and range to these functions, which may seem unusual to them since the domain is not numerical. This unusual aspect may help to deepen students' understanding of function as a process assigning a unique output to each input. Students also consider the question of whether a function has an inverse, and they find that neither function does because two different names can have the same number of letters or the same number of *Scrabble* points.

Perhaps more importantly, students are then asked questions that lead them to observe that number of points is not a function of number of letters, because two names with the same number of letters can have different numbers of points. Students are asked to produce a graph (scatterplot) of *Scrabble* points versus letters for the students in their class. Figure 1 presents these data for a sample of nine individuals. From this graph students can use the vertical line test to recognize further that number of points is not a function of number of letters.

Students continue with this example by calculating a new function: ratio of points to letters. They are asked whether this is a function of name or of points or of letters, and they should recognize that it is a function of name, but not of the other two. An instructor may also want to introduce the idea of a function of two variables with this activity. Students are also asked to identify the name that produces the largest ratio and whether that name necessarily has the most points or the most letters. Thus, some ideas and skills of working with ratios find their way into this activity.

Rather than stop there with the precalculus ideas, students can then explore more statistical aspects of these data, such as whether names with more letters *tend* to have more points than those with fewer letters. They respond in the affirmative, and then they are asked to identify a pair of names such that the one with more points has fewer letters. This part of the activity aims to introduce students to the important idea that statistical *association* between variables is not a deterministic rule, but rather reflects a *tendency*.

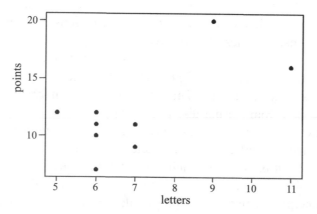

Figure 1. Scrabble points versus letters for a sample of nine individuals

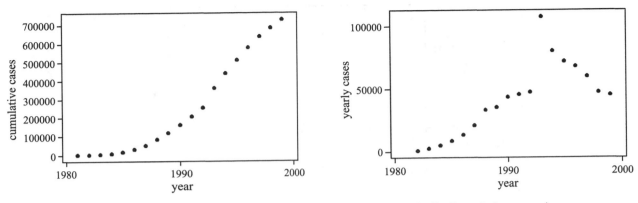

Figure 2. Cumulative and yearly AIDS cases (from Kaiser Family Foundation report)

While this statistical issue is not the focus of the activity, it would seem a shame not to delve into this aspect a bit since the earlier parts of the activity lead students' curiosity in this direction.

The next example moves from the fundamental question of "what is a function?" to considering properties involving the behavior of functions.

Example 2: AIDS Epidemic

For the twentieth anniversary of the AIDS epidemic, the Kaiser Family Foundation issued a report in 2001 that included data on cumulative numbers of AIDS cases and of AIDS deaths over the years (www.kff.org). Data on cumulative numbers of AIDS cases, as presented in the report, appear in the left graph of Figure 2.

These data provide opportunities for students to investigate properties of functions. First, they observe that the function is increasing since it records *cumulative* cases. Second, with a little prodding, they notice that the concavity changes around 1993 or so, indicating that the *rate* of increase slows down. Students then are asked how to calculate the number of AIDS cases per year rather than on a cumulative basis, which leads them to discover and calculate the *difference function*. The right graph in Figure 2 displays this difference function and therefore the yearly number of AIDS cases. Students then investigate how the difference function relates to the original function. They find that the difference function increases where the original function is concave up, that it reaches its maximum where the concavity changes, and that it decreases where the original function is concave down.

The next example asks students to apply their knowledge of power functions and illustrates that the discipline of statistics can provide interesting applications of functions.

Example 3: Margin-of-Error

One of the most important statistical concepts for educated citizens to understand is margin-of-error. Random variation leads survey results to vary from sample to sample, and the margin-of-error provides an indication of how much the sample result may deviate from the truth about the whole population. This is certainly not a precalculus concept, but it can be used to give students practice with the precalculus idea of a power function. Students are presented with the information in Table 18.1 on a survey's margin-of-error for several different sample sizes:

sample size	250	500	750	1000	1250	1500
margin-of-error	.063	.045	.037	.032	.028	.026

Table 1. Margin-of-error as a function of sample size

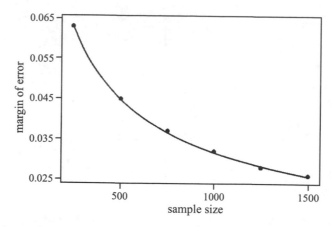

Figure 3. Margin-of-error as a power function of sample size

Students are asked to graph these margins-of-error as a function of sample size and then to find a power function that describes the relationship between sample size and margin-of-error. They should recognize that the decreasing function requires a negative power, and some trial-and-error should lead them to the negative one-half power (that is, margin of error $= 1/\sqrt{\text{sample size}}$) as the correct relationship, as shown in Figure 3. Students can go on to exercise their knowledge of exponents to address questions such as whether doubling the sample size of a survey serves to cut the margin-of-error in half.

The thrust of this example is not that students must learn statistics in their precalculus course. Rather, the point is that like physics and economics (to cite but two examples), statistics can provide applications in which students can enhance their study of functions. Just as learning some economics from supply-and-demand examples and some physics from distance-and-velocity examples is a fortuitous side benefit, so too is having students learn some statistics along the way.

The next example shifts from the concept of function to illustrating how data can motivate students to study the apparently mundane topic of properties of logarithms. It also serves as an example of using probability as an application area.

Example 4: Benford's Model for Leading Digits

Benford's model asserts that the leading digits of values for many variables follow a distribution predicted by the function $p(k) = \log(1 + 1/k)$. [The leading digit of a quantity is the first non-zero digit when reading from left-to-right.] This model has been used for detecting fraud in the accounting industry by identifying tax returns and other accounting documents where the leading digits fail to follow this model [8].

Students are presented with data summarized in Table 18.2, which reports the frequencies of leading digits in the populations of the 125 counties of Pennsylvania and California, as reported in *The 2000 World Almanac and Book of Facts*:

digit	1	2	3	4	5	6	7	8	9
freq.	38	17	21	16	8	8	6	6	5

Table 2. Frequencies of leading digits in 125 county populations

Students investigate how well Benford's model fits this data by calculating $\log(1 + 1/k)$ and comparing those probabilities to the relative frequencies in the data. They also use the property that the sum of the logs is equal to the log of a product in order to show that Benford's probabilities sum to 1. Students are then asked to find their own set of data, collect the leading digits, and analyze how well Benford's model applies to their data.

Data analysis principles for precalculus

The most prevalent use of data in precalculus books and courses is to illustrate fitting functional models to data. I suggest that while using data for this purpose, precalculus teachers should also seize the opportunity to acquaint students with sound principles of data analysis. In this section I present two examples of fitting functional models to data, both adapted from [10], and provide a list of data analysis principles along the way.

Example 5: Airfares

Students examine sample data on the cheapest airfare to fly to a destination and the distance to that destination, as reported by a local newspaper for flights from Baltimore, displayed in Figure 4.

One obvious fact that students recognize immediately is that a line will not fit this data exactly. This is in contrast to simpler examples that they may have seen, for example finding a line between two points or representing distance traveled as a linear function of time for a given (constant) speed. In fact, no simple function will model this airfare data perfectly. While this point is obvious, it suggests a first principle: *Variability is fundamental.* It is important for students to recognize that real data vary, so no functional model will provide an exact fit.

Students then fit a line to the data by eye and calculate its equation. They notice that their line differs from other students' lines, and they realize that some criterion is necessary for deciding which line "best" summarizes the relationship. This leads to a discussion of the least squares criterion and to a second principle: *Understand that least squares is just one criterion for selecting a best-fit line.* By applying this criterion to their hand-drawn line, students encounter an interesting and unusual function: sum of squared residuals (vertical deviations between the points and the line) as a function of the line drawn. Students can also work with other techniques for drawing lines. For example, Barrett et al. [2] has students work with the median-median technique starting with chapter one and does not introduce least squares lines until chapter five.

Students then investigate the impact of extreme observations on the least squares line: they change the airfare for the furthest city and find that it affects the line substantially, while changing the airfare for a city with a moderate distance has much less effect. The principle here is: *Recognize the potential influence of individual observations.* This is largely a consequence of the squaring of residuals with the least squares criterion, and the result is that a single observation could have undue influence on the line and cause it to poorly represent the pattern in the majority of the points.

Students are also asked to make predictions about airfares to other cities, based on their distances. The farthest destination represented in the data is 1500 miles away, but some of these predictions requested of students include cities much farther away. Students are asked to realize a fourth principle: *Be wary of*

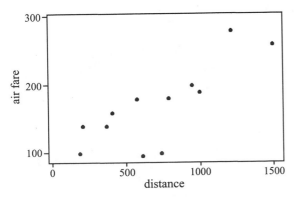 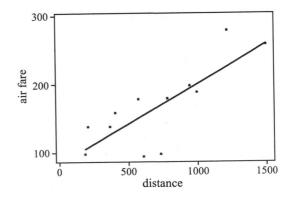

Figure 4. Airfare versus distance for a sample of twelve destinations

extrapolating the model beyond the domain of the sample data. This admonition also cautions students against interpreting the intercept of the line, for a distance of zero miles naturally lies outside the domain of the sample data.

Remembering the theme that variability is fundamental, students should also be taught not to regard predictions based on a model too precisely. Predictions should include not only a point estimate but also an interval to indicate the accuracy of the prediction. The principle to follow is: *Provide information about accuracy to accompany predictions.* The second edition of Barrett et al. [2] accomplishes this by asking students to construct prediction bounds that are two times the standard deviation of the residuals away from the line. For example, the least squares line predicts $171.31 for a flight to a destination 750 miles away. The standard deviation of the residuals is 36.07, so a prediction interval for the cost of that flight would extend from $99.17 to $233.45, providing an indication of the accuracy of the prediction.

Example 6: Televisions and Life Expectancy

The left graph of Figure 5 reveals the life expectancy for a sample of 22 countries versus the number of people per television set in those countries.

Students first note that there is a fairly strong negative association between the two variables. They are then asked whether this association means that sending televisions to countries with many people per television will cause their inhabitants to live longer. Students recognize the absurdity of this claim, and in the process they discover one of the most important principles of data analysis: *An observed association between two variables does not imply a cause-and-effect relationship between them.* This is a statistical point and certainly not part of the standard precalculus curriculum, but it is imperative for students to understand this principle whenever they are studying relationships between variables.

Students are then asked if the association between these variables can be modeled with a linear function, and the scatterplot clearly reveals that it can not. Another obvious, but probably the most important, principle of data analysis is: *Always look at a graph of the data first.* Doing so can help to prevent fitting inappropriate models to data, and it can also reveal unusual cases such as potentially influential observations. A classic illustration of this importance was provided by Anscombe [1], who devised four bivariate data sets with identical summary statistics, and therefore identical least squares lines, but with very different graphs so that the linear model is appropriate only in one of the four cases. Figure 6 displays the Anscombe data and their least squares lines.

Students are asked to examine transformations of the variables "people per television" and "life expectancy" that could linearize the relationship. It turns out that a logarithmic transformation on the "people per television" variable makes the relationship quite linear, as shown in the right graph of Figure 5. In this

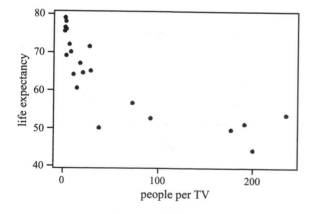 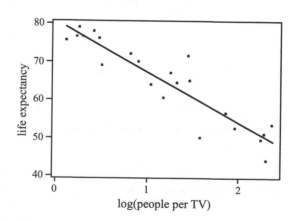

Figure 5. Life expectancy vs. people per television for a sample of 22 countries

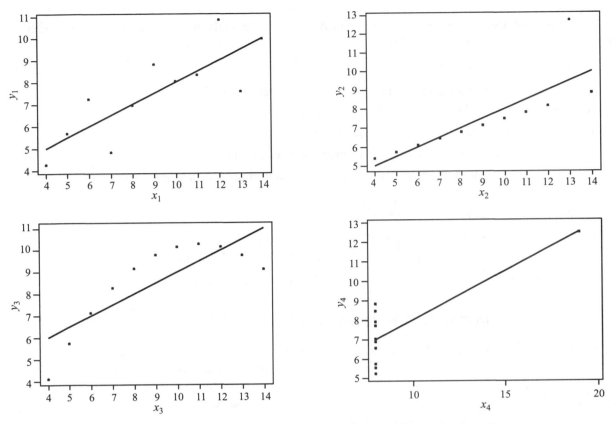

Figure 6. Anscombe data with identical least squares lines: $y = 3 + .5x$

manner students experience another data analysis principle: *Use re-expression of variables to try to make relationships linear.* After students fit a linear model for predicting life expectancy from this transformed variable, they practice their algebraic skills by making predictions based on this transformed model and re-writing the transformed linear model as a logarithmic one. Students are also asked a series of questions leading them to realize that the slope coefficient of the transformed model (-13.3) indicates that the predicted change in life expectancy is a decline of 13.3 years for a tenfold increase in the country's number of people per television set.

Students are also asked to examine residual plots (graphs of residuals versus x-values) for both the original and the transformed data. These plots can reveal departures from linearity that may be hard to see in the original scatterplot. The principle here is: *Analyze residual plots to assess how well the functional model fits the data.* Barrett et al. [2] emphasizes residual plots from the very first chapter, and Gordon et al. [7] also pays considerable attention to them.

In both the airfare and life expectancy examples, students use technology to calculate the correlation coefficient between the variables and also R^2, the proportion of variability in the y-variable explained by the model with the x-variable. They are cautioned about the principle: *Correlation and R^2 can be misleading about whether a model is appropriate.* For example, the correlation between the airfares and distances is 0.795, while the correlation between life expectancy and people per television (untransformed) is -0.804. These are very similar in absolute value, but the graphs reveal that the linear model is a reasonable one for the airfare data but not at all for the life expectancy data. This point is also reinforced by the Anscombe data, where all four datasets have the same correlation despite very different appearances. A very large value of R^2 does not necessarily mean that the model fits well, and R^2 should not be used as the sole basis for comparing different models. Residual plots are more effective for assessing model fit and for selecting among competing models.

A top-ten list results from re-stating these principles of data analysis that can be taught as students use functions to model data in precalculus:

1. Variability is fundamental.
2. Understand that least squares is just one criterion for selecting a best-fit line.
3. Recognize the potential influence of individual observations.
4. Be wary of extrapolating the model beyond the domain of the sample data.
5. Provide information about accuracy to accompany predictions.
6. An observed association between two variables does not imply a cause-and-effect relationship between them.
7. Always look at a graph of the data first.
8. Use re-expression of variables to try to make relationships linear.
9. Analyze residual plots to assess how well the functional model fits the data.
10. Correlation and R^2 can be misleading about whether a model is appropriate.

Data analysis in precalculus for general education

In his "Twenty Questions About Precalculus" appearing in this volume, Steen [13] poses many questions about the student audience for precalculus, which often includes students satisfying a general education requirement and who never take another mathematics or quantitative course. I believe that integrating data analysis throughout precalculus is even more important for these students than for those who go on to take calculus. Our contemporary world is awash in data; all citizens encounter data in their everyday lives. The need to make sense of data has become an important aspect of informed citizenship. Several calls for developing students' quantitative literacy, including Steen [11] and Steen [12], recognize the central role that data analysis plays in the effort. Considering the large role that precalculus plays in general education at many institutions, this concern provides further justification for including data analysis.

Conclusion

Data analysis can play an important role in enhancing students' learning experiences in precalculus. Genuine data often provide motivation and interest for students, and they reveal that concepts of precalculus do have application to analyzing data from a variety of disciplines as well as from everyday life. Beyond its usefulness for teaching precalculus ideas, data analysis also provides the opportunity for students to discover and explore some important statistical principles with which all educated citizens should be comfortable.

Acknowledgements. Most of the ideas presented in this article will appear in a text called *Workshop Precalculus: Functions, Data, Models*, by Nancy Baxter Hastings and Allan J. Rossman, to be published by Key College Publishing. This work was supported by National Science Foundation grant #9952483.

References

1. Anscombe, F., "Graphs in Statistical Analysis," *Journal of the American Statistical Association*, 27, 17–21. 1973.

2. Barrett, G. et al, *Contemporary Precalculus through Applications: Functions, Data Analysis, and Matrices*, Jansen Publications, Inc., Dedham, MA, 1992.

3. Cobb, G., "Teaching Statistics," in *Heeding the Call for Change*, L. Steen (ed.), Mathematical Association of America, Washington, DC, 1992.

4. Connally, E. et al., *Functions Modeling Change: A Preparation for Calculus*, John Wiley and Sons, New York, 2000.

5. Ganter, S.L. and W. Barker, (eds.), *A Collective Vision: Voices of the partner disciplines,* MAA Reports Series, Mathematical Association of America, Washington, DC, 2003.

6. Gordon, S., "Preparing Students for Calculus in the Twenty-First Century," in this volume.

7. Gordon et al., *Functioning in the Real World: A Precalculus Experience*, 2nd edition, Addison-Wesley, 2004.

8. Hill, T., "The Significant Digit Phenomenon," *The American Mathematical Monthly*, April 1995.

9. Moore, D., "New Content and New Pedagogy: The Case of Statistics," (with discussion), *International Statistical Review*, 65, 123–165. 1997.

10. Rossman and Chance, *Workshop Statistics: Discovery with Data*, Key College Publishing, Emeryville, CA, 2001.

11. Steen, L., (ed.), *Why Numbers Count: Quantitative Literacy for Tomorrow's America*, The College Board, New York, 1997.

12. ——, (ed.), *Mathematics and Democracy: The Case for Quantitative Literacy*, The Woodrow Wilson National Fellowship Foundation, 2001.

13. ——, "Twenty Questions About Precalculus," in this volume.

14. Swanson, T., J. Anderson, and R. Keeley, *Precalculus: A Study of Functions and Their Applications*, Brooks/Cole, 2000.

Theme 4. Student Learning and Research

One line of questioning that was repeatedly voiced at the conference *Rethinking the Preparation for Calculus* revolved around determining what works and what doesn't work. Participants asked: "How can we determine whether or not new pedagogical approaches or new curricular materials are effective? How can we measure the impact of a new approach on student learning?" In order to convince colleagues to adopt new pedagogies and instructional materials, we not only need data to show that there is a problem, we also need to measure the effectiveness of alternative approaches. In this section, Florence Gordon looks at the results of a study that compared student performance, success, retention, and attitudes in a reform version of precalculus and in a traditional version of the same course. Rebecca Walker reports on the experiences of college students who prepared to continue their study of mathematics in a reformed high school environment.

Florence Gordon reports that in the fall of 1999, when some mathematics faculty at NYIT decided to teach a reform version of precalculus, several other faculty members objected to the change. As a result, the department decided to compare the performances of students in two traditional and two reform sections of precalculus.

> For the last decade, the reform of mathematics education at the college level has been accompanied by the so-called "math wars." One dimension of this struggle has involved calls to prove that non-traditional courses are at least as effective as the courses they are designed to replace. From one point of view, this is certainly a reasonable request, especially as all of higher education faces pressures for accountability. From another point of view, I am not aware that anyone has ever been asked to prove anything about the effectiveness of the traditional courses.... In this article, we look at the results of a comprehensive, multifaceted study comparing student performance, success, retention, and attitudes in a reform version of precalculus to that in a traditional version of the same course. Another component of the study involved student performance, success, and retention in the follow-up calculus course based on the type of precalculus experience the students had....
>
> At the start of our experience at NYIT, some faculty expressed concern that the reform/modeling approach in precalculus would inflict irreparable damage to the students. In retrospect it is the traditional precalculus course that harms the students.
>
> *Assessing What Students Learn: Reform versus Traditional Precalculus and Follow-up Calculus*
> Florence S. Gordon

Based on the research she did for her doctoral thesis, Rebecca Walker reports on "the views of mathematics held by students who completed the four years of school mathematics using the Core-Plus Mathematics Project (CPMP)." She also reports on "case studies of the first semester college mathematics experiences of six of these students, with particular attention to the transition from this reform curriculum into college mathematics."

> A variety of questions arise while trying to rethink college precalculus. How will students react to a reform precalculus experience? Can a reformed precalculus experience help students develop a broader and more realistic perspective of mathematics? How successful will students be in calculus if they have a different type of preparation? Can a different learning environment promote deeper mathematical understanding? Is it

possible to create a classroom environment where students expect the mathematics to make sense and where they will struggle with complex problems? None of these questions is easy to answer but it is possible to begin to answer them by looking at the experiences of students who have prepared for college mathematics in a reformed environment.

Student Voices and the Transition from Reform High School Mathematics to College Mathematics
Rebecca Walker

19

Assessing What Students Learn:
Reform versus Traditional Precalculus and Follow-up Calculus

Florence S. Gordon
New York Institute of Technology

Introduction

For the last decade, the reform of mathematics education at the college level has been accompanied by the so-called "math wars." One dimension of this struggle has involved calls to prove that non-traditional courses are at least as effective as the courses they are designed to replace. From one point of view, this is certainly a reasonable request, especially as all of higher education faces pressures for accountability. From another point of view, I am not aware that anyone has ever been asked to prove anything about the effectiveness of the traditional courses. Most studies looking at the impact of a nontraditional approach to student learning have focused on the developmental mathematics/college algebra level and at the calculus level. For instance, Baxter et al [1] compared student performance in traditional versus reform calculus and the students' performance in subsequent science and engineering courses.

In this article, we look at the results of a comprehensive, multifaceted study, which we undertook in fall 1999, that compared student performance, success, retention, and attitudes in a reform version of precalculus to that in a traditional version of the same course. Another component of the study involved student performance, success, and retention in the follow-up calculus course based on the type of precalculus experience the students had.

Background

The precalculus offering at New York Institute of Technology (NYIT) is the standard college algebra and trigonometry course stressing traditional drill-and-skill techniques presumed to be necessary for calculus. Graphing calculators are used to augment the usual algebraic topics. For the science and engineering majors, this course has been followed by a reform calculus sequence based on the Harvard calculus materials [2].

In the fall of 1999, some faculty from the mathematics department decided to teach a reform version of precalculus based on mathematical modeling with an approach that emphasized conceptual understanding. Several other faculty members objected to this change, believing that such a course would inflict irreparable damage on the students in terms of their perceived weaknesses in algebraic manipulation. To resolve this issue, the department decided to offer two sections of the course in the traditional, algebra-oriented way and two sections using the reform/modeling approach and then to compare the results between the two groups. Four experienced faculty members, all of whom are considered excellent instructors, taught the four sections.

The students had no idea when they registered for precalculus that some of the sections would be reform/modeling and others traditional. Thus, their choice of section was basically random. Moreover, during the first week of the semester, the students in the reform/modeling sections were told that the classes were going to be very different from the traditional approach. These students were given the option of transferring to one of the traditional sections, but not one of them opted out.

While the core topics covered in both precalculus groups were the same (linear, polynomial, rational, exponential, logarithmic, and trigonometric functions), the emphasis, content, and sequence of topics were quite different. The traditional course, based on [3], was mainly lecture-based and stressed routine algebraic manipulations to improve student skills. The reform/modeling course, based on [4], stressed conceptual understanding of the mathematical ideas, problem-solving, and realistic applications. In the reform/modeling course, the algebraic manipulations arose only in the context of problem solving, not as long lists of drill-and-skill exercises. Students learned by applying the mathematical ideas and methods to real-world data, some of which they personally obtained on subjects of interest to them.

At the time, NYIT was a member of the NSF-supported Long Island Consortium for Interconnected Learning (LICIL) project, which was intended to foster greater interaction between mathematics and all other quantitative disciplines. With support from LICIL, a multifaceted assessment study was conducted by two external evaluators. One evaluator, who was responsible for LICIL's overall project evaluation, conducted an attitudinal study [5], and another evaluator was brought in specifically to assess the differences between the two groups in precalculus and in the succeeding calculus course [6].

The main components of the evaluation study included:

1. Comparison of placement test scores prior to the start of the precalculus course.
2. Comparison of student performance on a series of common problems on the final exams administered to the two precalculus groups.
3. Comparison of student attitudes in the two precalculus groups.
4. Comparison of student performance on weekly quizzes in the follow-up calculus course based on the type of precalculus experience the students had.
5. Comparison of student performance on class tests in the follow-up calculus course.
6. Comparison of student performance on the final exam in the follow-up calculus course.
7. Comparison of student retention and success rates in the follow-up calculus course.

Student backgrounds prior to precalculus

A department-generated placement test, which has been used for many years to assess algebraic proficiency, was administered to all students before the start of classes. On the basis of this test, students were admitted to the one-semester precalculus course or were directed either to a developmental course or to a slower two-semester algebra/trigonometry course.

For the 64 students who placed into precalculus, 37 were in the reform/modeling sections and 27 were in the traditional sections. The students in the reform/modeling sections scored a mean of 12.47 on the placement test with a standard deviation of 3.93. In comparison, those in the traditional precalculus sections scored a mean of 13.58, with a standard deviation of 3.36 (see Figure 1).

The standard deviations for both groups were very similar, indicating that there was very little variation in the placement test scores of both groups. To determine if the two mean scores were significantly different, a means/ANOVA t-test was conducted and the results indicate that, while the students in the traditional classes had better placement scores, the two means were statistically not significantly different: p-value $= 0.2512$, F-ratio $= 1.3426$.

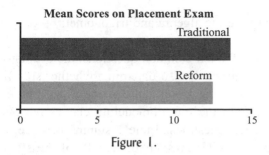

Mean Scores on Placement Exam

Figure 1.

Assessing student performance in precalculus

To comply with the departmental request for a comparison of student performance, all four faculty members teaching the course agreed to include ten common questions on the final examinations for both precalculus groups. These questions were primarily manipulative in nature. Note that the content of all of these questions was agreed to by the four faculty members who were confident that their students could handle the problems with relative ease.

The ten common questions were worth a total of 66 points. Students in the reform/modeling sections scored a mean of 49.69 with a standard deviation of 9.32, while those in the traditional sections scored a mean of 43.63 with a standard deviation of 12.03. A means/ANOVA t-test indicated that the two means were significantly different (p-value $= 0.0266$, F-ratio $= 5.1572$). Thus, the two groups differed in their performance on the common, algebraic manipulation questions on the final examinations with the students in the reform/modeling sections out-performing those in the traditional sections. (See Figure 2.)

This result is particularly striking when one considers that the students in the reform/modeling sections started with weaker algebraic skills on average (as measured by the lower mean scores on the placement exam), took a course that did not explicitly emphasize such skills, and ultimately outscored their peers on questions involving precisely those kinds of skills.

We can only speculate about the reasons for this. In a typical traditional course, the students are assigned many practice problems of the same type; the problems look indistinguishable and appear no different from problems that they've seen in previous math courses. Many students likely do not bother to work at many of the problems; they feel they've done them before. These students tend to equate familiarity with mastery. There is nothing in long assignments of routine problems that captures the students' interest.

Students in the reform/modeling sections were assigned primarily non-routine problems that did not resemble those they had seen before. The realistic contexts made the mathematics more interesting to the students; they cared about the answers. The algebraic manipulations were hidden in the contexts. The students were expected to create functions and use them to answer questions about the quantity being studied. The actual equations that arose were often much more complicated than those in a traditional course, because the numbers were realistic rather than being artificially contrived to make the answers work out evenly. In this environment, we conjecture that the students did more homework of a more

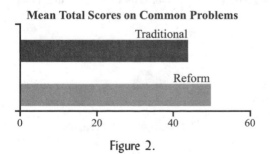

Mean Total Scores on Common Problems

Figure 2.

challenging nature. The relatively routine algebra and trigonometry problems on the final exam were then not difficult for them.

In addition to the total score on the ten common questions, the evaluator also analyzed student performance on each of the questions individually. To determine whether significant differences exist between the scores earned by the reform/modeling group and the traditional group on these questions, he used a MANOVA model. The students from the reform/modeling classes out-performed the students from the traditional group in seven of the ten questions. Table 1 summarizes the results and Figure 3 provides a visual display, where the bar on the left represents the results of the students from the reform/modeling sections for each question. The evaluator found that the differences were statistically significant for three of the ten questions (#1, #4, and #5).

The three questions where there were significant differences, and discussions regarding each of them, are as follows:

Question #1 *Brookville College enrolled 2546 students in 1996 and 2702 students in 1998. Assume that enrollment follows a linear growth pattern.*

(a) *Write a linear equation that gives the enrollment in terms of the year t (let t = 0 represent 1996).*

(b) *If the trend continues, what will the enrollment be in the year 2016?*

(c) *What is the slope of the line you found in part (a)?*

(d) *Explain, using an English sentence, the meaning of the slope here.*

(e) *If the trend continues, when will the enrollment reach 3500 students?*

Out of a total of 10 points, the students in the reform/modeling sections scored a mean of 9.14 with standard deviation of 1.38, while the students in the traditional sections scored a mean of 6.33 with

	Reform/modeling		Traditional			
Item	Mean	Std. Dev.	Mean	Std. Dev.	*F* Ratio	*p*-value
Q1	**9.14**	**1.38**	**6.33**	**3.71**	**17.8202**	**0.0001**
Q2	10.13	3.58	11.09	2.87	1.3156	0.2558
Q3	2.50	1.66	2.91	1.23	1.1639	0.2848
Q4	**1.00**	**0.97**	**1.81**	**0.56**	**15.2715**	**0.0002**
Q5	**7.20**	**1.77**	**4.44**	**2.69**	**24.4292**	**0.0001**
Q6	3.23	0.97	2.70	1.23	3.6470	0.0608
Q7	3.46	0.98	3.22	1.05	0.8597	0.3574
Q8	4.15	1.03	3.63	1.66	2.3867	0.1275
Q9	2.80	2.21	2.37	1.92	0.6484	0.4237
Q10	6.08	2.64	5.11	3.04	1.8510	0.1786

Table 1.

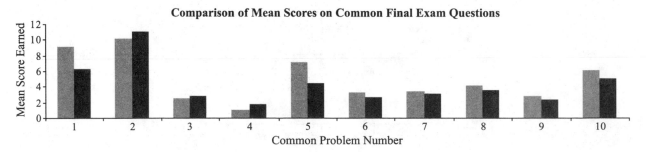

Figure 3.

standard deviation 3.71. A means/ANOVA t-test indicates that the two means were significantly different (p-value $= 0.0001$, F-ratio $= 17.8202$).

Question #4 Given $h(x) = 4(x^2 - 17)^{15}$, find two functions $f(x)$ and $g(x)$ such that $h(x) = f(g(x))$.

Out of a total of 2 points, the students in the reform/modeling sections scored a mean of 1 with standard deviation of 0.97, while the students in the traditional sections scored a mean of 1.81 with standard deviation 0.56. A means/ANOVA t-test indicates that the two means were significantly different (p-value $= 0.0002$, F-ratio $= 15.2715$).

Question #5 (reform/modeling sections) *The population of Peru was 24 million in 1995 and has been growing at an annual rate of 2.1%.*

(a) *Write a function for the population $P(t)$ after t years. (Let $t = 0$ represent 1995.)*

(b) *According to this model, what will the population be in the year 2000?*

(c) *Use any method to determine when the population will reach 30 million.*

Question #5 (traditional sections) *You deposit $8500 in a bank paying 4.25% interest.*

(a) *Write an expression for your balance $B(t)$ after t years.*

(b) *What will your balance be after 30 years?*

(c) *Use any method to determine when your balance will equal $15,000.*

The reason for these two different, though mathematically equivalent, problems is that the instructors in the traditional sections believed that their students would be more comfortable with the more standard money growth context that they had emphasized in class than with the population growth context.

Out of a total of 8 points, the students in the reform/modeling sections scored a mean of 7.2 with standard deviation of 1.77, while the students in the traditional sections scored a mean of 4.44 with standard deviation 2.69. A means/ANOVA t-test indicates that the two means are significantly different (p-value $= 0.0001$, F-ratio $= 24.4292$).

The most noticeable difference in the two groups is revealed by a comparison of the responses to part **(d)** of Question #1, which asked the students to interpret the meaning of the slope of the line. The complete set of responses appears in [7].

In summary, 35 of the 37 students in the reform/modeling sections gave a meaningful response, an indication of their understanding of the significance of the slope of a line. A typical response was: *"This means that for every year the number of students increases by 78."* The remaining two students wrote an appropriate statement for the meaning of slope, but calculated the slope as $\Delta t / \Delta y$.

In comparison, nine of the 27 students in the traditional sections were able to provide a meaningful response indicating an understanding of what the slope represents. Five left that portion of the question blank and three simply rephrased the algebraic formula $\Delta y / \Delta x$ for the slope in words. The remaining 10 students wrote statements that made no sense, including:

- *The point in which the # of students is increasing.*
- *The slope in this equation means the students enrolled in 1996. $Y = MX + B$.*
- *The slope is the average amount of years it takes to get 156 more students enrolled in the school.*
- *Since it is positive it increases.*
- *Its how many times a year it increases.*

Note that Question #1 was worth a total of 10 points on both final exams. Part (d), which asks for the interpretation of the slope, was worth only 2 points and so represents only a fraction of the difference in the overall mean scores (9.14 versus 6.33) between the two groups. Those students who had trouble interpreting the slope also had trouble using the equation of the line to answer the predictive questions posed.

It is interesting that the ability of both groups to calculate the slope of a line was comparable. However, any graphing calculator and many commonly available software packages, such as Excel, can do that also. What should be more valuable to our students is the ability to understand what the slope means in context, whether that context arises in one of their other courses in mathematics or courses in one of the quantitative disciplines or eventually on the job.

These results suggest that, unless explicit attention is devoted to emphasizing conceptual understanding of what the slope means, a majority of students are not able to create meaningful interpretations on their own. Lacking conceptual understanding, they are not able to apply the mathematics to realistic situations in new contexts or in other courses.

Many of us have heard complaints from colleagues in other disciplines about students who appear not to have learned key mathematical ideas and techniques, such as finding or using the equation of a line. In most other disciplines, linear functions do not arise in the form usually taught in a traditional course: *Find the equation of the line through the points* (1, 3) *and* (5, 11). Instead, one typically faces a collection of data relating two quantities that follow a roughly linear pattern and has to find and use the (regression) line that best fits the data. If students have such difficulty just giving meaning to the slope of a line, it is no wonder that they are unable to connect what they learn about lines and linear functions in their math classes to what they are expected to do in their other courses.

Moreover, if students are unable to make their own connections with an elementary concept like the slope of a line (which they have encountered previously), it is unlikely that they will be able to create meaningful interpretations and connections on their own for more sophisticated mathematical concepts, such as: What is the significance of the base (growth or decay factor) in an exponential function? What is the meaning of the power in a power function? What do the parameters in a realistic sinusoidal model tell about the phenomenon being modeled? What is the significance of the factors of a polynomial? What is the significance of the derivative of a function? What is the significance of a definite integral?

On the basis of this study, it is clear that we cannot simply concentrate on teaching mathematical techniques and skills. It is at least as important to stress conceptual understanding and the meaning of the mathematics. This can be accomplished by using realistic, contextual examples and problems that force students to think, not just to manipulate symbols. If we fail to do this, we are not adequately preparing them for successive mathematics courses, for courses in other disciplines, and for using mathematics on the job and throughout their lives.

Assessing student attitudes in precalculus

A student attitudinal survey was also conducted in precalculus the same semester by LICIL's external evaluator [5]. In this study, instructors A and B taught the traditional course; instructors C and D taught the reform/modeling course. All four instructors incorporated the use of graphing calculators. A set of 20 attitudinal questions was administered on the first day of class (pre-survey) and again on the last day of the semester (post–survey). These questions were designed to investigate the extent to which student attitudes changed as a result of each course. The questions covered several general areas. In analyzing the data, the evaluator combined the Strongly Agree and Agree responses, as well as the Strongly Disagree and Disagree responses.

One group of questions dealt with whether mathematics is an active, open-ended, discovery-oriented process or a passive, closed-ended, memory-based procedure. The results are shown as percentages of positive responses (percentage of students who Strongly Agree/Agree that mathematics is an active, open-ended discovery-oriented process) in Table 2. The pre-course responses were clustered close to the mean, with a range of only 4.5 and a standard deviation of 1.9, so that the four groups were comparable at the outset. The post-course responses, however, had a greater spread, with a range of 25 and a standard deviation of 12.1. Group C showed an increased level of positive attitudes and experiences toward mathematics,

	Pre	Post	Change
Traditional A	57.3%	47.2%	−10.1
Traditional B	56.0%	41.7%	−14.3
Reform C	58.0%	66.7%	+8.7
Reform D	60.5%	63.2%	+2.7

Table 2.

group D showed a slight increase, while both traditional Groups A and B indicated substantially more negative attitudes and experiences toward mathematics.

The second group of questions dealt with the usefulness of mathematics and whether the students viewed it as connected to situations beyond math courses. Some of the survey items were:

- *This course helped me to understand how to apply math to real world problems.*
- *In this course, I learned ways of thinking that are useful in situations outside of math.*
- *This course showed that math is useful in many non-math courses.*
- *This course made connections across disciplines.*

The corresponding percentage of positive responses for the four groups are shown in Table 3. The responses for the pre-course survey are again clustered, although not as closely as in the previous area, with a range 13.5 and a standard deviation of 6.1. Post-course results are scattered even more widely than in the first area, with a range of 62.3 and a standard deviation of 28.3. The combined pre/post means for the two reform/modeling classes moved from 55.4 to 53.2, while the combined pre/post means for the traditional classes fell from 58.75 to 9.25.

	Pre	Post	Change
Traditional A	65.5%	6.0%	−59.5
Traditional B	52.0%	12.5%	−39.5
Reform C	57.5%	68.3%	+10.8
Reform D	53.3%	38.0%	−15.3

Table 3.

Based on the responses to these two general areas, the evaluator concluded that students appear to respond much more positively to the reform/modeling approach than to the traditional approach. He hypothesized that perhaps differences in the student populations or in the instructors might account for the differences. Based on responses to other items, however, the evaluator rejected these hypotheses.

A third group of questions dealt with the importance of technology in learning mathematics. The results are shown in Table 4. The evaluator noted that the average for group C was the lowest in the pre-course survey, but increased to a 100% positive rating, while the percent of positive response of the other three groups all declined.

	Pre	Post	Change
Traditional A	78%	67%	−11
Traditional B	87%	70%	−17
Reform C	76%	100%	+24
Reform D	88%	79%	−9

Table 4.

The follow-up study in calculus

In the spring of 2000, a follow-up study was conducted on student performance and retention in the calculus I course for science and engineering majors based on whether the students had taken a traditional or the reform/modeling precalculus course. There were two sections of the calculus course taught by the same professor, and each section included students with both types of precalculus background. The course was based on the Harvard text [2].

Not all students who take precalculus at NYIT are required to take calculus I. Some take a business calculus course or an architecture calculus course. The precalculus course is a required, but terminal, course in certain majors, including physical therapy, occupational therapy, nursing, and communication arts. Some students do not immediately go on to take calculus I the semester after having completed precalculus.

The end result was that 13 of the 37 students (or 35%) who passed the reform/modeling precalculus course during fall 1999 started calculus I during spring 2000; five of the 27 students (or 18.5%) who passed the traditional precalculus course during fall 1999 started calculus I during spring 2000. In total, there were 52 students enrolled in the two sections of calculus I. Thirteen of them had a reform/modeling precalculus background while 39 had a traditional precalculus background. However, only five of those 39 students had passed traditional precalculus during fall 1999. The remaining 34 students had passed the prerequisite precalculus course in a variety of ways; some were transfer students, some had taken a traditional precalculus course during an earlier semester, some had taken traditional precalculus in high school, and some had completed a two-semester traditional precalculus track.

During the calculus study, the external evaluator compared students' performance on quizzes, class tests and the final examination, as well as student success rates, retention rates, and persistence in the course. Since only 13 students from the fall 1999 reform/modeling precalculus classes and only five students from the fall 1999 traditional precalculus classes enrolled in calculus I in spring 2000, the evaluator compared the 13 students with the reform/modeling background to all 39 students who had a traditional background (the full traditional group). In this article, we include the evaluator's analysis [6] for the full traditional group, along with accompanying graphs and tables. Wherever appropriate, we also add parenthetically the comparable performance of the subgroup of the five students who took the traditional precalculus course during fall 1999.

Assessing student performance and persistence on quizzes in calculus

As can be seen in Figure 4, the reform/modeling group consistently outperformed the full traditional group on all seven quizzes during the semester. (Note that the identical outcomes occurred when comparing the reform/modeling group to the subgroup of five students who came from the traditional background in fall 1999.) The quizzes on which the students with the traditional background scored closest to the students from the reform/modeling approach are the quizzes that were primarily algebraic in content (quizzes 4 and 6).

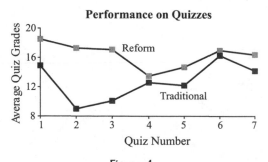

Figure 4.

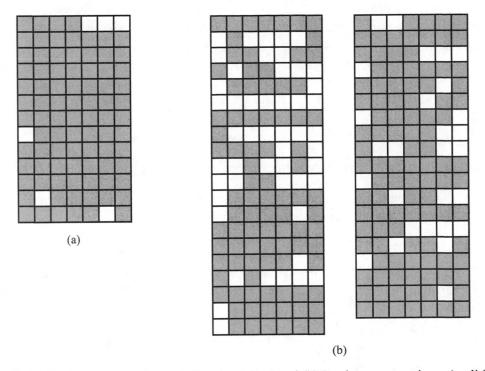

(a)

(b)

Figure 5. (a) Persistence on quizzes: (reform/modeling) and (b) Persistence on quizzes (traditional)

A closer examination of the quiz data reveals a striking pattern in terms of student persistence in calculus depending on the students' precalculus background. Figure 5a diagrams which students with the reform/modeling background took each of the seven quizzes in calculus. Each row represents an individual student and each column represents a particular quiz (#1 through #7). Each darkened box represents a quiz that a student took and each blank box represents a missed quiz. Similarly, Figure 5b illustrates the same information for the full group of students with a traditional background. Only one student with the reform/modeling background stopped attending calculus. It is evident that the persistence levels among the students with a traditional background were considerably lower and that a large component of this group gave up on calculus and stopped attending.

Student retention can also be seen in the graph in Figure 6 showing the percentages of students from the two groups taking each of the seven quizzes. The percentage of students with the reform/modeling background who took each of the seven quizzes was consistently higher than the percentage of students with a traditional background. (This was true for the full traditional group as well as for the subgroup.)

The comparison of student performance on the last few quizzes is distorted by the disproportionate number of students with a traditional background who stopped attending calculus. Since they were al-

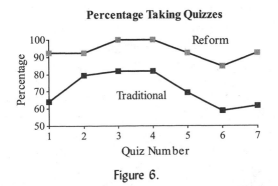

Figure 6.

ready doing poorly, their scores on the later quizzes, had they persisted, would undoubtedly have lowered the average for their group. That is, the lowest performing students were effectively removed from the traditional group when they stopped attending (both full group and subgroup), but the lowest performing students from the reform/modeling group completed the entire course. Had all these students persisted to the end of the semester, there would almost certainly have been an even wider discrepancy in the averages between the two groups.

Assessing student performance on class tests in calculus

Student performance on each of the three class tests in calculus was also compared. On the first class test, the full group of students coming from a traditional approach had a mean grade of 64.36 with a standard deviation of 19.23. (The mean grade of the subgroup was 59.0 with a standard deviation of 15.32.) In comparison, the students coming from the reform/modeling sections had a mean of 90.62 with a standard deviation of 8.96. See Figure 7. When the difference of means of the two groups is analyzed statistically, the associated t-value is $t = 4.71$ and the associated p-value is $p = 0.0000257$, so the difference between the groups is statistically significant.

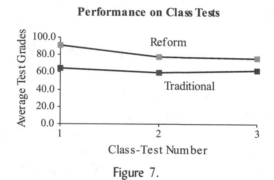

Figure 7.

The results on the second and third class tests in calculus demonstrate similar, though not as dramatic, patterns. On the second test, the full group of students coming from a traditional approach scored a mean grade of 59.58 with a standard deviation of 21.82. (The mean grade of the subgroup was 40.25 with a standard deviation of 7.15.) The students coming from the reform/modeling sections scored a mean grade of 77.42 with a standard deviation of 14.26. The resulting difference of means test gives $t = 2.61$ and $p = 0.01249$. The difference is statistically significant.

On the third class test, the students with a traditional background scored a mean grade of 62.3 with a standard deviation of 17.31. (The mean grade of the subgroup was 48.5 with a standard deviation of 16.5.) The students with the reform/modeling background scored a mean grade of 76.1 with a standard deviation of 18.13. The t-value for this difference of means is $t = 2.20$ and the associated p-value is $p = 0.035$. The difference is statistically significant.

From Figure 7, it appears that the results for the two groups are converging. Again, this is probably misleading because of the considerably higher percentage of students with a traditional background who gave up and stopped attending the course.

Assessing student performance on the final exam in calculus

On the final exam in calculus, the full group of students coming from a traditional approach scored a mean grade of 55.8 with a standard deviation of 23.09. (The mean grade of the subgroup was 11.5 with a standard deviation of 6.5.) In comparison, the students coming from the reform/modeling sections scored

Percentage of Students Passing Calculus

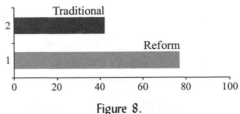

Figure 8.

a mean grade of 69.8 with a standard deviation of 26.21. The resulting difference of means test gives $t = 1.64$ and $p = 0.1$.

Note that 12 of the 13 students, or 92.3%, from the reform/modeling background who started the calculus course completed the course and took the final exam. In comparison, 24 of the 39 students, or 61.5%, of the full group who started the course took the final exam. (In the subgroup, three of the five students took the final exam.)

Finally, 10 of the 13 students from the reform/modeling background who started the course received passing grades in calculus, for a success rate of 76.9%. In comparison, 16 of the 39 students in the full group with a traditional background who started the course received passing grades in calculus, for a success rate of 41.0%. See Figure 8. (None of the five students in the subgroup passed calculus.) When the difference in proportions is analyzed, the associated z-value is $z = 2.24$ and the associated p-value is $p = 0.025$. Thus, the difference in the proportion of students who passed the course is statistically significant. (The corresponding values for the subgroup are $z = 2.94$ and $p = 0.003$, so again the difference in proportions is statistically significant.)

Discussion of results

The issue of student persistence and retention in calculus is perhaps the most critical factor for an institution such as ours. To summarize, only one of the 13 students from the reform/modeling background stopped attending the course, no other student from this group missed a single test or the final exam and, in fact, only two of the other students missed any quizzes. On the other hand, over 40% of the students from the traditional background stopped attending; six of them did not take the first test, eight of them did not take the second test, 16 did not take the third test, and almost half missed more than one quiz.

I can only speculate about the reasons for this. There is no doubt that the students with the reform/modeling background in precalculus were better prepared to handle the intellectual demands of a reform calculus course. They were comfortable with the need to understand the meaning of the mathematical concepts, not just to manipulate symbols by rote. They were used to non-routine problems, both conceptual and realistic, that required them to think and understand.

In contrast, many students coming from a traditional skills-focused precalculus experience may have brought with them higher levels of manipulative skills (even though the precalculus portion of this study suggests otherwise). But, they evidently were not prepared for a balanced emphasis on conceptual understanding, realistic applications, and algebraic manipulation.

As mentioned earlier, at the start of our experience at NYIT, some faculty expressed concern that the reform/modeling approach in precalculus would inflict irreparable damage to the students. In retrospect, it appears that it is the traditional precalculus course that harms the students.

References

1. Baxter, Judith Lee, Dibyen Majumdar and Stephen D. Smith, "Subsequent Grades Assessment of Traditional and Reform Calculus," PRIMUS, vol. VIII, 1998.

2. Hughes-Hallett, Deborah, Andrew Gleason, et al., *Calculus*, 2nd Edition, John Wiley & Sons, New York, 1998.

3. Larson, Roland, Robert P. Hostetler, and Bruce H. Edwards, *Precalculus Functions and Graphs: A Graphing Approach,* 2nd Edition, Houghton-Mifflin, Boston, 1997.

4. Gordon, Sheldon P., Florence S. Gordon, B. A. Fusaro, Martha J. Siegel, and Alan C. Tucker, *Functioning in the Real World: A PreCalculus Experience*, Addison-Wesley, Reading, MA, 1997.

5. Gafney, Leo, "Results of Student Attitudinal Study in Precalculus at NYIT" (unpublished).

6. Colley, Kabba, "Comparing College Students' Performance in Traditional and Reform Pre-Calculus and Calculus Courses: An Evaluation Report" (unpublished), <http://iris.nyit.edu/math/eval.doc>, 1999.

7. Gordon, Florence S., "What Does the Slope Mean?," *PRIMUS*, vol. XI, 2001.

20

Student Voices and the Transition from Reform High School Mathematics to College Mathematics

Rebecca Walker
Grand Valley State University

Introduction

A variety of questions arise while trying to rethink college precalculus. How will students react to a reform precalculus experience? Can a reformed precalculus experience help students develop a broader and more realistic perspective of mathematics? How successful will students be in calculus if they have a different type of preparation? Can a different learning environment promote deeper mathematical understanding? Is it possible to create a classroom environment where students expect the mathematics to make sense and where they will struggle with complex problems?

None of these questions is easy to answer, but it is possible to begin to answer them by looking at the experiences of students who have prepared for college mathematics in a reformed environment. Over the past ten years, a growing number of high school students have prepared for college mathematics by studying high school mathematics in a reform environment. The changes in high school mathematics came about in response to poor national and international test results [1–3] and the National Council of Teachers of Mathematics (NCTM) *Standards* documents [4, 5]. Beginning in 1992, the National Science Foundation (NSF) funded the development of several school mathematics curricula that would be in line with the vision of mathematics laid out in the NCTM documents. The curricula developed as a result of this funding are changing what is happening in many mathematics classrooms across the United States. One of the NSF curricula for grades 9–12 was developed by the Core-Plus Mathematics Project (CPMP) [6].

Students who study the CPMP four-year high school curriculum learn mathematics by first constructing mathematical meaning out of problem situations and then looking more formally at the mathematics involved. They work in situations and problem settings that were developed with the intent that students personally construct mathematical concepts and methods and make them their own. They study algebra from a modeling and functions perspective, and also study geometry, statistics and probability, and discrete mathematics each year. While learning mathematics using the CPMP curriculum, students are encouraged to value their informal knowledge and intuition and to try to link them to formal mathematical concepts. The objective is that, in addition to finding solutions to problems, students should be able to understand and to explain what they are doing. Thus, mathematics becomes a practice, something that they do, as well as a body of knowledge [7]. Through this learning process it is hoped that students develop, not only a specific knowledge, but also mathematical ways of thinking or habits of mind [8]. (For more information on the CPMP mathematics curriculum see [9].)

This paper first reports on a study of the views of mathematics held by students who completed four years of school mathematics, during the late 1990s, using the CPMP curriculum. Through the use of a Likert-scale survey, the Conceptions of Mathematics Inventory (CMI) developed by Grouws, Howald, and Colangelo [10], this study categorized students' conceptions of what it means to learn, know, and do mathematics before encountering college mathematics and again after one semester of college mathematics.

In addition, this paper reports on case studies of the first semester college mathematics experiences of six of these students, with particular attention to the transition from this reform curriculum into college mathematics. In particular, the case studies document describes the students' transitions in terms of placement testing, in-class experiences, out-of-class experiences, and college mathematics performance. Finally, each student's overall reflections about his or her preparation for and experiences in college mathematics is reported.

Design of the study

Students from eight high schools (five in Michigan, one each in Alaska, California, and Georgia) that were a part of the field test of the CPMP curriculum completed the CMI survey near the end of their senior year of high school and again after one semester of college mathematics. Whole classes of high school students ($n = 256$) completing the fourth year of the CPMP curriculum participated in the initial survey. All of the teachers of these students had been provided curriculum-specific professional development and ongoing support while implementing the CPMP curriculum and were aware of the goals of the curriculum. However, the success of the implementation varied from one school to another, and even within schools, depending on the classroom teacher. Two hundred fifty-six students completed the CMI at school during May of their senior year of high school and then were asked to complete it again in January after their first semester of college. One hundred thirty-two of the initial 256 students returned the follow-up survey after completing one semester of college. Of these 132 students, 38 of them did not complete the survey because they had not taken a mathematics course during their first semester of college. Two additional surveys could not be included in the pre- and post-testing because the students had not adequately completed the first administration of the CMI. This resulted in 92 students who completed both administrations of the CMI and were included in the analysis of the pre- and post-CMI data. The CMI data was analyzed by considering measures of center and measures of spread. Two-tailed paired t-tests were used to determine whether or not the mean responses in each dimension changed between May and January.

The case study participants were chosen from among volunteers in the classrooms that were surveyed. They came from five different high schools and attended one of two large midwestern universities. Each student was recommended by his or her high school mathematics teacher. These recommendations were based upon three criteria: the student being responsible enough to complete the entire set of five interviews, the student being articulate enough to explain what he or she is thinking, and the student enrolling in either precalculus or calculus during the first semester of college. Data for the case studies was collected from a variety of sources: two interviews and three problem-solving sessions with each student (conducted by the author), review of student work on tests and quizzes, classroom observation, and instructor interviews. The interviews occurred in August before the students went to college and in January after completing the first semester of college mathematics. The problem-solving sessions were evenly spaced throughout the semester and involved students in think-aloud problem-solving sessions where they solved problems similar to those they were working on in their classes. All interviews and problem-solving sessions were transcribed and coded for analysis. Classroom observations and instructor interviews were used to help determine the type of college mathematics classroom environment that each case-study student was in.

Analysis of student beliefs and conceptions of mathematics

The Conceptions of Mathematics Inventory (CMI) was used to measure students' conceptions of mathematics [10]. The CMI is a 56 item, 6-point Likert-scale instrument, with response choices of strongly disagree, disagree, slightly disagree, slightly agree, agree, and strongly agree to the statements. The inventory

---

Composition of Mathematical Knowledge

Knowledge as concepts, principles and generalizations   vs.   Knowledge as facts, formulas and algorithms

Structure of Mathematical Knowledge

Mathematics as a coherent system   vs.   Mathematics as a collection of isolated pieces

Status of Mathematical Knowledge

Mathematics as a dynamic field   vs.   Mathematics as a static entity

Doing Mathematics

Mathematics as sense-making   vs.   Mathematics as results

Validating Ideas in Mathematics

Logical thought   vs.   Outside authority

Learning Mathematics

Learning as constructing and understanding   vs.   Learning as memorizing intact knowledge

Usefulness of Mathematics

Mathematics as a useful endeavor   vs.   Mathematics as a school subject with little value in everyday life or future work

---

Figure 1.

measures students' conceptions in seven different dimensions: Composition of Mathematical Knowledge, Structure of Mathematical Knowledge, Status of Mathematical Knowledge, Doing Mathematics, Validating Ideas in Mathematics, Learning Mathematics, and the Usefulness of Mathematics. The conceptions in each dimension are measured on a continuum with respect to two poles. The dimensions and their associated poles are provided in Figure 1.

The conceptions in each dimension are assessed using eight statements. Of the eight statements, four are written to reflect one pole and four to reflect the other pole. For the purposes of this study, the pole in the first column of Figure 1 will be referred to as the positive pole.

In order to facilitate interpretation of the results, the responses to all statements were scored so that a mean score of six indicates complete agreement with the positive pole in each dimension. The positively worded statements were scored from 6 for strongly agree to 1 for strongly disagree and the negatively worded statements were scored 1 for strongly agree to 6 for strongly disagree.

After this coding had been completed, the mean score for each student in each dimension was calculated. This mean is referred to as the dimension rating for the student in that dimension. Box plots showing the dimension ratings for the participants in May and in January are given in Figure 2.

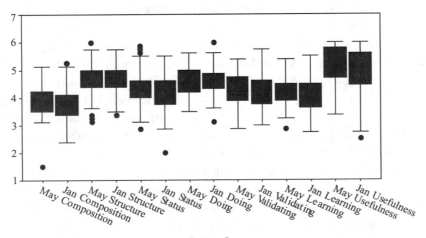

Figure 2.

	N	May Mean	May SD	Jan Mean	Jan SD	p-value
Composition of Mathematical Knowledge	92	3.8900	.5361	3.7745	.5394	.048
Structure of Mathematical Knowledge	92	4.6671	.5388	4.6943	.5046	.643
Status of Mathematical Knowledge	92	4.3356	.5541	4.2228	.6424	.093
Doing Mathematics	92	4.5666	.4894	4.6128	.5055	.394
Validating Ideas in Mathematics	92	4.2649	.5475	4.2500	.5962	.802
Learning Mathematics	92	4.2106	.5024	4.0870	.5456	.060
Usefulness of Mathematics	92	5.1889	.6873	4.8832	.8977	.000

Table 1.

The box plots indicate that the median dimension ratings were above 4.0 for all dimensions except the Composition of Mathematical Knowledge. This indicates that in general these students viewed mathematics as being a useful, coherent, and dynamic system of concepts and ideas, where learning is accomplished by sense-making and can be validated through logical thought. At both times the conceptions regarding the usefulness of mathematics were stronger than any other conception. The plots further indicate that the changes in conceptions between May and January were not large.

To further explore the stability of these conceptions, the May and January scores of the 93 students who adequately completed both CMI administrations were analyzed. The sample mean scores were analyzed using paired t-tests to determine whether or not the mean response in each dimension changed between the first and second completion of the CMI.

Table 1 gives the means and standard deviations for the May and January survey results and the p-values for the paired t-test.

There were two dimensions for which the changes in the individual conceptions were significant: Composition of Mathematical Knowledge and Usefulness of Mathematics. The conception that mathematical knowledge is composed of concepts, principles and generalizations decreased in strength. After a semester of college mathematics, students believed that facts, formulas, and algorithms played a greater role in mathematics than at the end of high school. It should be noted, however, that there was still a majority that believed that mathematics was more about concepts, principles, and generalizations than about facts, formulas, and algorithms.

After a semester of college mathematics, students' conceptions of the usefulness of mathematics had also decreased. They moved toward seeing mathematics as a school subject with little value in everyday life or work.

The change in the ratings for the learning mathematics dimension approached significance ($p = .06$), which indicates that the students' conceptions did change some in this dimension. After one semester of college mathematics, these students believed that memorization of intact knowledge played a larger role in learning mathematics than they did at the end of high school. It should be noted that, at both times, their conception averages were slightly more toward the positive pole of learning as constructing understanding.

Without further research, it is not clear why these students' beliefs changed as they did. However, based upon responses to a brief survey about the college mathematics classes that the 92 students were enrolled in and the information gathered through the case studies, it is possible to make some conjectures. Most of the students reported that the majority of the class time in their college mathematics classes was spent taking notes while the instructor lectured to them about the new material. They were given sample problems and were expected to be able to work similar problems. There was very little discussion or interaction during the class sessions. It also may have been the case that explanations were not valued in the student work that was assessed. This combination of things may have led students to believe that learning how to do mathematics (learning the algorithms) was more important than knowing why they were doing it (learning the reasoning behind what they were doing.) This may help explain the changes in the composition of mathematical knowledge and the learning mathematics dimensions.

One hypothesis about why the student beliefs about the usefulness of mathematics changed is also based upon the brief course descriptions provided by the students. These students were enrolled in college mathematics classes that did not seem to develop the mathematics out of contextualized situations. In fact, many of the students indicated that they did very few problems that were in a context at all. This could make it hard for students to identify how the mathematics that they were studying could be used outside of the mathematics classroom and lead to a decrease in the usefulness of mathematics dimension.

These results should be considered in light of the fact that students' conceptions about what mathematics is, what it means to do mathematics, and how one goes about learning mathematics all influence how students approach mathematical situations and what they take away with them from such situations [10–13]. Learning mathematics in this reform environment helped students to develop conceptions that should support a deeper learning of mathematics, a better disposition toward problem-solving, and the development of more autonomous learners of mathematics. If students were to go into calculus with these beliefs about the nature of mathematics, and what it means to do and learn mathematics, it might make it easier to get students to truly engage in mathematics at the calculus level and beyond.

Transition to college mathematics

The paper will now discuss how the six case-study students navigated the transition from a reform high school curriculum through the first semester of college mathematics. Although the six case study students each had different collegiate experiences, it is helpful to step back and look in general at their experiences and try to assess what went smoothly for them, what did not go smoothly for them, and what, if any adjustments they made along the way. Sometimes their experiences were similar enough that they are considered as a whole. At other times it is important to split them into two groups of three students each: the students who attended Southern University (Ann, Cathy, and Rita) and those who attended the Northern University (Randy, Sally, and Ted). (Both student and university names have been changed to conceal their identities.) This separation is necessary because the students at Southern University experienced a reform calculus curriculum, using the Harvard calculus materials [14] that included a variety of teaching and assessment methods and those at Northern University learned mathematics through a traditional curriculum, using either *Calculus* by Thomas and Finney [15] or *Precalculus* by Bittinger, Beecher, Ellenbogen, and Penna [16] with traditional teaching and assessment methods. When considering the stories of these six students, it is important to remember that they all were recommended for this study by their classroom teachers and that they all had been successful in their high school mathematics classes. (For more detailed information about these six students, see [17].)

College mathematics placement

The first experience that many students have with college mathematics is the placement test that determines which mathematics course to take. One student, Ted, did not have to take a placement test because his ACT score allowed him to enroll directly in calculus I at Northern University. The other five students all completed the placement test for their respective universities. The three students at Southern University all placed into and enrolled in calculus I. The remaining two Northern University students, Randy and Sally, placed into and enrolled in precalculus. The placement process went smoothly for all five students. The only complaint was that the tests were given during the summer after the students had been away from mathematics long enough to be out of practice with the details necessary for completing some of the problems that were on the test. Once the students began attending their college mathematics classes they encountered a variety of classroom expectations, some that were different from those with which they had become familiar during high school.

Experiences in the classroom

Because the in-class mathematics experiences of the students who attended Northern University differed from those who attended Southern University, each group of students is discussed separately.

Northern University

All three students at Northern University were in classes where the primary means of delivering new mathematics was through instructor lectures. None of these students found this to be a difficult way of learning mathematics. They all indicated that they had taken non-mathematics classes during high school for which lecture was the primary teaching method and so they were all able to adjust easily to that pedagogy.

In their college courses, the general approach to developing mathematics was also different from what they had experienced in high school. In high school, these students had developed mathematical concepts out of contextual situations and then considered the formal mathematics. The opposite was true in their college mathematics classes. They first were taught the necessary skills and then, in some cases, considered applications of those skills. Ted indicated that math had not been difficult for him in high school and so working in groups and developing the mathematics out of contexts worked for him there. But he thought that at college, because the material was more difficult, he learned better by not focusing so much on the applications and by doing lots of problems. In the final interview, when asked about how working in contexts might have influenced his high school learning he said:

> Sometimes I was kind of agitated, because you go through like all this writing down and it's really not cut and dry what you actually did. When I go and do the problems over and over again, I can go back and see where I'm really going wrong. With numbers that are on the paper. It's not just explanations about why you got something.

Ted did not feel he needed to supply explanations because he believed that if he could do the problems correctly, then he understood the mathematics.

My conversations with Sally focused on what she did to construct her understanding of the mathematics. She mentioned that, during high school, when she was working through a problem that the mathematics just "snuck up on her" as she worked. But in college "it was faster and just like do it, do it, do it, and then if you don't understand then you are lost." Despite this statement about either understanding it or not, Sally indicated that she thought it was easier to just be shown how to do the math rather than trying to develop it on her own. In the last interview she said:

> In high school they wanted us to go try this before they taught it. Which I mean, it makes sense for some people if they are going to actually try it and think about what they already learned. But I think when you have a lot of other classes going on and you look at it once and you go, well I don't know how to do that, and you put it away. So I mean this semester was easier, like, thinking wise: Because I just wrote down my notes and followed along, instead of having to go out and figure it out myself.

Throughout the semester, Sally indicated that, in general, she only did as much as she needed to in order to do well in her classes.

Randy identified the differences between his high school math classes and his college math class as follows: "The differences, a faster pace. This class is faster paced, more amount of work. We learned a lot, we learned more. The other classes were story-problem based, this is just straight math." Although he could identify the difference in the two approaches to mathematics, he did not have an opinion on which one worked better for him. He was able to be highly successful in both settings.

Another change for the students at Northern University was the limited, if any, use of calculators in their college classes. In their high schools, graphing calculators were tools that were used to help students understand the mathematics they were learning and to solve problems. Technology was not used at all in Ted's calculus class and he was not allowed to use a calculator on any of his tests. This did not bother Ted at all and did not seem to affect his overall performance. During the last interview, Ted said the following in response to questions about not using a calculator: "I prefer not to use it. I can use it, but I just know that it will slow me down and you get lazy at times. I like to do long multiplication in my head if I can. Just think it through." This belief that using a calculator makes you lazy came up several times during conversations with Ted. It may be this belief that made the transition into a class that did not allow calculator use go as smoothly as it did for Ted.

In the precalculus class that Randy and Sally took, a graphics calculator was required, but was not fully integrated into the curriculum for the course. The professor would use the calculator to help motivate solution methods, but most of the time would require that the students know how to solve the problem symbolically. He encouraged them to use the graphical and numeric capabilities of their calculators to check their solutions. This resulted in some frustration for Sally, but did not bother Randy. Sally knew that she could often find the correct solution using her calculator, but was not able to solve the problem symbolically. She did not understand why she needed to know both methods; after all, she could get the answer. Randy had no problem with not being able to rely on calculator solution methods and did make good use of the calculator to verify his work.

In addition to three large lecture sessions each week, Randy and Sally experienced two recitation sections. Similar to their high school classes, they were expected to work in groups during the recitation period and they did so up to a point. They both worked primarily with one other person. In the last interview Randy and Sally provided the following descriptions of their work in groups during the recitation periods.

Randy: Me and three other people. We worked [the problems] pretty separately, but me and one girl in particular, worked a little closer. Like through the problems, we would make sure we were both on the same track. Then we all compared our answers once we were done basically.

Sally: I would just work with a partner and we did the workbook assignments. And we never had enough time. We always were like scrambling to get the workbooks done. Because it seemed like there were a lot of problems for the 20 minutes that they gave us. If we didn't get something we asked the other groups that were sitting around us. And then we would end up just copying to get it done, because we wanted our points. Maybe it would have been different if I could have stayed or if we just had more time to work on the workbook. I would have tried to understand better.

Although both Randy and Sally were working with other people in their recitation sessions, by both their descriptions and my observations, it seemed as though neither of them was really working collaboratively with the other students. They were primarily working independently and verifying their answers at the end. There did not seem to be much discussion about solution procedures.

Southern University

The three students who attended Southern University did not experience transition difficulties related to what happened in their mathematics classrooms. They were all enrolled in reform calculus classes. For these students, the day-to-day routine and expectations of the classroom were not very different from what they had experienced in high school. The mathematics was almost always tied to some context and interpretation of the context was integral to the class. They were expected to learn why they were doing things in addition to how to do them. It was expected that the students had already studied the appropriate material, through reading text material and completing some related exercises, so the primary purpose of the class time was to further develop understanding through working problems. The instructors spent varying amounts of time at the beginning of each class providing a short lecture or explanation of the material. Thus the amount of time students actually spent working on problems varied from one section to another: Cathy spent the least amount, Rita was in the middle, and Ann spent the most. During the class meeting times, all three of the instructors expected that students would actively participate in class by working problems with other students. Students were also expected to understand concepts behind the problems. They needed to know in what settings and in what ways the concepts might be used. Most of the time it was not sufficient for students to provide only an answer to a problem. They were expected to provide a complete solution and to explain their reasoning.

At the beginning of the semester, Rita and Cathy were pleased with their classes and thought that class time was helpful to them. This changed for Rita after one month. After the first test, on which she did quite well, Rita indicated that she sometimes skipped class because she didn't always find it very helpful. She said that it wasn't that helpful because the professor just went over the section in the book and she felt that she could do that on her own and in less time. However, she did not do nearly as well on the second test and so after the second test she resumed regular class attendance. Rita felt like she could determine if she understood the material and she did not think it was very helpful to have someone telling her the mathematics that she needed to know.

From the beginning of the semester, Ann thought her instructor should do more explaining and fewer problems. During each of our problem-solving sessions, Ann said something about how she wished her instructor would explain more about the mathematics behind what they were doing and why they were doing it and go through step by step. It is hard to know exactly what Ann wanted and she was not able to clearly explain it. Based upon my class observations, the instructor tried hard to get the students to think about what they were doing, why they were doing it, and what mathematical concepts they were using. Rather than just telling the students what to do, he tried to get the students to talk about the concepts and how they might be able to use them to accomplish the task at hand. Although the instructor was not standing at the board explaining to the students each mathematical concept and how it could be used, there certainly was an emphasis on developing conceptual knowledge rather than solely on getting the problems done. But it was clear that Ann did not feel she was understanding what she needed to. In an effort to get the explanation that she wanted, Ann went to the tutor lab. She indicated that she found the assistance she received there to be somewhat helpful to her.

During the final interview, in order to help these students think about what, if anything, about the transition was hard for them and to get them to reflect on the semester, I asked them what their instructor could have done differently to help them learn calculus. Rita and Cathy talked about working problems and going through explanations.

Rita: Well I would say maybe just do more problems. But I think that's just for me. When I see examples I can like use those, but he did a lot of problems. I wish he did more… Yeah, it (working on problems in class) was helpful; because it was just like doing group work again, more group work. But it was better because he was there to guide you through it.

Cathy: Can I compare it to the class I am in now [Calculus II]? I think I'm learning a lot more in the class that I'm in now, like getting better in terms of he's, I mean, he'll lecture for a little bit. It'll be like sample problems in class and he'll explain in like general stuff and we'll take notes on it. But he'll also have people go up to the board and spend a lot of time in class and they'll put their answers to the homework problems on the board. And he'll go through it and he's really picky about things, which in one way is bad but in another way is really good, because it forces you to remember to put those things down to remember what to do. And I think that's helping me a lot. It also helps me to see how other people think, and I could have done it that way, instead of straight by the book. That seems to be helping me more.

Both of the above comments imply that these students wanted to get more explanations about why and what they should be doing. Although each of these participants had suggestions about how her calculus class could have been more helpful, neither of them complained much about how things were going during the semester.

Ann felt differently about what would have helped her be more successful. As previously stated, she felt that her instructor was doing too many problems and not enough explaining. In the final interview she said:

> I think instead of doing all those problems in class spending a little more time going over the sections instead of the problems. Like teaching the concepts once and showing how to apply it in problems. 'Cause I think once you understand the concept, no matter what problem you do, it will be easier to do any problem if you have a stronger grasp of what the concept is about. It's easier when somebody's showing you how to do it and you're doing it with them step-by-step, like while they're teaching you. You don't teach the whole concept in like one example, you should do it step-by-step so it's easier, it's broken down.

This statement indicates that Ann would have liked her professor to show her step-by-step how to do things and in that process carefully explain each step. Since it was not clear to me what Ann meant by understanding or learning the concept, I asked her. Her reply was:

> Like learning the integral and derivatives, like learning how to do a certain problem. Understanding what the problems are asking you, I guess that is what I mean. I don't know myself... I think it is important to know what the derivative is telling you, but at the same time you have to understand the basic rules in order to solve it.

From this statement, it seems that Ann wants to know something about what the big picture is and why she would do each step. However, in the problem-solving sessions that we had, she rarely was able to provide reasoning and rather seemed focused on getting to the correct solution. Whether this was because she did not really believe that understanding the concepts and knowing why was important or whether she was just too far behind in the class was not clear. In terms of what happened in class, the transition to college mathematics was probably more difficult for Ann than for any of the other case-study participants. This may have been due to the fact that quite early in the semester Ann got fairly far behind because of extenuating circumstances and spent most of the semester trying to catch up. As a result, she was frustrated that she did not understand the mathematics being taught in class.

Experiences outside of the classroom

What regularly happens in the classroom is only part of a student's experience with college mathematics. Another large part of a college mathematics class is the work students are expected to do outside of class, which includes their work on assignments, other daily preparation for class, and preparation for tests. All of the students indicated that they needed to work much harder and spend more time studying for their mathematics classes at college than they did in high school. They also all said that the pace of the class was much faster than they had experienced in high school; they were responsible for learning much more material in a given time. These challenges seemed to be more about the transition from high school to college in general than they were about the transition from a reform high school mathematics program to college mathematics. The manner in which students confronted the need to learn more, and to learn it faster, and other issues that are related to the work students did outside of the classroom will be discussed in this section. The expectations placed on students at the two universities differ enough that separate consideration is again given to the students based upon the university they attended.

Northern University

The students at Northern University were all told that they should read the upcoming section of the text before going to class. This expectation was not present during high school because the CPMP textbooks do not contain material designed to be read by students outside of the class. However, despite the explicit directions, none of the three students read their textbook on a regular basis. For the first two-thirds of the semester, Sally said that she only used the book to get the homework problems and to copy the things from the green boxes when studying for a test. Randy also primarily used the book for the homework problems. Both of these students indicated that they did read the book when they were studying trigonometry, especially trigonometric identities. They used the book more at this time because they were having a hard time with the material and believed that they could solve some of their confusion by reading the text. Ted said that he used the examples in his book to help him decide what he needed to do to get the homework done properly. Despite not following the recommended process, the approach to textbook use adopted by each of these students did not cause any difficulty.

Although these students did not spend much time outside of class reading their textbook, they did spend time on their daily homework assignments and were concerned about the correctness of their work. Unlike high school, homework was not collected in the precalculus class. This did not bother Sally or Ted and they both completed their assignments on a regular basis. Some of their motivation for this was the opportunity to ask questions about the homework problems during recitation and that, approximately once a week, they had either a test or a quiz. If either of them had fallen behind in their work, it would have been reflected in their quiz and test grades.

Homework assignments were collected weekly in Ted's calculus class. Despite this, he did not complete his homework assignments as regularly as did Randy and Sally. He said that he often did not work on his calculus assignments until the night before they were due. He believed that as long as he was able to do the work and solve the problems on the test, he was "ok." Thus, he did the work when it was convenient for him, not as daily preparation for class. His homework grades were satisfactory and so it appears that this did not cause a transition problem for him.

A third area of work outside of class is preparation for tests. Ted had to prepare much more for tests in college than he needed to in high school. He indicated that in high school he just looked over "stuff" to make sure that he knew it. He also said that in college he spent much more time actually working problems in preparation for tests. Part of the reason for this difference is that his college instructor provided specific review problems to complete and that had not been the case in high school. Although the review sheet contained a wider variety of problems than were on the test, it contained problems that were similar, most often in form and content, to all problems on the test. He worked on the review sheets by himself and asked friends who were in calculus II for help when he couldn't get the correct answer. He also indicated that he looked back over his completed homework assignments. To complete the review sheet for the final exam, Ted said that he worked with a group of people. But this was the only time that he mentioned working with other people in any extended fashion.

Sally and Randy also were given suggested problems to complete as review for each test. Sally did not mention working the review problems in preparation for tests. Instead she went through her notes and made a review sheet of important facts and algorithms that she wasn't sure of. She also looked through the book to see what was in the boxes. Sally indicated that this was helpful to her because the more times she wrote something down, the more likely she was to remember it. In studying for the final exam she just made sure that she knew how to do the problems that were on her previous tests. This was different from what Sally had done during high school. She was able to use her notes and her textbook on all tests in her high school mathematics classes and so felt that she did not need to study anything. In college, she felt as though she had to memorize more because she was not allowed any references. While this was an adjustment that Sally had to make, it did not seem to present a big hurdle as she made the necessary changes without complaint and did well in the class.

Randy did not work the review problems for the first two tests. However, after not doing as well as he wanted to on the second test, he decided that he needed to work all the review problems in preparation for the remaining tests. But that was the only preparation he did for these tests. Randy reviewed his old tests in preparation for the final. He indicated that he did not do anything different in studying for tests in college than he had done in high school.

In college, both Sally and Ted needed to adopt new ways of studying for tests. However, this did not cause any problems for either one of them. All three of these students were able to adequately adjust to different expectations regarding what they should do outside of the classroom. These students had no difficulties with this part of the transition.

Southern University

The calculus classes at Southern University were structured such that students were expected to read each section and to work some basic problems before going to class. The instructors assumed that students had at least some familiarity with the topics for the day and that the goal during class was to help the students deepen their understanding of the mathematics. Although this was different from high school, all three of these students recognized these expectations, and at the beginning of the semester, they seemed to have been diligent about getting the reading done before going to class. However, as the semester progressed, none of them regularly got the reading done in advance. They all indicated that they eventually did read the book, but not always on time. They further indicated that this resulted in some frustration during class sessions. Because they had not read the section, they were not able to follow along as well during class. The reason they provided for not getting the reading done before class was that they gave other tasks more priority and ran out of time. Although the expectation that they read the textbook before each class period was a change from high school, none of these students felt that the expectations were unreasonable.

Besides reading the book, these students were expected to complete group homework assignments. Each instructor assigned students to groups and groups were changed at least once during the semester. The homework assignments consisted of four to six relatively difficult problems to be completed by the group as a whole. The paper that was turned in was supposed to include complete solutions with detailed explanations about the mathematical reasoning used. The first few assignments seemed to be completed by each group as a whole, but after that it was more common for the groups to split up the problems and then come together to write up the assignment. The amount of discussion that occurred during this write-up time seemed to vary greatly.

These students all talked favorably about the group-homework experience. Additionally, they all felt that it was a better experience when the group worked together on the problems, rather than splitting them up and just coming together to write them up. Representative comments from Rita and Ann include:

Rita: Group homework helped a lot. Group homework was really hard, it was very difficult. I don't think I'd be able to do it on my own. I could see how other people figured stuff out and that helped me in the way I could figure stuff out. That helped a lot.

Ann: Well like in my other group we did almost all the problems together, but it seems that in this group that everybody just does one of the problems and goes over it with everyone the night before we do the homework. Actually, even the night before we usually don't even meet because we just go over it in class before we hand it in. So it's not as productive as it was with my first group. Because now we just do the problems, and like if I have a problem and I don't know how to do it, I'll just go to the math lab and they will help me. But I don't understand the other problems because I never got to do them with the whole group.

All three students saw the group homework as very much like the in-class group work they had done during high school. First, they indicated that it provided an opportunity to talk with other students about mathematics and to see how others were solving problems. They also indicated that the expectations about how solutions were to be written were very similar to what they had experienced during high school. They further indicated that the idea of carefully explaining oneself in mathematical situations was foreign to many of their classmates. They felt they were much better at this than other students in their groups and thus were able to serve as resources for their groups. The transition to group homework was not a transition for these students.

These students did not indicate that they did anything differently while studying for tests at college than they had done in high school. As a part of the course packet that students purchased at the beginning of calculus I, each student received a copy of all tests and the final for the previous two semesters of calculus I at Southern University. All of the students worked these problems as part of their preparation for each test and for the final exam. Cathy and Rita said the following about studying for tests:

Cathy: I went over a lot of like older things and stuff to see what I got wrong and like try to work through them and figure out how to do them the right way. Like I said I did a lot of review problems and just tried to make sure that I knew the basic process of how to do all kinds of problems.

Rita: I knew I was ready because I did the previous exam. I did them and I understood them all. Like a lot of it was like explain it in no more than two sentences. And so I actually did it using the proper terms and correct grammar and everything. It makes sense, and I knew I was right because they're also doing it in a big group and I would explain it to them why I was right and they're like, oh yeah, I remember that from class. Or if I was wrong they would explain it to me what I was doing wrong. And then I understood.

For these students there did not seem to be any transition issues related to preparation for tests.

College mathematics performance

One final portion of the transition that these students experienced, is the formal assessment that they encountered in their mathematics courses. As in high school, all of the tests or exams that these students completed in college were composed primarily of problems for which they were asked to show their solutions. There were portions of the precalculus final exam at Northern University and the calculus midterm exams and final at Southern University that did not allow for partial credit, but in general, all of the students were expected to provide their reasoning and were given partial credit for what they did correctly.

The majority of each student's final grade was determined by his or her grades on tests given during the semester and a final exam. The students at Southern University had two tests and a final exam. They were required to have a graphing calculator and were allowed one note card with reference material. All three students did well on the first test. On the second test, Ann and Rita experienced major decreases in their test scores, while Cathy's score only decreased by a small amount. The scores on the final exam were somewhere between those of the first two tests. The final grades for these three students were: Rita, B-; Cathy, C+; and Ann, D+. In the final interview in January, I asked them how they felt about their grades. They all indicated that they did not do as well as they hoped that they would and gave the following reasons:

Rita: I just didn't have enough time because I had to write a lot of papers last semester. Because we went section by section, I figured I'll just put it off one more day and I'll just have two sections to read. It's no big deal. And so I guess in a way I really got behind and so I wouldn't do it, I'd be like, well, I mean it doesn't take that long to read the section and do the problems. So I figured I'd just do it the next day or the next day and never got around to it. But like in high school it was way different. If you didn't do it, it was okay; you'd do it the next day.

Cathy: (in response to what she would do differently) I guess that I really need to keep up with the reading before I go to class. Because a few times I waited to learn things in class and it didn't make much sense as when I'd read it before. Doing extra problems, not just the ones that were assigned. I guess taking the initiative and asking other people questions instead of relying on them to explain things to other people that didn't understand them. Asking questions in class.

Ann: I think a lot of it has to do with taking it first semester, like in college, and stuff like that, just adjusting. I think I can do better. I was not happy with the grade, but I'd been pretty much behind for a while, and I couldn't catch up in enough time to be able to master the stuff to get a better grade... The material was difficult, but yet it was attainable to understand. You could get to understand it after a while. It was pretty fast, the pace of the class, but I think a lot of it had to do with being behind.

It is interesting to note that in all of the above comments each student could identify what she might have done differently and did not blame anyone but herself for the grades she received.

The students at Northern University were all pleased with their course grades. The grades they received were: Ted, A; Randy, A; and Sally, B. Sally and Randy had six tests during the semester and a final exam. Except for the second test, which covered polynomial and rational functions, Randy's grades were consistently about 90%. Randy did not have any difficulty with the course and actually said that it was easy for him. Sally's test scores were in the 80s except for the test on trigonometric identities, on which she received a grade in the 50s.

Ted had three tests and a final exam in calculus. His scores consistently increased throughout the semester from the first test grade of 74 to a final exam grade of 98. Because of the class policy that students would not receive a final grade lower than their grade on the final exam, Ted got an A in calculus I. He was pleased with his grade and did not have any comments about the tests or his overall performance.

Students' final reflections on their preparation

Each preceding section has separately considered a different portion of the transition to college mathematics for these six students. In an effort to get the students to reflect holistically on their experiences, I asked them to evaluate their overall preparation and to indicate if they would recommend CPMP mathematics to other students. Ted and Randy said they would not recommend CPMP; Cathy and Ann were not sure; and Rita and Sally said yes.

Ted's response reflects several of his beliefs. He believes that facts, formulas, and algorithms are very important in mathematics. He further believes that hard work is the best way to learn.

Ted: I would not recommend it. I don't think it gives you enough skills, really intellectual ones. It doesn't really test your mind at all. You can get by real easily by just doing the homework. You don't have to apply yourself. I know that I didn't do much work, 15 minutes, or so. You can do it in 10 minutes before you go to class.

His other big concern about having used the CPMP curriculum was that he didn't take calculus I in high school and could not go directly to calculus II. Ted's not taking calculus I in high school probably had more to do with his being in a very small school (limited number of course offerings and one track for all students) than it did with his learning mathematics using the CPMP curriculum. He finished our conversation with the following statement, "I guess I don't have very many complaints. I'm going into calc II here in college, so I guess I could just see myself being one semester ahead of where I am now."

Randy's reason for not recommending CPMP was that he thought it might make the transition to college harder for some students and that he didn't get through calculus in high school.

Randy: In terms of going to college, what you learned in the traditional math track is better geared towards how they teach in college. The way they teach integrated, if you are not understanding it, and getting it there, it's a lot harder to come to college and do it. It's taught differently.

He then added:

I think I learned what I needed to learn. I think that most people that don't do well in integrated wouldn't do well in the other track either. So I don't think it was all that bad, really. I learned a lot more than I thought I learned. I'd still have rather been in the other one. I wish I could have gotten through precalc and calc in high school.

However, Randy had not taken an algebra class in eighth grade and so it is unlikely that he would have taken calculus in high school had he been in the traditional math track. In the end, Randy was not unhappy with his preparation, but still held some belief that the traditional track would have been better for him. He was not able to really be specific about why he felt that way.

Cathy and Ann were not sure about whether or not they would recommend CPMP to other students. They both said that the reasoning and problem-solving skills that they learned in CPMP were very valuable to them, but they weren't sure that those skills were the most important for college. What they really wanted was a combination of both CPMP mathematics and a traditional curriculum.

Cathy: I wish there was something in between. Because like they're (students in CPMP) getting like how to think about it in a good way. I think in Core-Plus you learn how to think about things. In the other courses, I think they are more. I don't want to say mathematical knowledge because that's too general, but you are learning more like concrete problems like tons of different kinds of algebra problems, geometry, more than I think we did in Core-Plus. I just wish I had gotten a stronger foundation through the other path of mathematics, but kept the same way of thinking from CPMP. It just seems like, I don't think one is more important, but I just think like in terms of what people expect you to be able to apply is more of the other path, that's just what I've come across so far.

Cathy somehow wanted a curriculum in which she learned both the thinking that was in the CPMP and the algebraic skills that she felt she had not fully developed. She was clear that she didn't want to give up one for the other but wasn't sure how one could do both because this might be too much material.

Ann also said that she learned to think in CPMP. She said that she didn't have to worry about skills because she had taken skill-oriented summer courses in algebra and geometry. She thought that students need skills to get a good grade in college and thinking just because it was important. When I asked her what about CPMP was important she said, "Like the whole process of learning math. I don't know how to explain it, but like the way I was able to go step-by-step in learning something." Whether Ann would recommend CPMP seemed to depend on what your goals are: If you wanted good grades in college, take the traditional; if you want to really understand mathematics, take CPMP.

Rita and Sally had no reservations about recommending CPMP to other students. Rita felt that the focus on sense making and understanding that she developed during high school was extremely valuable to her.

Rita: I would tell them to stick with Core-Plus. I would say that Core-Plus had helped me a lot. Just with being able to explain. Like Core-Plus is all about the student understanding the concept. Whereas in calculus it's not about that. What I mean to say is in calculus they expect you to understand without making you understand. In Core-Plus the teacher makes you understand. And I think that Core-Plus has helped me a lot in the sense of like I can, not only can I look at a problem and understand, or look at a concept, for example not only do I understand, like what I am doing, I understand why I'm doing that specific thing.

Sally on the other hand, was not able to talk about the mathematics that she learned in high school. She said that she would recommend CPMP because she really liked her math classes. She liked working in groups and felt she had learned what she needed to know. She wanted to tell students in CPMP that they wouldn't "always get to use notes and their book on everything, so just have good study skills and don't get behind in your work." Sally's recommendation fits her very well. She wants to enjoy life and doesn't want to work too hard. The social part of the CPMP classroom appealed to her and since she had done well in college, she was pleased with her high school mathematics experiences.

Summary

These case studies allow for an initial look at the transition from a reform high school mathematics curriculum to college mathematics. They also provide an opportunity to begin to identify issues that might arise between a reform precalculus class and future mathematics courses. Based upon the reports of the participants and my observations, none of them had any real problems making the transition from this reform high school curriculum to college mathematics. All of the students had to make small adjustments, but they were aware of what they had to do and made necessary adjustments without difficulty. The only concern voiced by several students was the need for more algebraic skills. This came out in our conversations about what they thought would help make the transition easier for future students. The biggest potential for a transition problem was in the calculus class at Northern University. However, because Ted's conceptions about mathematics were similar to those that were present in the course, he did not have difficulty with the transition.

The case-study analysis provides evidence that students who study mathematics using the NCTM Standards-based CPMP high school mathematics curriculum can make the transition to college mathematics without difficulty. In fact, the beliefs that they develop during high school about the importance of communication and the value of reasoning and justifying makes part of the transition go quite smoothly for them. They are accustomed to working through problems on their own, and so when they confront the need to learn mathematics without the teacher having already explained it to them, they are able to do so. They also are able to explain their thinking to their peers and instructors. This supports the research findings [18-22] that students who learn mathematics in active, sense-making ways tend to be better able to think mathematically and will try to make sense out of new situations.

Limitations of the study and possibilities for further research

The research described here has several limitations. As in all studies that depend on mailed surveys the response rate must be taken into consideration. Although over 50% of the students returned the second CMI, it is not known why the others chose not to return the survey. Thus the picture drawn about the students' conceptions of mathematics is necessarily incomplete.

As with any survey data, what the students were thinking when responding to the CMI was not known. It is possible that students interpreted statements in ways other than intended when the inventory was written. This limitation is always present when gathering data that does not involve conversations with the participants.

Also, the case study students were not chosen randomly. Rather, they were all students who had been successful in high school mathematics and were considered by their teachers to be good students. The very fact that they were competent and responsible young people may have been enough for them to adjust to the new situations with relative ease.

This study provides an initial assessment of the conceptions held by students who have completed four years of high school mathematics using a reform curriculum and how those conceptions impact a student's college mathematics experiences. Further research is needed to more deeply understand the complex relationships that exist among curricula, pedagogy, and the conceptions that students develop and how those conceptions influence students' actions in mathematical situations. Further research might consider the following:

1. This research did not take into account the precise manner in which the Core-Plus Mathematics Project curriculum was implemented. Studies involving students from classrooms where a reform curriculum was well implemented would provide information about how a reform curricula can influence student conceptions of mathematics.

2. This research suggests that conceptions in the seven dimensions identified by the CMI were not independent. The degree to which conceptions in one dimensions are correlated with conceptions in another dimension was not explored. Studies that investigate how conceptions in the different dimensions correlate with each other would be helpful in building a more complete picture of student conceptions of mathematics.

3. Because there is not previous research about the conceptions of mathematics held by students upon graduation from high school, it is not possible to determine exactly how the conceptions of the students in this study would differ from those of students who completed a more traditional curriculum. Studies that allow for comparisons of conceptions between students using different curricula and/or pedagogy would be helpful in determining the impact of curriculum and/or pedagogy on student conceptions of mathematics.

4. As an increasing number of students at all levels are learning mathematics using reform curricula and varied teaching methods, it is important to continue investigating how these changes affect both the attitudes of students towards mathematics and the amount of mathematics that students are learning.

References

1. Dossey, J.A., I.V.S. Mullis, M.M. Lindquist, and D.L. Chambers, *The Mathematics Report Card: Are we Measuring up?,* Educational Testing Service, Princeton, NJ, 1988.

2. Crosswhite, F.J., J.A. Dossey, J.O. Swafford, C.C. McKnight, and T.J. Cooney, *Second International Mathematics Survey: Summary Report for the United States,* National Center for Educational Statistics, Washington, DC, 1985.

3. National Research Council, *Everybody Counts,* National Academy Press, Washington, DC, 1989.

4. NCTM, *Curriculum and Evaluation for School Mathematics,* NCTM, Reston, VA, 1989.

5. NCTM, *Professional Standards for Teaching Mathematics,* NCTM, Reston, VA, 1991.

6. Coxford, A.E., J.T. Fey, C.R. Hirsch, H.L. Schoen, G. Burrill, E.W. Hart, A.E. Watkins, M.J. Messenger, and B. Ritsema, *Contemporary Mathematics in Context,* Everyday Learning, Chicago, IL, 1997.

7. Hirsch, C.R., A.F. Coxford, J.T. Fey, and H.L Schoen, "Teaching Sensible Mathematics in Sense-making ways with the CPMP," *Mathematics Teacher,* vol. 88, 1995, pp. 694–700.

8. Cuoco, A., E.P. Goldenberg, and J. Mark, "Habits of mind: An Organizing Principle for Mathematics Curricula," *The Journal of Mathematical Behavior,* vol. 15, pp. 375–402, 1996.

9. Schoen, H.L. and C.R. Hirsch, "Responding to calls for change in High School Mathematics: Implications for Collegiate Mathematics," *American Mathematical Monthly,* Vol. 110, 2003, pp. 109–123.

10. Grouws, D., C. Howald, and N.Colangelo, "Student Conceptions of Mathematics: A Comparison of Mathematically Talented Students and Typical High School Algebra Students," presented at American Educational Research Association, New York, 1996.

11. Schoenfeld, A.H., *Mathematical Problem-Solving,* Academic Press, Inc, Orlando, FL, 1985.

12. Oaks, A., "The Effects of the Interaction of Conception of Mathematics and Affective Constructs on College Students in Remedial Mathematics," University of Rochester, 1987, p. 468.

13. Silver, E.A., "Research on Teaching Mathematical Problem-Solving: Some Underrepresented Themes and Needed Directions," in *Teaching and Learning Mathematical Problem-Solving: Multiple Research Perspectives,* E.A. Silver, (Ed.), Lawrence Erlbaum, Hillsdale, NJ, 1985, pp. 247–266.

14. Hughes-Hallett, D., A. Gleason, D. Flath, P.F. Lock, S. Gordon, D. Lomen, D. Lovelock, W. McCallum, D. Quinney, B. Osgood, A. Pasquale, J. Tecosky-Feldman, J. Thrash, K. Thrash, and T. Tucker, *Calculus,* Second ed., John Wiley & Sons, Inc, New York, 1998.

15. Thomas, G. B. and R. L. Finney, *Calculus and Analytic Geometry,* 9th Edition ed., Addison-Wesley Publishing Company, Reading, MA, 1996.

16. Bittinger, M.L., J.A. Beecher, D. Ellenbogen, and J.A. Penna, *Precalculus: Graphs and Models,* Addison Wesley Longman, Inc, Reading, MA, 1997.

17. Walker, R.K., "Students Conceptions of Mathematics and the Transition from a Standards-Based Reform Curriculum to College Mathematics," unpublished dissertation, Department of Mathematics and Statistics, Western Michigan University, Kalamazoo, MI, 1999.

18. Schoen, H.L., C. R. Hirsch, and S.W. Ziebarth, "An Emerging Profile of the Mathematical Achievement of Students in the Core-Plus Mathematics Project," presented at Annual Meeting of the American Educational Research Association, San Diego, 1998.

19. Silver, E. and M. K. Stein, "The QUASAR Project: The 'Revolution of the Possible' in Mathematics Instructional Reform in Urban Middle Schools," *Urban Education,* vol. 30, pp. 476–521, 1996.

20. Lampert, M., "Knowing, Doing, and Teaching Multiplication," *Cognition and Instruction,* vol. 3, pp. 305–342, 1987.

21. Lampert, M., "When the Problem is not the Question and the Solution is not the Answer: Mathematical Knowing and Teaching," *American Educational Research Journal,* vol. 27, pp. 29–63, 1990.

22. Boaler, J., "Open and Closed Mathematics: Student Experiences and Understandings," *Journal for Research in Mathematics Education,* vol. 29, pp. 41–62, 1998.

Theme 5. Implementation

In the six papers in this section, we learn from the experience of others who have implemented changes at their institutions, who have developed new curricular materials and designed new courses, or who are utilizing emerging technologies. We also revisit the impact of education reform on the transition from high school to college and the appropriate placement of students. In particular, Robert Megginson offers some suggestions for successfully implementing a new curriculum from the faculty standpoint, while Judy Ackerman offers suggestions from an administrator's standpoint. In support of Zalman Usiskin's earlier claims about the importance of placement, Sheldon Gordon argues that implementing new pedagogies and new curricula necessitates rethinking the way students are placed into college courses. Lawrence Moore and David Smith discuss the impact of technology on the way students learn. Al Cuoco describes designing new instructional materials that are based on students' "habits of mind," instead of being topic driven. Many participants at the conference, *Rethinking the Preparation for Calculus*, felt that the problem with college algebra and precalculus is that they are trying to serve too many audiences. Bonnie Gold describes how at her institution, they solved the one-course-does-not-fit-all situation by dividing their college algebra course into several different courses "each with a clear mission and a separate clientele."

The University of Michigan at Ann Arbor initiated some major curricular revisions in its precalculus and introductory calculus courses in the early 1990s.... The changeover from more traditional to reformed courses in the first-year program required an intensive and, at times, exhausting effort, but actually went remarkably smoothly considering the size of the undertaking and the controversies that reform efforts elsewhere have sometimes faced. We did encounter a number of practical and political issues along the way that had to be dealt with to assure a successful transition, some of which we anticipated and others which caused us to scramble a bit when they arose. The purpose of this paper is to describe some of those issues and provide suggestions for handling them.

Some Political and Practical Issues in Implementing Reform
Robert E. Megginson

It is easier to effect change in precalculus courses if the dean is on board than without the dean. The dean can be an ally when faculty recognize the need for change, or the dean can be the instigator when the mathematics department resists change. Be prepared to educate your dean about the issue and make sure that you understand the current state of affairs with respect to students in precalculus at your college. In this way there is a chance that you can make an effective case for how a reformed precalculus course will improve things for students.

Implementing Curricular Change in Precalculus: A Dean's Perspective
Judy E. Ackerman

In large measure, the problems with mathematical transitions are due to the rapidly growing reform movements in mathematics education at both the secondary level and the college level. NCTM's efforts to promote a school curriculum based on their *Standards* documents are bearing fruit around the country. Instead of the relatively uniform secondary curriculum that most of us went through, many schools across the country have implemented a variety of reform curricula that provide students with very different content and very different teaching and learning environments.... The smooth transition from school to college mathematics is breaking down.... However, the transition problems involve considerably more than differences between school and college mathematics offerings. Perhaps the most significant, yet often overlooked, aspect of transition is the issue of placement—the interface between the two.

The Need to Rethink Placement in Mathematics
Sheldon. P. Gordon

Students with notebook computers connected to a campus backbone by wireless cards are increasingly common. Extensive use of communication technology such as NetMeeting is less common but should be the norm in a couple of years. And, if this were an interactive, online article, we could provide a live link to video of students working through our module. If our scenario is an accurate glimpse of the future—and we believe this future is almost upon us—what are the issues for student learning? ... Technology is changing the way students approach learning. Increasingly, they will conceive of their work in terms of interactive learning materials, computer algebra systems, spreadsheets, and Web-based cooperation—with occasional use of pencil and paper. Learning how to learn in this environment is as important as learning about the mathematics itself.

Changing Technology Implies Changing Pedagogy
Lawrence C. Moore and David A. Smith

Some very useful "modes of thought" in mathematics are given short shrift in high schools (and especially in precalculus courses): hardly showing up at all are reasoning about algorithms, combinatorial thinking, and using the linearity of certain maps on the plane. Furthermore, even for students who go on to calculus and advanced mathematics, the emphasis on traditional precalculus skills and methods is misplaced. Calculus instructors have long complained that the real stumbling blocks for their students are the hard *ideas* in the subject: notions like limit, approximation, convergence, and error estimation. Organizing curricula around these *mathematical habits of mind* provides an alternative to topic-driven design.

Preparing for Calculus and Beyond: Some Curricular Design Issues
Al Cuoco

A cornerstone of the American democracy is that all children should be given equal opportunity. As a result, the standard school mathematics track leads to calculus. While this may be a reasonable policy at the school level, by the time students arrive at college, they have become unequal in many ways. Some have been stimulated by their school mathematics, while others have been crippled by their early mathematical experiences. Some have a clear interest in a mathematically intensive discipline, while others are clearly focused on the humanities, business or social sciences and others are still undecided. One size no longer fits all (if it ever did), in college mathematics courses.

Alternatives to the One-Size-Fits-All Precalculus/College Algebra Course
Bonnie Gold

21

Some Political and Practical Issues in Implementing Reform

Robert E. Megginson
Mathematical Sciences Research Institute

The University of Michigan at Ann Arbor initiated some major curricular revisions in its precalculus and introductory calculus course in the early 1990s with which the author has been closely involved as a member of the Michigan mathematics faculty. A number of political and practical issues had to be addressed to help assure the success of the efforts. The purpose of this paper is to describe some of those issues and provide suggestions for dealing with them when they arise in other implementations.

Introduction

In the fall semester of 2001, about 3000 students enrolled in the three courses considered to be part of the University of Michigan's first-year mathematics program, namely, differential calculus, integral calculus, and Michigan's one precalculus course. About 700 of those students were in precalculus, most of whom were taking the course specifically to get ready for courses in calculus for which placement information indicated they were not yet fully prepared. All three of these introductory courses are taught by methods commonly called "reformed" featuring the appropriate use of technology, texts [1], [3] that support the pedagogical emphasis in the courses, and various forms of cooperative learning and other teaching methods not based exclusively on lecture to take advantage of different student learning styles.

These curricular reforms, in essentially their current shapes, have been in place since 1992 in the case of calculus and 1993 for precalculus, with the precalculus reform following hard on the heels of that for calculus so students would not experience a sudden change in the look and feel of the courses when passing from precalculus to the first calculus course.

The changeover from more traditional to reformed courses in the first-year program required an intensive and, at times, exhausting effort, but actually went remarkably smoothly considering the size of the undertaking and the controversies that reform efforts elsewhere have sometimes faced. We did encounter a number of practical and political issues along the way that had to be dealt with to assure a successful transition, some of which we anticipated and others which caused us to scramble a bit when they arose. The purpose of this paper is to describe some of those issues and provide suggestions for handling them. Though this paper appears in a volume on curricular changes in precalculus, in practice the same issues can arise in any effort to reform introductory mathematics courses, and so are addressed here in that more general context. The first is one that can quickly doom a nascent reform effort if colleagues get the idea that *they* are considered to be the biggest problem that needs to be addressed.

213

Show respect for your colleagues' teaching styles

At all costs, one must avoid sending the message to colleagues that those involved in a reform effort have found the secret to good teaching, and those who do not use the methods must therefore be bad teachers. Frankly, some of the rhetoric from curricular innovators in the early days of reform sent this message quite loudly. Faculty who have been caring teachers doing an excellent job using traditional lecture-oriented methods, but often with some truly innovative twists to those methods, quite rightly resent the implication. Because of this, whenever I give a talk about the Michigan reformed precalculus and calculus programs at another institution, it almost always happens that someone in the audience asks why "we" (those of us who have been involved in reform projects) believe "they" are all bad teachers. "We" certainly do not—at least those of us do not who remember how much we learned about teaching from watching some superb teachers who use traditional methods—and we need to say so. A reform effort will not succeed without support from our colleagues, and we cannot expect to have that if we do not respect what they have accomplished in the classroom, even if we believe the pedagogical methods associated with reform will often be more effective with more students.

More generally, implementers of a mathematics reform project should do everything possible to avoid creating an "us" versus "them" division in the department over the project, and this means listening closely to colleagues' concerns about the effect of the curricular revisions and addressing them where possible. To be able to do this, one must make sure colleagues actually know what is changing and why.

Keep your colleagues in the loop

Make certain that colleagues understand from the beginning the full extent of your reform effort, and how its pieces fit together. In particular, make sure they understand that reform is not just the selection of the textbook, or the use of technology. In part, this is to assure that your colleagues do not feel blindsided when they later see that the changes went beyond a new textbook or the introduction of calculators into a course. You may also find that there will be more support for your effort if you can demonstrate how its pieces complement each other, with the textbook, technology, and pedagogical changes working together to enhance student learning. A seminar or two about your intentions and the problems that will be addressed, with some hands-on work with a few difficult and interesting exercises from the textbook you will be using, can do much to reassure your colleagues about the likely results of the curricular changes being planned. It is particularly important to make sure that the persons most likely to be resistant to the changes attend these sessions; invite them personally.

Get the backing of senior faculty

It is important to have senior faculty who are part of the power structure of the department buy into your reform effort at a very early stage, preferably by taking a direct part in it. Our precalculus and calculus reform efforts were aided greatly by the strong support of the chairs of our department, D.J. Lewis and B.A. Taylor, during the implementation phase, and administrative support for the programs has remained strong since then. The calculus reform effort that preceded and laid the groundwork for our precalculus reform was directed by a respected senior faculty member, Morton Brown, with the help of another faculty member, Patricia Shure, who is well known at the national level for instructor preparation and educational innovation. Though most of the rest of our faculty in the early 1990s were not really familiar with the issues that calculus and precalculus reform were addressing, most did know that there was already some controversy surrounding reform. However, with senior departmental personnel supporting the reform efforts, the rest of the faculty were willing to give the curricular revisions a chance to prove themselves, and in many cases to teach the revised calculus courses to see for themselves how they had changed. Unfortunately, but not unexpectedly, senior faculty involvement in teaching the reformed courses has been almost exclusively in calculus, with only three having taught precalculus at the time of this writing[1]. However, the reforms in the precalculus course are very similar to what we did with calculus, and the faculty know that. In any case, almost all faculty who have actually taught the courses are now convinced of the value of the changes. With broad, continuing support from departmental administration and faculty, the curricular changes are now institutionalized.

Because Michigan's reform programs are well known nationally, graduate students and postdoctoral faculty who have taught in the programs are sometimes sought by other departments wishing to implement their own reform programs. A job candidate in this situation needs to check out carefully where the support for reform in that department actually lies. If there is a strong commitment from the departmental administration and a substantial collection of senior faculty for a change in the courses, and a solid understanding of the issues that reform is supposed to address, then well and good. On the other hand, if that department seems not to be sure why they might want to reform their courses in the first place, but wishes to try an experiment by bringing in someone from outside to conduct a few sections of a reformed course and see what happens, then there is a great potential for professional disaster for the job applicant. Junior faculty already in a department who wish to undertake a substantial reform effort without obtaining the backing of respected senior faculty, but instead assume that the changes will automatically prove themselves, should have similar concerns.

Get the backing of client departments

Client departments should be brought on board from the beginning. They can be a great source of support with the higher administration, as well as within your own department when faculty from the client disciplines can help reassure uneasy faculty from your own that the changes in the courses should have a positive impact on preparing students for study in other mathematically-based fields. It may turn out that the client departments are concerned with the same pedagogical issues that you wish to address, and will be quite supportive if they understand what it is you are doing and why you are doing it.

At the beginning of our reform efforts at Michigan, we had extensive meetings with the science and engineering departments about our intentions and to get their advice on how our revisions could better prepare students for courses in those departments. The result has been generally strong support by those departments for the program and its goals. One piece of anecdotal evidence of this occurred in a joint meeting of the curriculum committees of our liberal arts and engineering colleges attended by the author of this paper. A faculty member who is not in one of our usual client disciplines, but who had read an anti-reform article in the popular press, initiated a discussion about whether our reform efforts had compromised Michigan's introductory mathematics program. The most vocal supporter of our efforts in that meeting turned out to be a physics professor who said that, based on the early discussions between mathematics and physics about the goals of the reform programs, his department had decided to try assigning exercises in the introductory calculus-based physics course that would require students not just to be able to compute integrals mechanically, but to understand more conceptually what integrals really represent. It was discovered that the students could actually do those exercises, which our supporter from physics was confident would not have been the case prior to our reform efforts. He closed by stating that he would not want mathematics to go back to our previous way of presenting the material. His words in that meeting had a far more positive impact than just about anything a member of our own department could have said to defend our program.

Get the backing of academic counselors

It is important to explain to your institution's academic counselors the reasons for the reform and how it will affect their advising of students, and obtain their support for the program. Many students taking

---

[1]Editor's note: The original version of this paper was submitted in February, 2002.

college precalculus have already done well in high school mathematics; about a quarter of our precalculus students in the fall 2001 term had previously taken a high school *calculus* course. When faced with a reformed course as an introduction to college mathematics, such a student may become very concerned that someone has changed the rules under which the student had done well in the past, particularly after a bad first quiz or homework grade.

When this happens, the student is likely to head straight for an academic counselor for advice. It is important that the counselor be able to explain to the student the reasons for the curricular revisions and their ultimate benefits, and provide suggestions for improving performance that might be a bit different from those appropriate for a more traditional course. The advisors will be able to do this if they have been brought into the loop early in the planning stages of the project and have had any concerns of their own about the curricular changes addressed, and in this case they can end up being some of your strongest supporters. If this is not done, then the advisors will be at a loss about what to tell students who are in trouble in the course. They also might end up advising students away from it at registration time, and spreading the word on campus that there seems to be something strange going on in your introductory program.

The next recommendation addresses one very reasonable concern that colleagues in your own department are almost certain to bring up, and that you might also hear from colleagues from other departments and academic counselors if the issue is not dealt with from the beginning.

Make sure skills are learned

In the early days of the current mathematics reform efforts, quite a bit was said about the need to increase the emphasis on concepts and understanding in introductory college mathematics courses, with correspondingly less stress placed on algebraic and computational skills. Though the de-emphasis on skills was in most cases never as great as some of the rhetoric on both sides of the reform issue would have one believe, this issue became a hot-button item for many persons worried about the early direction of mathematics reform. The concern by colleagues that students might not learn needed skills in courses that are supposed to prepare them for more advanced study has brought down fledgling reform efforts in more than one department. To help allay fears about this issue at Michigan, and, more importantly, to make sure our students really were getting required skills from our precalculus and first-year calculus courses, we implemented gateway examination programs in those courses.

A gateway examination is a test of a student's mastery of important basic skills, such as applying fundamental differentiation and integration rules quickly and accurately, that need to be part of a student's personal mathematical toolkit even though computer algebra software or calculators can do the computations. In most implementations of gateway testing, including ours, students may continue taking different versions of a gateway examination over a particular set of skills without penalty so long as the test is finally passed by some deadline, which allows the student to shore up shaky skills between attempts. However, in trade for being allowed the multiple attempts, the skill level required to pass is high. For example, on Michigan's eight-question differentiation test containing some quite difficult derivatives covering all of the basic differentiation rules, the student is allowed to miss only one question, and errors the student might think are small, such as an omitted set of parentheses, are not forgiven.

The effort involved in implementing and maintaining a major gateway-testing program should not be underestimated. So that a student will get fundamentally different versions of a test on successive attempts, Michigan's gateway tests have been computer-generated from the beginning of the program in the early 1990s. However, before the 2001–02 academic year most of the tests were given in paper form, mostly in a testing center rather than in the classroom, and were proctored and graded primarily by undergraduates; (see [4] for a description of the early days of the program). The logistics involved in administering, grading, and returning thousands of tests in a timely fashion each semester were formidable, so almost from the beginning we sought ways to mechanize as much of the process as possible. By the late 1990s, computer-based testing systems were becoming sophisticated and stable enough to handle our requirements, and at the time of this writing Michigan is in the midst of a two-year effort to convert all of these tests to be administered in a computer laboratory designed specifically for this purpose, using testing software originally developed by John Orr and collaborators at the University of Nebraska and currently marketed by John Wiley & Sons under the name eGrade. This conversion and a further extension of the program has been made possible by the support of the National Science Foundation, through grant DUE-0088264.

Know who takes the courses

It can happen that a successful program at one institution will not work well at another because of fundamental differences in the student populations. When problems arise because of a bad fit between program and students, it might be possible to make matters better with a few modifications, but it might also be too late if the project is already perceived to be a failure. For this reason, it is particularly important for implementers of reformed introductory mathematics courses to consider the nature of the population to be addressed by the reforms.

At the University of Michigan, the "typical" precalculus student is a first-year student in the liberal arts college, with the goal of preparing for calculus rather than taking a liberal arts mathematics course, and has not ruled out the possibility of a major or minor in mathematics. Here is one example of how such information affects the conduct of the course. A major feature of the program is group homework, with extensive exercise sets due weekly from students who solve the exercises together outside of class in teams of size three or four. We would have been more concerned about the logistical issues students would face in arranging meetings to do the group homework if we did not know that a sizable majority of them, as first-year students, live in residence halls that are located in four clusters on campus. Where possible, the initial assignment of students to homework groups is made so that the students in each group live near each other. If one were to attempt to transplant the Michigan model without modification to another institution where most of those taking precalculus were nontraditional students living at home, then scheduling meetings outside of class to work on group homework could cause major problems.

Prepare instructors for a changed classroom

At Michigan, there is a weeklong professional development program before the fall term starts that is required of all instructors who are going to be teaching the reformed courses for the first time. This is followed up with weekly meetings in precalculus and differential calculus where further pedagogical issues are addressed, often as they arise in the classrooms. Visits are also made to the classroom of each instructor new to the program, usually twice during the term.

This instructor training model may not be practical for institutions where only a few instructors would require the training each year, but it is still important for those instructors to learn what they will need to do in a classroom that may be radically different from those in which they learned mathematics. One good resource for such instructors, whether or not a full-blown instructional training program is available to them, is [2]. Both authors of that volume are former Michigan instructors who helped with the coordination of the reformed courses and instructor development program.

Finally, and perhaps most important . . .

Do not underestimate the total impact that a curricular change can have on the department. Because teaching precalculus and calculus is such a large part of the role of almost every mathematics department, a serious curricular change in the first-year courses will affect the entire operation of the department. If a commitment to smaller class sizes is made, then there is an obvious impact on the hiring of faculty and the support of teaching assistants. Reformed courses taught in multiple sections tend to require more attention from a course coordinator than more traditional courses, particularly when the instructors have not taught such courses before, in which case the reward system in the department may need some modification to assure that the coordinators are appropriately rewarded, both monetarily and professionally, for their efforts. When many instructors will be teaching the courses, training the instructors and following up with classroom visits can consume substantial resources. All of these issues, as well as others specific to the implementation, will quite likely require an increase in resources for the department as a whole and a reprioritization of resources within the department.

The personal effort required from someone coordinating part of a reformed introductory mathematics program can also be substantial. There are occasions on which each of us involved in the Michigan program would go home quite exhausted, or occasionally would not go home at all; more than once I watched the sun set from my office window while working on some problem involving the coordination of precalculus or calculus, and then saw it rise again before leaving. However, the effort is worth it. Michigan's students are now getting better courses from instructors who are better prepared to address differing student learning styles, and that is paying dividends for both the institution's own programs and those of the institutions that Michigan's teaching assistants and postdoctoral faculty ultimately make their academic homes.

References

1. Connally, E., D. Hughes-Hallett, A.M. Gleason, et al., *Functions Modeling Change: A Preparation for Calculus*, John Wiley & Sons, New York, 2000.

2. DeLong, M. and D. Winter, *Learning to Teach and Teaching to Learn Mathematics: Resources for Professional Development*, MAA Notes #57, Mathematical Association of America, Washington, DC, 2001.

3. Hughes-Hallett, D., A.M. Gleason, W.G. McCallum, et al., *Calculus: Single Variable*, 3rd edition, John Wiley & Sons, New York, 2002.

4. Megginson, R.E., "A Gateway Testing Program at the University of Michigan," in *Preparing for a New Calculus*, MAA Notes # 36, A.E. Solow, ed., Mathematical Association of America, Washington, DC, 1994.

22

Implementing Curricular Change in Precalculus: A Dean's Perspective

Judy E. Ackerman
Montgomery College

Introduction

Mathematics departments have not been overly enthusiastic about rethinking precalculus courses despite changes in calculus, changes in K-12 mathematics that have resulted from the NCTM Standards, and an increased emphasis on accountability. In four-year colleges and universities, some faculty equate precalculus with precollege mathematics or at best as the one mathematics course that students take to meet their graduation requirement. However, in the two-year colleges, precalculus often serves as a true pathway to calculus and to majors that require a significant amount of mathematics.

For many years, calculus reform was the rallying point for mathematics faculty around the country. It involved much more than the addition or deletion of topics from the calculus curriculum. Rather, it initiated the fundamental questioning of what was really important for calculus students to know, particularly in the light of the increasing availability of technology in the form of computers, graphing calculators, and computer algebra systems. The balance of depth versus breadth, applications, and theory was questioned. Calculus reform was much more than just curriculum reform since it also demanded significant change in pedagogy and assessment. Today, even so-called "traditional" mathematics courses and textbooks reflect elements directly attributable to calculus reform. With the history of collegiate calculus reform, why isn't precalculus reform being embraced by the mathematics faculty?

Initiating curricular change

What will it take for significant change to take place in the collegiate precalculus course? Who needs to get on board for it to happen? Although the literature is relatively silent about a dean or administrator's role in curricular reform there are a few suggestions that indeed there is a role and informed deans can be advocates for change. In *Crossroads in Mathematics: Standards for Introductory College Mathematics Before Calculus* it is suggested that although the faculty have the primary responsibility for implementing educational reform, deans can facilitate reform by providing leadership, resources and incentives [1].

A few years ago, when I was the chair of a mathematics department in which there was a limited amount of interest in implementing curricular change in some of the courses, I handed my dean an article by Lynn Steen. Steen articulated twenty questions that deans should ask their mathematics department if they wanted to improve mathematics instruction on their campus [7]. Since the dean had not seen the article, this was a way to initiate a long overdue dialogue between the dean and the mathematics department in order to accomplish change that would benefit our students. Steen's questions were independent of course, pedagogy, technology, or type of higher education institution, and are still applicable. Today, as we grapple with the issue of fundamental change in precalculus, the following questions based on Steen's earlier questions might be particularly relevant to the discussion:

- Who are the students at your institution and what mathematics preparation do they come with?
- What do your students achieve in your precalculus course and in each of your other mathematics courses?
- Do you know what happens to students after they leave your precalculus course?
- Is technology used extensively and effectively in mathematics courses?
- Are the mathematics faculty aware of the national discussion concerning the NCTM Standards, AMATYC Crossroads, Quantitative Literacy, and MAA's work on the first undergraduate mathematics course?
- What steps has your department taken to be sure that faculty are well-informed about curriculum studies and research on how students learn?
- What resources are required to achieve the objectives that will result in change in your precalculus course?
- How well do department priorities match institutional priorities?

Of course, since each institution is different, the answers to these questions will differ and should inform how each college addresses the issue of change in precalculus.

Deans should not sit around indefinitely waiting for mathematics departments to initiate needed improvements in their courses. There are a number of trigger points that should signal to a dean that the mathematics department needs to take a close look at what is going on in their precalculus course. If the department doesn't raise the issue of change, then the dean should raise it when one or more of these is present:

- The success rate in precalculus is significantly lower than for other introductory college level mathematics courses.
- The success rate in calculus I for students who complete precalculus at the institution is low.
- The number of students who successfully complete precalculus and go on to calculus I is small.
- Departments that offer courses with a prerequisite of precalculus are complaining about students' mathematics preparation for these courses.
- The pattern of student complaints about precalculus is different than for other introductory college-level mathematics courses.

One of the issues regarding precalculus reform is that there is not a well-defined definition as to what is meant by precalculus. In fact, at the national workshop held in October 2001, Rethinking the Preparation for Calculus, participants were talking about precalculus with a big "P" being different from precalculus with a small "p." An additional source of confusion comes about because in some colleges, college algebra is the precalculus course. Since it's pretty clear that those of us in the mathematics community have some difficulty defining what is meant by a precalculus course, how can we expect those from outside of the mathematics community to understand the distinction between "Precalculus," "precalculus," and "college algebra" and advise students appropriately? So, to clarify discussion on precalculus course reform, I recommend that we come up with better names for these courses that clarify the intent of each of them. Then we can proceed on the task of reforming all three of these so called precalculus courses.

One of the courses, that today is often called "Precalculus," is for students who plan to continue on through a rigorous calculus sequence. The name "Precalculus" might even be reserved for this course.

Another course is for students who expect to take a limited number of additional mathematics courses that might include applied calculus and/or statistics. Today such a course might be known as either "precalculus" or "college algebra." Finally there is the mathematics course that is frequently called "college algebra," that students take as their last college mathematics course. There are usually administrative policies that require this course to be called college algebra, but many different types of courses come under this name. For example, in Maryland, the Maryland Higher Education Commission (MHEC) initially planned to issue regulations defining the statewide general education requirement in mathematics for all two-year and four-year college graduates as college algebra. Mathematics faculty from around the state's two-year and four-year public colleges got together and proposed a modification of the wording to "at or above the level of college algebra." Although this policy is not totally problem-free, introductory college level mathematics courses can be identified with a meaningful name that describes the actual scope of the course.

Let's start with the assumption that the purpose of the Precalculus course is to prepare students for a calculus I course. Does it make sense today, to offer the same type of manipulative-oriented, skills-driven, precalculus course that was offered in the past in which we assumed that most of the students in the class were going to be math majors or majors that required a significant amount of mathematics? Instead, shouldn't we be considering changes in pedagogy, content, type of applications and use of technology that are consistent with changes that have already been made in most calculus courses and in the K–12 preparation of the students who come directly from high school? This same argument can also be made for precalculus and college algebra courses.

Implementing curricular change

Case studies on curricular change in higher education are few and usually relate to changes in general education programs. Sandra Kanter, Director of the Doctoral Program in Higher Education Reform at the University of Massachusetts–Boston, suggests that change not be viewed as a one-time occurrence, but rather as a series of incremental happenings. She further asserts that "successful implementation of curricular changes required the energies and talents of many faculty members. To the degree that the process was open and collaborative, it built trust and good will among and between faculty and administrators, and only this ensured that faculty felt committed to the eventual outcome" [3].

An article by Alison Schneider [6], contrasting what happened in the overhaul of general education at two universities, points to the need for political savvy, considerable time spent in anticipating objections, and the active participation of the dean. At the university where the dean promised to provide the resources necessary to implement the new plan the revamped curriculum succeeded, whereas it did not at the other university.

Robert Diamond suggests that in many institutions of higher education the faculty promotion, tenure and reward system doesn't recognize significant time and energy devoted to improving courses and curricula [2]. If this is the case, there is actually a disincentive for faculty to make changes to precalculus courses or any other course.

The Long Island Consortium for Interconnected Learning reported in its progress report for year one [4] and year two [5] on how one of the deans from a member institution said that evidence of instructional innovation would be required for promotion to full professor in his college. This is an example of a dean taking on a leadership role and providing incentive for faculty participation in instructional innovation and curricular reform.

Making a case to the dean

Faculty who hope to initiate change in one of the precalculus courses at their college cannot assume that their dean is familiar with the issues surrounding the course. They need to be prepared to make a reasoned argument that makes the case as to how students at their college will benefit from the proposed change(s). A clear understanding of who enrolls in precalculus and what they take next is crucial. One of the most powerful arguments to make to a dean is that a reformed precalculus course will increase student success and satisfaction. You should be prepared with knowledge of the current status at your college of the course and its outcomes for students, in addition to being acquainted with the existing literature that documents the need for this type of change in precalculus courses. Be ready to suggest new models for the course.

An implementation plan needs to include the goals to be accomplished by a reformed precalculus course, specific strategies to carry them out, and anticipated outcomes resulting from the implementation. It also needs an evaluation component so that the effectiveness of the change(s) can be documented. Don't forget that important outcomes may also be in the affective domain. Course objectives usually don't state that students completing a precalculus course will be interested in enrolling in an additional, non-required mathematics course, or that students completing a precalculus course will consider majoring in mathematics. In most other disciplines these are goals, so why shouldn't they be goals for mathematics courses too?

Finally, the dean needs to be given a realistic estimate of the resources necessary to implement a reformed precalculus course. Consider the resources needed to start the project as compared to those needed to maintain a project. It helps if you have considered alternative ways to fund the project. Are the proposed changes such that partial funding might be available from the National Science Foundation to develop a new course, or to adapt an existing, reformed precalculus course?

Carefully consider how much faculty development will be necessary to implement a reformed precalculus course. In every implementation of a reformed mathematics course I have been involved with, faculty development has been the underestimated component of implementation. Plan for time to prepare for implementation and to anticipate all of the potential difficulties that might be encountered. Request support for a project coordinator who can stay on top of the implementation process. Require each faculty member teaching the reformed course to participate in regular course meetings during their first semester teaching the course. Depending on the teaching load at your institution, you may want to provide alternate time for faculty for this activity too.

Pilot first

A pilot implementation of a reform precalculus course provides the opportunity to observe the intended and unintended effects of the reformed course and make necessary adjustments. For colleges with a large number of students and many different instructors, I would highly recommend this approach. This is also recommended when a mathematics department is not in general agreement as to the nature of a reform precalculus course.

The proposal for a pilot implementation of a reform precalculus should specify the length of time of the pilot, the number of sections to be included in the pilot, and the criteria to be used in deciding whether or not to go from the pilot to full implementation. For example, early on when our mathematics department piloted the use of graphing calculators in our precalculus course, there were those who were sure that students using graphing calculators would not perform well in calculus I. During the pilot we learned that students using graphing calculators did as well as those who did not use them even when they went on to take a traditional calculus I course. We learned other things as well including the fact that over half of our precalculus students did not take a calculus course with us during the two years following their successful completion of either version of the precalculus course. At the conclusion of the pilot, graphing calculators were required in all sections of precalculus.

Some of the criteria that you might consider to evaluate a pilot of a reformed precalculus course would be student success in the reformed precalculus course, change in student attitude towards mathematics, student success in their next mathematics course, and student enrollment patterns in an additional mathematics course. Although cost of a reformed course might not be a major consideration for faculty, this is something that deans need to consider. As a dean, I have chosen to go with an instructional choice that is somewhat more expensive when there has been a positive impact on student success. It is important to be open to changes or results that were not anticipated.

Support for faculty implementing change

Professional development is an important component of implementing curricular and instructional change, particularly for the faculty who did not initiate the project. After all, the faculty who support change are already knowledgeable about the change while the others who will be teaching the course need to be brought on board. In a department where the majority of the faculty who teach precalculus are adjuncts or TAs, this can be a problem. If incentives can be provided, this is the place. They may be in the form of travel to a conference, alternate time for project activities, or in the case of adjuncts, extra pay.

Although deans are not usually involved in the day-to-day implementation of curricular reform, it is a good idea for the project leadership to keep the dean informed about how things are going and to alert the dean to what unexpected things are happening. There should also be opportunities for informal discussions of the project between the dean and the rest of the project team. Finally, encourage and help the dean to understand the student perspective. The best way to do this might be to invite the dean to sit in on the course that is being changed. Whenever I sit in on a class, I always try to ask the students about the course at the end of the class period. The student perspective is important and may help shape a better course.

Summary

It is easier to effect change in precalculus courses if the dean is on board than without the dean. The dean can be an ally when faculty recognize the need for change, or the dean can be the instigator when the mathematics department resists change. Be prepared to educate your dean about the issue and make sure that you understand the current state of affairs with respect to students in precalculus at your college. In this way there is a chance that you can make an effective case for how a reformed precalculus course will improve things for students.

References

1. Don Cohen, Editor, *Crossroads in Mathematics: Standards for Introductory College Mathematics Before Calculus*, American Mathematical Association of Two-Year Colleges, Memphis, 1995.

2. Robert M. Diamond, "Broad Curriculum Reform is Needed If Students Are to Master Core Skills," *The Chronicle of Higher Education*, August 1, 1997, p. B7.

3. Sandra Kanter, "Reflections on Reform," *Peer Review*, Summer 2000.

4. Long Island Consortium for Interconnected Learning, "Progress Report for Year One," 1996 (http://www.licil.org/licil/Progress.htm).

5. Long Island Consortium for Interconnected Learning, "Progress Report for Year Two," 1997 (http://www.licil.org/licil/Progress98.htm).

6. Alison Schneider, "When Revising Curriculum, Strategy May Trump Pedagogy: How Duke Pulled Off an Overhaul while Rice Saw Its Plans Collapse," *The Chronicle of Higher Education*, February 19, 1999, p. A14.

7. Steen, Lynn Arthur, "Twenty Questions That Deans Should Ask Their Mathematics Departments," *Bulletin of the American Association of Higher Education*, 44:9 (May 1992), pp. 3–6.

23

The Need to Rethink Placement in Mathematics

Sheldon P. Gordon
Farmingdale State University of New York

Several years ago, Richard Riley, secretary of education in the Clinton administration, challenged the mathematics community to address the problems of articulation in mathematics education between high schools and two- and four-year colleges. Riley called for this national initiative, through the National Research Council, because of the growing breakdown in the once smooth transition between high school and college mathematics, as well as the differences between mathematical experiences in different colleges when students transfer from one institution to another.

In large measure, many of the problems with mathematical transitions are due to the rapidly growing reform movements in mathematics education, both at the secondary level and at the college level. NCTM's efforts to promote a school curriculum based on their *Standards* documents are bearing fruit around the country, as described in other articles in this volume. Instead of the relatively uniform secondary curriculum that most of us went through, many schools across the country have implemented a variety of reform curricula that provide students with very different content and very different teaching and learning environments.

- There is a major emphasis on conceptual understanding, not just routine manipulation;

- There is an emphasis on realistic problems, not just artificial template problems whose solutions are to be memorized and regurgitated;

- There is an emphasis on mathematics via discovery, not mathematics as a collection of facts and procedures to be memorized;

- There is an emphasis on the use of technology;

- There is an emphasis on writing and communication and working collaboratively.

Most of these themes are also part of the reform movements in collegiate mathematics. However, the extent to which these changes have permeated school mathematics is considerably more extensive than the extent to which they have affected collegiate mathematics.

Thus, the smooth transition from high school to college mathematics is breaking down. In particular, we have the following four scenarios:

- a traditional high school preparation leading to traditional college offerings.

- a traditional high school preparation leading to reform college offerings.

- a *Standards*-based high school preparation leading to traditional college offerings.

- a *Standards*-based high school preparation leading to reform college offerings.

224

The first of these scenarios should present no major *transition* problems, either to the students or to the institutions. Students are placed into courses offered in the same spirit as their high school experiences and the level of the courses should be comparable to the students' level of previous accomplishment. The fourth scenario should likewise present no major transition problems. (Of course, students can still encounter significant mathematical problems, but that is another issue altogether.)

However, the second and third scenarios can present significant transition problems, especially to the students. In one case, students arrive on campus, presumably with strong manipulative skills, and suddenly they are faced with the expectation that they have to think deeply about and fully understand the mathematics, and that they cannot succeed just by memorizing procedures by rote. In the other case, the students arrive on campus expecting to expand on their understanding of mathematical concepts, to apply mathematics to more sophisticated realistic problems, to use technology, and to work collaboratively in teams. When they are faced with courses that focus almost exclusively on skills and the expectation that they need to memorize procedures by rote, the effect is comparable to running into a brick wall.

Unfortunately, in practice, things are not quite this clear cut. Very few institutions can be selective enough to choose students with any single type of mathematical background. Thus, most schools need to think through how to deal with students having all sorts of different mathematics background, but few are doing so.

However, the transition problems involve considerably more than differences between school and college mathematics offerings. Perhaps the most significant, yet often overlooked, aspect of transition is the issue of placement—the interface between the two. What are the usual placement tests that decide how much students know and what courses they are placed into? There are several widely used, standardized placement tests, which are all based on the traditional school curriculum and are designed to assess students' ability at algebraic manipulation. Also, many mathematics departments use home-grown tests, which likewise typically focus on the traditional high school curriculum. All of these placement vehicles are fine for the first scenario listed above, but what of the other three scenarios?

For instance, one of the two national placement tests typically starts with a component measuring a student's ability in algebra. Students who do well are automatically moved on to a higher level component that tests college level (precalculus) mathematics; those who do poorly on the algebra level are automatically moved down to a lower level component testing arithmetic and introductory algebra ability. The algebra portion of this test covers 12 topics in an adaptive manner:

1. Square a binomial.
2. Determine a quadratic function arising from a verbal description, e.g., area of a rectangle whose sides are both linear expressions in x.
3. Simplify a rational expression.
4. Confirm solutions to a quadratic function in factored form.
5. Completely factor a polynomial.
6. Solve a literal equation for a given unknown.
7. Solve a verbal problem involving percent.
8. Simplify and combine like radicals.
9. Simplify a complex fraction.
10. Confirm the solution to two simultaneous linear equations.
11. Traditional verbal problem—e.g., age problem.
12. Graphs of linear inequalities.

Now picture what happens to students who have come through a *Standards*-based high school curriculum. Such a student has likely developed an appreciation for the power of mathematics based on understanding the concepts and applying them to realistic situations, as illustrated in some of the lovely examples and problems described in several of the accompanying articles in this volume, such as Dan Teague's or Eric Robinson and John Maceli's. But, this type of traditional placement test clearly ignores much of what they have learned in the way of non-manipulative techniques, of conceptual understanding, and of contextual applications. So, what happens when such students sit down to take a traditional placement test, which is designed only to determine how many manipulative skills the students have retained? Is it surprising that many such students end up being placed into developmental mathematics offerings because their algebraic proficiency is seemingly very weak? This is certainly unfair to students if they were never exposed to some of those skills, or if the emphasis on those particular skills was lower than in the past to make time for more important mathematics or if the students' experience in mathematics has led them to think of mathematics as something considerably more important, more practical, and more intellectually demanding than squaring a binomial. The result is that many students are placed one, two, or even more semesters behind where they likely should be placed based on the amount of mathematics they took in school.

Furthermore, the standardized tests and most of the home-grown tests deny students' use of technology, even though that had been an integral part of their mathematical experience in high school. (Supposedly, some of the national placement tests will soon allow students to use any standard calculator, including most graphing calculators.)

It certainly seems unreasonable to take students who have completed two, three or even four years of high school mathematics and place them into low level developmental courses because their algebra skills are weak. That weakness is perhaps because those skills may not have been emphasized or perhaps because those skills have grown rusty due to a long lay-off since the last math course in high school. All too often, both courses and textbooks assume a blank-slate philosophy, presuming that the students have never seen anything previously. That is not likely the case and will be less the case in future as the reported percentages of students who continue on to successive mathematics courses in high school increases. (Historically, the drop-out rate was on the order of 50% each year; recent evidence indicates, for instance, that the drop-out rate from first year algebra to second year algebra is now on the order of 10–15%. For additional data, please see "High School Overview and the Transition to College," by Zal Usiskin, in this volume.) It seems that a better solution would be for departments to rethink some of the "remedial" courses they offer to see if they are reasonable based on the overall mathematical backgrounds of today's students.

Now picture what can happen with students who took traditional mathematics courses in high school and who are going into reform courses. On the basis of these traditional placement tests, the students' level of manipulative skills may well be assessed as high enough to place them into courses that are well above the level of their conceptual abilities. If they have never had to understand the mathematics they have apparently mastered and have never been expected to read a mathematics textbook, these students may well be overwhelmed by the intellectual expectations of a reform course. (We would not dream of putting a student coming out of elementary algebra into a course in linear algebra; although the student might have the necessary skills, he or she would need to develop a much higher degree of conceptual ability.)

To illustrate just how bizarre these issues can become, consider the situation in New York state. Over 20 years ago, the State Education Department implemented the Sequential Math curriculum, whose *content* is much in the spirit of the NCTM *Standards*. (Effective in 2001, the state began to implement a new version of this program, a pair of courses called Course A and Course B.) However, apparently not a single college in the state has changed its mathematics offerings to reflect what their in-state students are actually taught in the Sequential Math curriculum, nor the nature of the mathematical experiences that the students came through. Moreover, most of the colleges in the state use the standardized, national placement tests that are based on the old syllabus. Some use home-grown tests, but they are typically as traditional in what they seek to assess.

For instance, the author's school and two neighboring institutions all use the same national placement test, which is designed to assess what students learned from a traditional curriculum that has not been offered in New York for over 20 years. So countless students are being declared "remedial-level" and being penalized for not knowing things they were never taught. Moreover, the mathematics curricula at these three neighboring institutions differ markedly. The curriculum at one school is totally traditional, mirroring the old New York state curriculum, so the students are being squeezed through a filter that has little validity for their backgrounds. At a second school, the curriculum is reform from precalculus up, while at the author's school, the entire curriculum is totally reform starting at the development math level. Thus, at the latter two schools, the students not only are being squeezed through a filter that has little validity for their backgrounds, but also they are being squeezed through a filter that has little validity for mathematics courses they are about to take. Our department has been trying to address the placement issue, but has encountered resistance from the placement office, which does not want to implement a new test, and we have been unable to identify a computer-administered test that reflects our philosophy and needs.

To illustrate just how poorly these tests can assess what students have learned in high school, some 15 years ago, when the author was on staff at one of these neighboring two year colleges, the school first adopted and implemented one of the two national placement tests. Just as the fall classes were about to begin that year, the then-department head discovered that more than 140 entering students who had taken some calculus in high school had been placed into developmental arithmetic by this placement test. The test just kept finding the weaknesses in the students' mathematical ability and eventually traced them all the way down to things like problems with manipulating fractions. To avoid this issue subsequently, the department simply re-normalized the results of the placement test. That is, the bar was significantly lowered—the cut-off scores needed for placement into the various courses were lowered sufficiently to assure that appropriate numbers of students would be placed into each course.

Reportedly, the test-makers such as ETS (Educational Testing Service) have been under pressure to develop a new generation of national placement tests that are more aligned to *Standards*-based courses. That would certainly be a huge step in easing the transition problems. However, the process of developing, testing, and validating such tests is a long-term undertaking and we probably cannot expect to see such products available in the immediate future. Unfortunately, departments that depend exclusively on such tests—most likely because of the ease of administering them to large numbers of students—probably can't do much until then.

However, there are some adjustments that can be made rather simply in terms of placement. For instance, some departments have a placement scheme that utilizes the number of years of high school mathematics that a student has taken and his or her ACT or SAT score in conjunction with a placement test to decide on the appropriate course. Other departments take the number of years since the student's last math course into account in placement decisions. In fact, the author is aware of one large scale study conducted some 10 or 15 years ago at a large two year college where about 18 different factors, including placement test score, SAT or ACT score, age, last math course, and years since last math course were all studied in terms of being effective predictors of student performance. They found that about 12 of the factors were statistically significant and so developed a multivariate regression formula for prediction based on all the relevant factors.

There is one other factor that may be particularly relevant today in terms of the new emphases in reform courses. The greater stress on conceptual understanding, on real-world problems, and writing and other communication skills requires a significantly greater level of verbal ability on the part of the students. As such, it is reasonable to link the score on the verbal/English portion of a placement test with the mathematics score. For instance, the author's department has considered ways to add extra points to a student's math score based on high levels of performance on the verbal portion of the placement test. We believe that the verbal ability will likely compensate, to some degree, for relatively low math scores achieved by some students on such a traditional test.

In the meantime, there is much that departments that give their own placement tests can do to help alleviate many of these problems. The first step is to recognize that they will likely need two different placement tests, one for students coming out of a traditional high school program and another for those coming out of a *Standards*-based program. (Alternatively, such departments might try to develop a single placement test that covers both sides and is designed in such a way that the faculty can interpret the results based on their own needs.) The key is to find ways to identify which student is which; it is unlikely that most students will be able to identify the kind of program they went through.

The second issue is to determine the appropriate mix of problems that are mechanical in nature versus those that are conceptual in nature. In a department offering reform courses, just what are the key manipulative skills that are necessary to succeed in those courses? Is it necessary, for instance, to be able to add or divide relatively complicated fractions, say $\frac{5}{24} + \frac{3}{16}$ or $4\frac{3}{4}/1\frac{5}{8}$, or should students be allowed to convert such expressions to decimals and get the answer using a calculator? Is the ability to get the right answer as important as the ability to look at the second expression and estimate that the value is about 3? (Or should the inability to perform such operations relegate students to a course in remedial arithmetic despite their having successfully completed three years of high school mathematics?)

Then there is the reverse issue. What are the key conceptual skills that are necessary to succeed in a reform course, especially for a student coming out of a reform curriculum in high school? How do you determine if a student truly understands the notation, say for a function, or can only move the symbols around mechanically? How do you measure whether a student has the verbal ability to handle the emphasis on mathematical concepts? Perhaps it would be desirable to use the score on a verbal or English placement test in conjunction with the score on a math placement test.

On the other hand, if a department offers only reform courses, how should it assess the skills of a student who has undergone a traditional high school preparation? If a student lacks key conceptual skills because they were not stressed in high school, but has extremely strong manipulative skills, is a reform college algebra or precalculus course the appropriate solution? Must each such course begin with a review of fundamental concepts that students are expected to know?

And, finally, if a department is offering only traditional courses, how should it assess the non-traditional skills that students may bring to the courses? If a student lacks facility with algebraic manipulations because they were not stressed in high school, but has a relatively deep understanding of the mathematics, is a standard remedial course the appropriate solution? Similarly, if a student has a much broader mathematical experience that includes, for instance, an understanding of statistics, data analysis, and probability, does a standard remedial course make sense? If the answer to these questions is "no," how should such courses be redesigned to build on what such students have learned? Can courses be created that emphasize the development of algebraic skills that take advantage of some of the relatively sophisticated knowledge and experience such students bring instead of treating them as individuals who have never mastered any mathematics? Certainly, if such courses can be designed, they would have a much better impact on the students in terms of both motivation and morale.

Clearly, if we can ease the mathematical transitions of the students, we would make things better for all of us. The students will be better served when they arrive on campus; enrollment in "remedial" courses may actually diminish because many of the students being placed there may not really need remediation; enrollment in college-level mathematics offerings might even increase. The students will be happier, the faculty will be happier, and the administrators will be happier.

Acknowledgement: The work described in this article was supported by the Division of Undergraduate Education of the National Science Foundation under grants DUE-0089400 and DUE-0310123. However, the views expressed are not necessarily those of either the Foundation or the projects.

24

Changing Technology Implies Changing Pedagogy

Lawrence C. Moore and **David A. Smith**
Duke University   *Duke University*

Introduction

Sam looked up from the stack of orders on his desk and glanced at his watch. 3:30, time to work on his project with Andrew. He pushed the orders to one side and turned to his computer. No picture this time, but Andrew's voice came through with sounds of students playing frisbee in the background.

Sam, 28, was a non-traditional student, fitting his course work around his work schedule. Andrew was a traditional first-year student. The two had been partners now for four weeks—though they had never met in person.

The background rock music ceased, and Sam heard Andrew's voice, "Hey Sam! What's it like in the real world today?"

"Usual thing, pushing paper... Better watch out for wild frisbees," Sam replied.

"Right. But it's too nice to stay inside," Andrew explained.

"You kids have a soft life," Sam teased. "OK, let's get started."

The instructor's discussion of the project flashed up on the screen. It was just text—Professor Rodriguez was not much for adding voice descriptions. Not like Sam's political science prof, who always added a video stream with her verbal instructions.

Sam proposed a plan of work. "OK, we need to find a picture of a cross-section of a chambered nautilus, then construct a model of the shell's spiral curve. And then we compare it with the real thing. Why don't you search the Web for a good picture, while I look through our class notes for the right formulas?"

Andrew agreed, and his end of the connection went dead. Sam entered a search query, refined it, and found what he wanted. He opened a computer algebra worksheet, made some notes, copied in code, modified it, and produced a test graph.

Andrew's voice returned, and a great picture of a shell appeared in the communication window. "I've put in the x- and y-axes. You can see. And here is a table of coordinates that I pulled off the picture with that cursor widget. ... Oh great, you are all set with the modeling function. Right, exponential growth. I worked through that lesson last week. Now how do we match that up with the coordinate data?"

After another fifteen minutes of trial and error and a return to the class notes, the graph of the model function fit well—except for a stretch near the center that just wasn't the same as the rest of the spiral growth pattern.

"Let's ask Rodriguez about this center stuff," suggested Andrew. They quickly drafted a question, attached the picture and the worksheet, and e-mailed the lot to their instructor.

"Once we hear back from Rodriguez, we need to write up the report. Should be able to wrap this up in another hour," summarized Andrew.

Sam heard the rock music resume and then the dull thunk of a wayward frisbee catching Andrew in the head as he bent over his wireless notebook. The communication screen vanished just as Sam's boss showed up at his desk with another pile of orders.

Our scenario is only partly fanciful. For over five years we have had students working on a project similar to this—albeit in a classroom environment with help available from the instructor. The team project in the scenario could have come from our Equiangular Spiral module [5] with some minor changes. (For example, we continue the project into its calculus implications, and we supply the picture.) Indeed, if Andrew did a Google image search for "chambered nautilus," he would have found over 200 great pictures, one of which is the one we used. And if he searched for "spiral" at the math.duke site (not a likely choice, to be sure), he would have found about 20 images, one of which is the picture on which we ask students to do their measurements.

Note in particular the following features of the scenario:

- Assigned group work

- Remote collaboration

- Use of the World Wide Web as an integral part of the project

- Traditional and non-traditional students working together in real time

- Time on task outside of classroom hours, but with (asynchronous) contact with the instructor

Students with notebook computers connected to a campus backbone by wireless cards are increasingly common. Extensive use of communication technology such as NetMeeting is less common but should be the norm in a couple of years. And, if this were an interactive, online article, we could provide a live link to video of students working through our module.

If our scenario is an accurate glimpse of the future—and we believe this future is almost upon us—what are the issues for student learning? We will discuss the following issues in this paper:

1. Learning and working in an increasingly rich technological environment

2. Making sense of mathematical information—using technology to check

3. Student-to-student interactions

4. Creation of interactive learning materials

5. Intellectual demands of these new forms of learning

Learning and working in an increasingly rich technological environment

Technology is changing the way students approach learning. Increasingly, they will conceive of their work in terms of interactive learning materials, computer algebra systems, spreadsheets, and Web-based cooperation—with occasional use of pencil and paper. Learning how to learn in this environment is as important as learning about the mathematics itself.

Of course, technology has changed how we work and think about work in many ways. Let us illustrate with an example. Suppose you are thinking about writing a paper. You have a couple of ideas; possibly you jot them down on a pad. Then you want to expand them, so you make some more notes, circle them, and draw an arrow to the spot where they should be inserted. Reading the change to be inserted, you realize that other sentences need to be changed as well, and so on. Soon you have several sheets covered with words, lines, loops, and arrows that look more like an abstract painting than a draft of a paper. You quickly abandon paper and resort to a word processor to straighten things out. The point is not so much that you eventually used the technological tool, but that right from the beginning you were framing your thoughts about the paper with the use of the word processor in mind. Technology has changed the way you conceived of the task, as well as the way you carried it out.

Just as technology has changed the way that most of us approach a writing task, it also is changing the way students think about mathematical activities and carry out mathematical investigations. Graphs are now easy to display and can guide an investigation rather than just be an end product of a difficult calculation. With a symbolic calculating system, long trial calculations are also relatively easy and can also serve to guide an investigation. Similarly, data can be gathered, plotted, and compared. Now the important issues become what calculations, graphs, and data to display and how to interpret them.

Making sense of mathematical information—using technology to check

While it is true that technology will enable students to work with their favorite mathematical representations —symbolic, graphical, numeric—it is even more true that students will need to learn how to work and think productively, using many different modes of representation. Indeed, learning how to work and think in multiple representational modes may be one of the most important learning goals of mathematics courses in the age of technology.

In the old pencil-and-paper days, each calculation was likely to be long and subject to errors. Checking, if it was done at all, was likely to consist of performing the same calculation over again—probably making the same error. Now, complicated calculations are easy, and, more importantly, many new ways of checking are readily available. One can compare the symbolic derivative with a difference quotient calculation, a symbolic integration with a numeric integration, or a model function with data. Indeed, modeling provides a strong incentive for students to check their work and correct their mistakes. A student who is not bothered by a pencil-and-paper calculation of a negative volume is much more unsettled by a graph of a model function that does not lie anywhere near the data.

Since students have less emotional attachment to a short computer algebra system calculation than to a long pencil-and-paper one, they are more willing to check the result. They are not looking at the possibility that another 15-minute calculation will have to be repeated. With the pain of checking largely mitigated, the teacher is free to make checking a requirement—and to build checking strategies into the content of the course. Think of the consequences: Getting a confirmed right answer every time will be a normal expectation for both teachers and students. That means we will have to abandon the bell-shaped grading curve—which was never a scientifically sound idea anyway. But it also means—if we have the will—we can eliminate high withdrawal/failure rates and turn mathematics into a subject in which students expect to succeed.

The National Research Council study *How People Learn* ([1], [3]) identifies self-monitoring as one of the key findings from research about successful learning. Specifically [3], p. 13, "A 'metacognitive' approach to instruction can help students learn to take control of their own learning by defining learning goals and monitoring their progress in achieving them." The concept of confirming every mathematical calculation is a local implementation of this principle, since most students start most assigned tasks with the goal of getting the right answer.

This self-monitoring function, known to be important for learning in general, takes on added importance with ubiquitous Web access. Using the Web, a student may find many others who have already dealt with the problem under consideration. How can one know which calculations or conclusions to trust? The ability to evaluate information, to decide what is reasonable, what is correct, is vital to making intelligent use of Web resources.

One illustration of this is [8], a page of lecture notes for a mathematically oriented biology course. This page contains a lot of apparently correct and useful information, but it draws an incorrect mathematical conclusion—one that is obvious to a mathematician but that would easily fool a student. Specifically, Sugg analyzes the classical Lotka-Volterra predator-prey model (in its differential form, not a discrete model) and concludes that the model is inherently unstable. Nowhere on the page is there any hint of the population cycles that are the correct trajectories of the differential system—and also the observations from nature that motivated both Lotka and Volterra.

Another related issue, one that is general across the sciences and engineering, is created by simulations that eliminate the need to perform real physical experiments. This becomes increasingly important in mathematics as modeling becomes a central part of mathematics courses. The issue is not just the accuracy of the simulation, but also the student's conception of the physical world. What models and data are being used to create a given simulation? How reasonable is it that the simulation accurately represents the aspect of the world under consideration? How can one check?

Student-to-student interactions

A particularly important challenge of this new environment will be designing learning experiences that support cooperative work and the development of a class-wide community of learners. One way to go about this is described in [7] in the context of a differential equations course–but the same principles could be expected to work with lower-level courses as well.

As we imagined in our opening scenario, there will be great opportunities for productive cooperative work—even for students with little or no opportunity for face-to-face contact. In addition to Microsoft's *NetMeeting* (http://www.microsoft.com/windows/netmeeting), many other ways to accomplish real-time collaborations are now available. Some other examples include

- Blackboard (http://www.blackboard.com)
- WebCT (http://www.webct.com)
- Netopia's *Timbuktu* (http://www.netopia.com/en-us/software/products/tb2/)
- AT&T's *Virtual Network Computing* (http://www.uk.research.att.com/vnc/)
- Interwise's *Enterprise Communications Platform* (http://www.interwise.com/)

The capabilities of these products are all different from one another, as are their prices, but each enables collaborators to share work in real time via the Internet.

In the other direction, there is a tendency for technology to provide the individual with a personal learning environment, insulated from contact with others. With headphones delivering a stream of background music and individual hand-held computing devices replacing workstations that can accommodate two people, the individual student may retreat from any significant learning interaction. It will be important for both curriculum developers and instructors to focus on this issue.

Creation of interactive learning materials

What are the implications of technology for developers of learning materials? In the recent past, individual faculty have been creating interactive class materials shortly before they were needed in class. Then, more often than not, the materials were left alone until the next time the author-instructor was teaching the same course. Even if an author did more work, it was unlikely that the materials were ever "finished" in any reasonable sense. In some ways, this is comparable to the period in the 1970s when many individuals wrote their own word processing programs. After a short transition period, users came to expect more from a word processing program than most individuals were willing or able to produce. Now most of us use one of the common commercial programs.

For learning materials, there are currently two trends. One is for teams of individuals to work together to produce materials that include sophisticated interactions delivered in a setting that is easy to use and very flexible. The other trend is similar to the phenomenon of open-source software. Authors cooperate in a loose federation that combines compatible learning components in different ways as necessary and leaves the product for further development by others.

In the old textbook-oriented model, a small group of authors, working very intensely, produced most of the major text material. The individual faculty member's responsibility was to create a syllabus around the published text. Now, regardless of the interactive materials used, the instructor is going to be much more closely involved, often adapting the materials for his or her own use. Beyond that, many more instructors will be part of the design and development of the materials. However, if it is done well, the development of learning materials that incorporate technology will take extensive time and effort. How are authors to be rewarded? The rewards will probably not be royalty income so much as scholarly recognition. So far, this sort of recognition has been slow to develop.

Intellectual demands of these new forms of learning

Finally, we need to be clear that students will be expected to do more challenging tasks than in the past—particularly in precalculus and calculus courses. In the past, just deciding on a symbolic calculation algorithm and executing it with care represented a satisfactory learned response. Now the student will need to recall and evaluate the usefulness of and connections among a variety of representations and computations. This is a higher-order intellectual activity—one that will allow learning at a deeper level. Fortunately, reforms such as the NCTM *Principles and Standards* [6] have paved the way for this change.

It is no longer acceptable to assess student learning by asking them to solve calculational problems because computer algebra system (CAS) capabilities are widely available to almost everyone. For example, the Texas Instruments TI-89 (about $150) provides powerful algebra and calculus capabilities (with 2-D and 3-D graphics) in a handheld calculator. Many schools and colleges provide site-license access to *Maple®* or *Mathematica®*. *StudyWorks* (essentially a fully functional version of *Mathcad®*) is available from Mathsoft for about $40. And there are a number of free or inexpensive online services that will accept a problem input and provide the output from, say, *Maple®* or *Mathematica®*. One example is *The MathServ Calculus Toolkit* at Vanderbilt University [2], which includes a number of precalculus topics as well. Simply forbidding the use of any of these tools is about as effective as sticking a finger in a crumbling dike.

In fact, it never did make sense to assess student understanding of mathematics solely or primarily by their ability to do unaided symbolic calculations. At best this ability is a poor proxy for understanding, as anyone can learn simply by asking students to explain what they are doing as they carry out a calculation. And generations of students have come to believe that the calculations are what mathematics *is*. Worse, reserving the rewards for those who are proficient at calculations in a timed, closed-book, no-technology test setting has denied success to many other students who are quite capable of understanding mathematical concepts—as we have learned by teaching those students in technology-rich environments. Whatever the limitations on our profession in the past, we are not condemned to repeat failing practices forever.

On the positive side, a recent analysis and synthesis [4] of research on the use of technology in mathematics instruction at all levels has documented strong support for welcoming technology as a component of our pedagogical practices. One of us co-authored the calculus chapter [9] in this volume, which includes among its conclusions the following:

- "Technology integrated intelligently with curriculum and pedagogy produces measurable learning gains...."
- "There is evidence that using tools such as Mathematica and Maple for conceptual exploration ... leads to conceptual gains in solving problems that can transfer to later courses. In comparison, students following traditional courses tend to use more procedural solution processes."
- "Technology enables some types of learning activities (e.g., discovery learning) and facilitates some others (e.g., cooperative learning) that are harder or impossible to achieve without technology."

Of course, the completed research all refers to technologies that have been available in the past. The technologies becoming available to us now hold promise for even more exciting gains—if we can keep up with the intellectual challenge of adapting our pedagogies to the realities of the world in which our students live.

References

1. Bransford, J. D., A. L. Brown, and R. R. Cocking, eds. *How People Learn: Brain, Mind, Experience, and School*, National Academy Press, Washington, DC, 1999.

2. Crooke, P. S., and Tschantz, S. The MathServ Calculus Toolkit, http://www.math.vanderbilt.edu/~pscrooke/toolkit.html (Retrieved September 11, 2002).

3. Donovan, M. S., Bransford, J. D., and Pellegrino, J. W., eds. *How People Learn: Bridging Research and Practice*, National Academy Press, Washington, DC, 1999.

4. Heid, M. K., and Blume, G. W., eds. *Research on Technology and the Teaching and Learning of Mathematics*, Information Age Publishing, Greenwich, CT, in press.

5. Moore, L. C., Smith, D. A., Mueller, W. "The Equiangular Spiral," *Connected Curriculum Project*, http://www.math.duke.edu/education/ccp/materials/mvcalc/equiang/, 2000 (Retrieved August 28, 2002). Also published in *Journal of Online Mathematics and its Applications* 3(1), 2001, http://www.joma.org/vol1-3/modules/equispiral/ (Retrieved August 28, 2002).

6. National Council of Teachers of Mathematics, *Principles and Standards for School Mathematics*, NCTM, Reston, VA, 2000.

7. Smith, D. A. "The Active/Interactive Classroom," pages 167–178 in Derek Holton, ed., *The Teaching and Learning of Mathematics at University Level: An ICMI Study*, Dordrecht: Kluwer Academic Publishers, 2001. Also available at http://www.math.duke.edu/~das/essays/classroom/ (Retrieved August 28, 2002).

8. Sugg, D. W. "Predation," Lecture notes for an Ecology class at SUNY Geneseo, http://darwin.bio.geneseo.edu/~sugg/Classes/Ecology/Lectures/Lecture_20.htm, October 9, 1996 (Retrieved August 28, 2002).

9. Tall, D. O., Smith, D. A., and Piez, C. "Technology and Calculus," Chapter 7 in Heid and Blume [4] (in press).

25

Preparing for Calculus and Beyond: Some Curriculum Design Issues

Al Cuoco
Center for Mathematics Education–EDC

This paper outlines an alternative to the topic-driven design principle that is the basis for most precalculus courses, arguing that the real power of mathematics lies in the methods used to produce results as much as in the results themselves. It describes a fourth-year high school course that adopts this design, with examples and student work.

Introduction

Curriculum design in US precollege mathematics is largely topic driven; a course is defined by the topics it treats. The major criteria for including a topic in any particular course include:

- does it review and deepen important ideas from previous courses?
- is it a prerequisite for likely subsequent courses?
- did it fall through the cracks in earlier grades?
- does it appear on high stakes tests?

As one moves up the grades, the effects of this design principle compound. By the time one reaches the fourth year of high school, we end up with 18-chapter, 800-page compendia of topics that range from trigonometry to data analysis to complex numbers. These monster texts all go under the name "precalculus," which is therefore defined as everything from trigonometry to data analysis to complex numbers.

Of course, there's much more in these texts than what one needs as preparation for any of the current calculus offerings. Indeed, it's a well-known fact among high school teachers that one can only finish slightly more than half of these chapters in a given year, and yet many students who go on to calculus from such experiences have what they need to get respectable grades.

But in addition to being too big, these courses are, at a deeper level, too small. There has been a growing consensus among all involved in secondary mathematics education that this topic-driven curriculum is not serving our students well. Today's high school graduates enter a highly technological world in which mathematics plays an essential role. However, the widespread usefulness and effectiveness of mathematics in fields outside the discipline come not just from mastering specific skills, topics, and techniques, but more importantly, from developing the ways of thinking used by scientists, mathematicians, engineers, and others. National reports and standards documents ([1], [2], [3], [4], for example) have articulated a sentiment that is widely held by people in the mathematics education community: essential student outcomes from a modern mathematics curriculum should include skills like:

- finding and analyzing patterns,
- designing and conducting experiments,
- describing and communicating,
- tinkering and inventing,
- visualizing and conducting thought experiments,
- conjecturing and guessing,
- theorizing and abstracting, and
- making logical connections and explanations.

See ([6], [7]) for an elaboration on this list. Over a decade ago, *Everybody Counts* [5] described it this way:

> Mathematics offers distinctive modes of thought which are both versatile and powerful, including modeling, abstraction, optimization, logical analysis, inference from data, and use of symbols. Experience with mathematical modes of thought builds mathematical power—a capacity of mind of increasing value in this technological age that enables one to read critically, to identify fallacies, to detect bias, to assess risk, and to suggest alternatives [p. 31].

Some very useful "modes of thought" in mathematics are given short shrift in high school (and especially in precalculus courses): hardly showing up at all are reasoning about algorithms, combinatorial thinking, and using the linearity of certain maps on the plane. Furthermore, even for students who go on to calculus and advanced mathematics, the emphasis on traditional precalculus skills and methods is misplaced. Calculus instructors have long complained that the real stumbling blocks for their students are the hard *ideas* in the subject: notions like limit, approximation, convergence, and error estimation.

Organizing curricula around these *mathematical habits of mind* provides an alternative to topic-driven design. It provides another criterion for including or excluding a particular topic, and, what's more important, it has a great influence on how topics are developed: Explicit attention is given to the methods behind the results.

In mid-1996, NSF convened a group of curriculum developers, mathematicians, and educators to discuss possible directions for fourth-year high-school mathematics. At the time, NSF had invested heavily in comprehensive curricula for the first three years of high school. It was time to look at what belonged in the bridges between high school, postsecondary programs, and the workplace. At that meeting, participants expressed a need for problem-based and student-centered materials that:

- build on and make use of the rather different backgrounds that students would develop in any of the three-year "standards based" programs,
- identify and formalize "big mathematical ideas" developed during the first three years of high school (for example, a general notion of function or proportional reasoning),
- use mathematics itself as a context for developing mathematical ideas, and
- treat topics and thinking processes that are prerequisites for postsecondary education and for school-to-work transitions (trigonometry or algorithmic thinking, for example) in the spirit of reform without losing their essentially technical nature.

It was clear at this meeting that there were many different topics, organizing principles, and viable directions that could be taken in such a curriculum for the latter part of high school. Everyone agreed that the last years of high school, more than any other part of the K–12 curriculum, called for a multiplicity of approaches and options, not only for the preparation of future scientists and engineers, but also for developing informed citizens.

My colleagues and I at EDC's Center for Mathematics Education decided to apply our design principle—putting mathematical thinking at the center of curriculum development—to the creation of a fourth year course for high school seniors. More precisely, we wanted to develop a program that would:

- give students a sense of what mathematics is about and the experience of what it is like to do mathematics,

- help students develop the habits of mind used to create the major results of modern mathematics, and

- prepare students for future work in mathematics and science, should they choose to pursue those fields.

This paper describes that attempt, an attempt that evolved over the course of several years by adding constraints imposed by the field, making compromises imposed by the nature of schools, and incorporating the brutal and enlightening feedback one gets from field-test teachers and (especially) students. The result is a course for high school juniors and seniors, appropriately named *Mathematical Methods* (or M², for short) [11], that looks quite different from traditional precalculus courses, in spite of all the compromises we made.

Goals for the program

While mathematical thinking was our primary goal, there were several others at play. Some are deeply rooted in our approach to mathematics education, while others are more focused on the particular audiences for this course. Like all projects just starting out, we spent some time creating a list of lofty (and rather poorly defined) goals for the work:

1. The materials should center around mathematical thinking.

2. The materials should be accessible to our intended student and teacher audiences.

3. We should set high expectations and help people meet them.

4. Each chapter should be easy to start and should take students farther than they dreamed possible.

5. The chapters should form a web of interconnected ideas.

6. We should use the most effective technology available for helping students develop the habits that we want to foster.

7. The materials should help students see the value and importance of mathematics.

8. We should work to enlarge students' and teachers' notions of reality to include mathematics.

9. Graduates of our course should love mathematics.

10. The benchmarks for choosing a topic include:

    - it provides an opportunity to develop mathematical ways of thinking

    - it is of historical importance

    - it is useful

    - it is beautiful

    - it prepares students for future mathematics

    - it contributes to a broad picture of mathematics

11. The materials should be useful across the wide range of upper high school courses.

12. The materials should be faithful to mathematics as a discipline, emphasizing major historical themes, results, and problems.

Looking back, these seem grandiose, subjective, and a little naive. Some of them seem to pull in opposite directions. Seven years later, there's a story to tell about what happened with each of them. We'll tell a few of these stories in the last section of this paper.

What habits? What topics? What students?

There's an old saying that you can't think about thinking unless you think about thinking about *something*. A course organized around mathematical thinking would be little more than a mathematics appreciation course if it contained chapters on "doing experiments" or "using linearity." Our intention was to use mathematical topics as vehicles for certain ways of thinking. But, of course, the topics needed to serve other purposes as well—the ones laid out in item 10 above.

And what was the audience for the course? As other papers in the proceedings have documented so well, there is wild variation in the reasons students take a fourth year of mathematics. At one extreme are the students who aren't interested in (or lack the grades for) a college-prep precalculus or calculus course. These students typically take courses, with titles like Senior Topics, that are just watered down versions of precalculus, with some SAT review thrown in. At the other end are students who've taken all the AP the school has to offer and want an advanced elective. These students often find haven in a computer science course, a directed study with a willing teacher, or a college level course at a local college. In the middle are most of the college-bound students, taking precalculus, calculus, or one of the AP calculus courses.

The design therefore turned into a three-dimensional effort:

- What students will we serve?
- What mathematical habits do we want students to develop?
- What topics are good vehicles for developing these habits and serving the needs of students after they finish M^2?

Based on the belief that the vast majority of students are capable of serious mathematics, we decided to develop materials that could be used across a wide spectrum of fourth year courses, from the topics courses to the advanced electives. This led to some early decisions:

- M^2 would be designed for students who have either completed one of the comprehensive high school curricula or a more traditional Algebra 2 course.
- We would aim the materials at:

  – secondary students who intend to work in a technical field;

  – college-intending students who are looking for an alternative fourth year of mathematics;

  – students who wish to take a mathematical elective during their junior or senior year (either instead of or in addition to precalculus/calculus sequence).

We also wanted to address the needs of students who may not be planning to continue their mathematical studies.

To reach such a broad audience we borrowed a structure from an earlier curriculum effort [14]: We would write a small number of large chapters around big themes. Each chapter would start with an extremely simple entry point that would be tractable for almost every student and then would carry the development to levels to challenge the most advanced students. The idea was that each chapter would have certain jumping off points, so that teachers could customize the materials for their students in a way that, instead of forgetting the last half of the book, allowed them to help all students to experience the important themes, develop the central mathematical habits, and dig into the major results in the course. This decision was by no means unanimous. As one staff member put it, "It sounds to me like we're willing to take on students of widely different backgrounds and prepare them for very different future mathematical careers. I'm not sure just how feasible this really is." None of us was sure, but we decided to see how far we could take this approach.

As for mathematical ways of thinking that we wanted to foster, we decided on "the big four":

- algebraic thinking

  – designing and using algorithms

    – reasoning about numerical and algebraic calculations

- linearity

    – reasoning by linearity (in the sense of linear algebra)

    – using linear approximations

- combinatorial methods

    – recognizing isomorphic combinatorial problems

    – counting without explicit enumerations

- analytic thinking

    – reasoning by continuity

    – making successive approximations

Our benchmarks for choosing a topic, our choice of habits of mind, and this "no threshold, no ceiling" decision put some strong constraints on our choices for chapters. For example, not every mathematical topic has easy entry, and school mathematics is notorious for introducing topics that don't go very far. Our first cut contained four chapters:

- Solving Equations (including polynomial and trigonometric equations)
- Coordinates and Linear Methods (including the use of matrices)
- Counting/Recursion/Sequences and Series (including limits and infinite series)
- The Complex Numbers (building on and using everything that comes before)

Lists are seductive, and we had created several. I was convinced that we had a simple plan for a lean course that would give students a real taste of what mathematics is all about—the rest of the work ahead seemed straightforward. Then we took these ideas out for a spin, in classrooms, with a local advisory board of teachers, and with a national advisory board of high school and college teachers, mathematicians, and mathematics educators.

The Evolution

After consulting with our advisors, drafting sample activities, and piloting the drafts in local schools, it became apparent that the design had to be refined. There were missing topics (that *had* to be there, of course), there were inter-chapter dependencies that couldn't be resolved, teachers were worried that the chapters were *too* big (covering too much material), and the easy entry and useful benchmarks seemed to be violated in some of the drafts—we needed some better hooks. The outline passed through several iterations. We stayed faithful to the habits we wanted to develop and the audience we wanted to reach, but the outline at each iteration began to look a little more like a precalculus course, with more and shorter chapters.

And there was a problem with technology. The writers were convinced that, in addition to a graphing calculator, a computer algebra system was essential for some of the topics, methods, and ideas we wanted to develop. The CAS of choice was one installed on a computer, something like Mathematica or Maple. It was especially important that the system we use be *extensible*, so that students could build computational models of their own for algebraic objects (sequences of polynomials, for example) and so that they could express in the CAS abstract relationships (of *their* invention) between mathematical structures. However, it became clear that M^2 would not be used, except by the wealthiest districts, if we insisted on a computer-based CAS. Indeed, actual use of desktop or laptop computers in mathematics classes is on the decline among the teachers in our advisory group. Instead, teachers are using calculators. The CAS system on calculators lacked some of the functionality and power that we thought we needed. I was especially worried about the interface, memory limitations, and notational conventions built into these hand-held machines, seeing many places where students could develop *bad* habits of mind through their use. But it was clear that we needed to adapt to the constraints or give up hope of using a CAS. It turned out that the system on the TI-89 and 92 allowed us to do most of what we wanted (especially with the wonderful support we got from the TI development staff), was extensible enough to get across the idea that one could build computational models for mathematical objects, and had an interface that posed very little difficulty for students. And in at least a few cases, the memory limitations of the machine turned out to be a perfect pedagogical device for developing important methods like mathematical induction (see [9] for details).

After several iterations, we ended up with a book containing seven chapters—more than we wanted but less than the norm. We had to make some hard decisions about what to leave out. The linearity of rotations and summations remains, but a more general discussion of linear transformations fell out. We didn't get as far into difference equations as I had wanted. I had hoped that we'd take a much more structural approach to complex numbers. And we just touched on ideas that I wish could have been given better play: Applications of complex numbers to the theory of regular polygons, and the mathematics behind public key cryptography. Many of these decisions to leave topics out would have been made differently had we been using topic-driven design, but the key focus on mathematical habit became the arbiter that caused many beautiful topics to be put aside. Other decisions came about from purely practical issues. For example, in the last months of the development, we did a major rearrangement of the *order* of the chapters, because teachers said that, especially in senior classes, technical material (like trigonometry) should never come near the end of the year. And, of course, we had to make sure that students would be prepared for calculus.

Here's a brief description of the seven chapters:

(1) Tables, Patterns, and Rules.   This chapter asks students to find functions that agree with input/output tables. Students generate closed form and recursive rules for these functions, and then learn about mathematical induction as a way of showing that two seemingly different functions agree on the non-negative integers (see [10] for an elaboration of the approach). Advanced topics include methods for finding genuinely *different* functions that agree on a table and Lagrange interpolation. The focus is on algorithmic thinking, finding and describing patterns, and algebraic thinking.

(2) Polynomials.   This is a chapter on advanced algebra. There's a dual focus: Polynomials as formal objects (algebraic thinking and reasoning about calculations) and polynomials as representations for certain continuous functions (reasoning by continuity and analytic thinking).

(3) Complex Numbers and Trigonometry.   The typical way to establish the geometric interpretation of complex numbers is to use the addition formulas for sine and cosine. This chapter goes the other way: By establishing the linearity of rotations in the plane, one can use the geometry of complex multiplication to derive the addition formulas for sine and cosine (as well as most other trigonometric identities). This has several advantages:

• It shows the essential connection between geometry and the algebra of complex numbers without the heavy machinery of trigonometry.

• It provides an *application* of the algebra of complex numbers: Students develop a general purpose machine for proving trigonometric identities.

• It provides some coherence to the topic of identities, showing how most depend on the invariance of the unit circle under rotation about the origin and reflection in a diameter.

(4) Count It Up.   This is a chapter about combinatorics and combinatorial thinking (predicting the outcome of an enumeration without having to carry it out explicitly). Combinatorial formulas are developed in special cases by solving problems and are then generalized. After developing results about permutations, combinations, and binomial coefficients, the chapter culminates with one large problem—the "Simplex Lock" problem—described at: http://www2.edc.org/makingmath/mathprojects/simplex/simplex.asp

(5) Add It Up. The idea here is to build on Chapter 1, looking at techniques for summing series. This can all be done in an informal way, using the kind of mathematical induction developed earlier. Topics include:

- arithmetic sequences and series,
- geometric sequences and series,
- sums of squares and cubes, and
- telescoping sums and how to build them.

Students use their CAS to investigate and prove properties of the $\sum$ operator, many of which have parallels in calculus with $\int$. This chapter is about algebra and algorithms, reasoning about and transforming calculations.

(6) The Ideas of Calculus. This chapter tells the story of how mathematics evolved to solve the problem of finding areas enclosed by curves. Another message is that functions can't always be given by algebraic formulas. Using techniques for summing powers from Chapter 5, students can tackle the problem of finding areas bound by graphs of polynomial functions. We also develop Fermat's brilliant method for finding the area under *every* curve of the form $y = x^n$ ($n \neq 1$) using geometric series. Again, techniques from Chapter 5 allow students to recreate Fermat's solutions.

(7) Algebra and Cryptography. The chapter starts by looking at some simple and historically important encryption schemes. For example, students look at linear functions $x \rightarrowtail ax + b$, applied to a letter's position in the alphabet. This introduces the structure of Z/26Z (because if a has a common factor with 26, you make a bad cipher). More general ciphers are defined by 1–1 functions on Z/26Z. From here, we will investigate some elementary number theory, including:

- solving equations in modular systems,
- units and zero divisors, and
- Fermat's Little Theorem.

The chapter culminates in a treatment of public key cryptography.

In the next section, I'll describe a student activity that illustrates more precisely how the design principles play out across the chapters.

An example

The course opens with students trying to find rules that agree with tables. As many people who've tried this know, students naturally gravitate to recursively defined rules. So, asked to find a function that agrees with this table:

Input	Output
0	3
1	5
2	7
3	9
4	11
5	13
6	15

students notice both the closed form rule $n \to 2n + 3$ as well as the recursively defined rule

$$f(n) = \begin{cases} 3 & \text{if } n = 0, \\ f(n-1) + 2 & \text{if } n > 0. \end{cases}$$

While many students notice the recursive pattern, the transition between a verbal description and the above mathematical notation is not easy. Modeling the recursive rule in a CAS helps many students make the transition. In the TI-89 system, the model looks very much like standard mathematical language:

```
: f(n)
:       Func
:           if n = 0 then
:               return 3
:           else
:               return f(n-1) + 2
:           endif
:       endFunc
```

The interplay between recursive and closed form models for functions is a theme that runs throughout the course. For example, early in the year, with no formal machinery behind them, students are presented with the following problem:

Suppose you want to buy a car. You don't have much money, but you can put $1,000 down and pay $250 per month. The interest rate is 5%, and the dealer wants the loan paid off in three years. What price car can you buy?

This leads to the question, "How does a bank figure out the monthly payment on a loan?"

A recursive approach lets students experiment using their CAS. They begin with a simpler model: one in which there is no interest. Suppose the original price of the car is $10,000. If $b(n)$ is the balance owed at the end of n months, the model looks like this:

$$b(n) = \begin{cases} 9{,}000 & \text{if } n = 0, \\ b(n-1) - 250 & \text{if } n > 0. \end{cases}$$

Students gradually refine the model to include interest. They translate the following verbal description:

What you owe at the end of the month is what you owed at the start of the month, plus 1/12 of the yearly interest on that amount, minus your monthly payment.

into a mathematical function:

$$b(n) = \begin{cases} 9{,}000 & \text{if } n = 0, \\ b(n-1) + \frac{.05}{12}b(n-1) - 250 & \text{if } n > 0. \end{cases}$$

Students can then experiment, adjusting the "250" until they make $b(36) = 0$. In fact, many students add another input to their balance function, so they can change the monthly payment on the fly. A TI-89 model looks like this:

```
: b(n,m)
:       Func
:           if n = 0 then
:               return 9000
```

```
:          else
:              return (1 + 1/12*.05)* b(n-1,m) - m
:          endif
:      endFunc
```

Students love to get the monthly payment down to the penny. In fact, we go up one level of abstraction and ask them to calculate monthly payments on several loans, seeing how the payment changes with the cost of the car:

Pick an interest rate and keep it constant. Suppose you want to pay off a car in 36 months. Investigate how the monthly payment changes with the cost of the car.

(1) Make a table like this:

Cost of car (in thousands of dollars)	Monthly payment
10	
11	
12	
13	
14	
15	
16	
17	
18	

(2) Describe a pattern in the table. Use this pattern to find either a closed form or a recursive rule that lets you calculate the monthly payment in terms of the cost of the car in thousands of dollars. Model your function with your CAS and use the model to find the monthly payment on a $26,000 car. Check your result with the original approximation method.

Each entry in the table is calculated by a series of approximations, and then the entire table is treated as a new data set in which students find a surprising relationship. A snapshot of one student's work can be found on the next two pages.

Michelle notices a linear relationship between the cost of the car and the monthly payment. She doesn't yet have the background to prove that the observed relationship is, in fact, linear, but she has evidence for a conjecture. Indeed, Eric Karnowski, a colleague at EDC, was working through this problem and found an extremely beautiful way to use the full features of the CAS on the TI-89 to get an explicit formula for the monthly payment with no need for successive approximation. The balance at the end of 36 months with a monthly payment of $250 can be obtained by entering b(36,250) in the calculator (we get $764.92 as a balance). Eric thought of m (the monthly payment) as a *variable*, and he wanted to find the value of m that makes b(36, m) output 0. So, he asked the CAS to simplify b(36, m). Rather than assigning m a value, he left it as an indeterminant and entered b(36,m) in the TI-89. The calculator outputs an expression in m—it gives

$$10453.250082011 - 38.75333552005m.$$

But we want this expression to be 0. So, we enter

```
solve(10453.250082011 - 38.75333552005*m = 0,m)
```

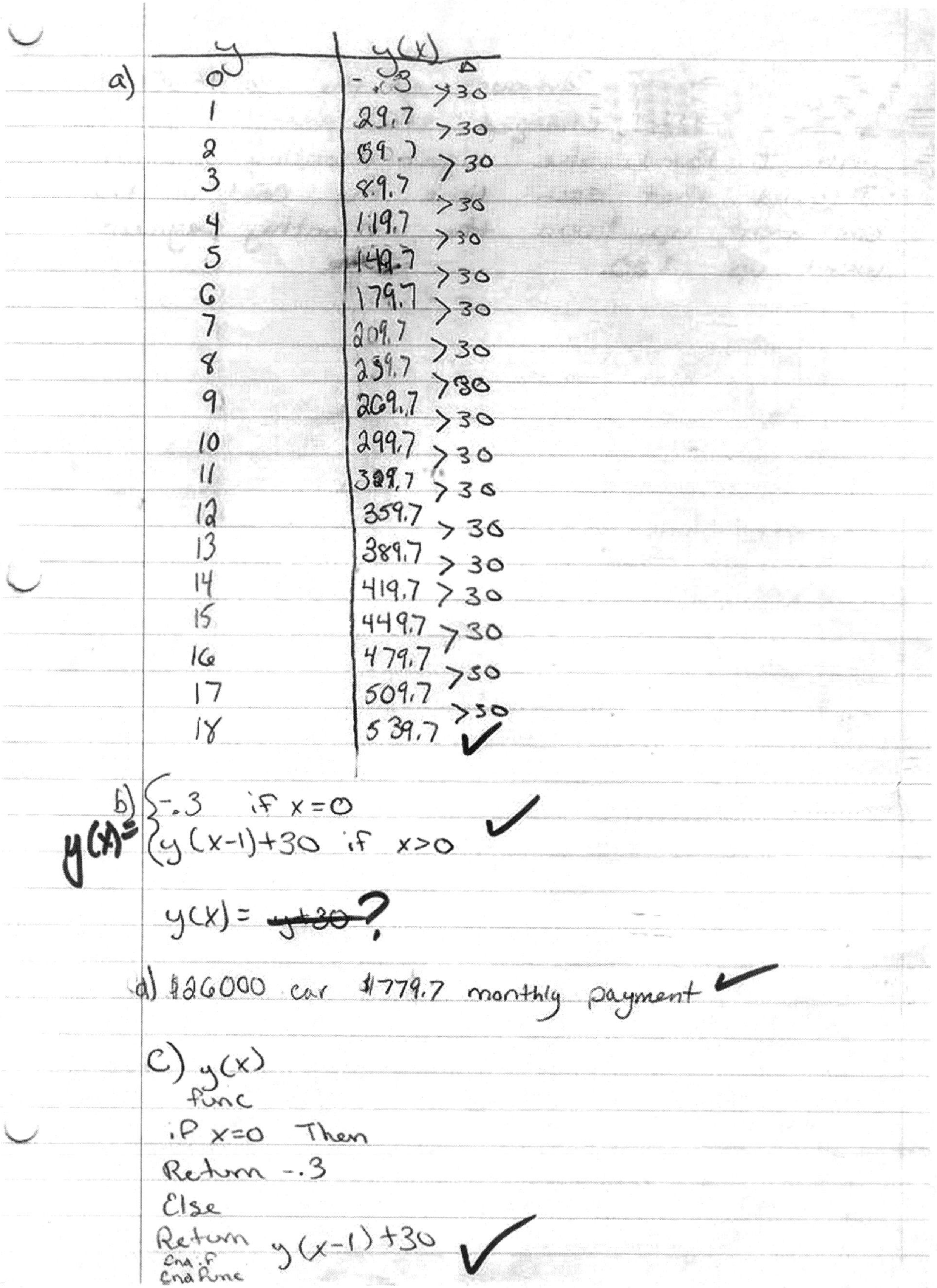

a)

y	y(x)
0	-.3	> 30
1	29.7	> 30
2	59.7	> 30
3	89.7	> 30
4	119.7	> 30
5	149.7	> 30
6	179.7	> 30
7	209.7	> 30
8	239.7	> 30
9	269.7	> 30
10	299.7	> 30
11	329.7	> 30
12	359.7	> 30
13	389.7	> 30
14	419.7	> 30
15	449.7	> 30
16	479.7	> 30
17	509.7	> 30
18	539.7	✓

b)

$$y(x) = \begin{cases} -.3 & \text{if } x = 0 \\ y(x-1) + 30 & \text{if } x > 0 \end{cases}$$ ✓

$$y(x) = \; \cancel{+30}\, ?$$

d) $26000 car $779.7 monthly payment ✓

c) y(x)
   func
   iP x=0 Then
   Return -.3
   Else
   Return y(x-1)+30 ✓
   Ena if
   Ena Func

Figure 1.

• I changed the amount of the cost of the car then I changed the monthly payment until I found the right monthly payment.

• I found that each time the cost of the car went up $1000 the monthly payment went up $30.

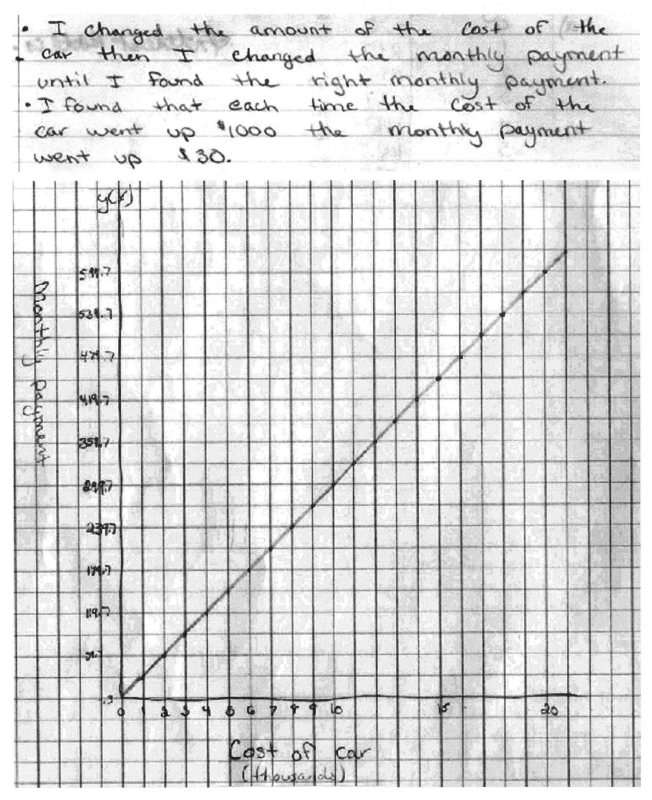

Figure 2.

and find that

$$m = 269.738\ldots.$$

A monthly payment of \$269.74 will do the trick. For more examples like this, see [8].

Much later in the year, when studying series, students develop the technique of unstacking a recursive definition to express it as a summation. They apply this technique to the problem of finding an explicit formula for monthly payments. The development goes something like this:

Suppose you borrow \$12,000 at 5% interest. Then you are experimenting with this function:

$$b(n,m) = \begin{cases} 12000 & \text{if } n = 0, \\ (1 + \frac{.05}{12}) \cdot b(n-1,m) - m & \text{if } n > 0. \end{cases}$$

Notice that:

$$1 + \frac{.05}{12} = \frac{12.05}{12}.$$

Call this number q. So, the function now looks like:

$$b(n,m) = \begin{cases} 12000 & \text{if } n = 0, \\ q \cdot b(n-1,m) - m & \text{if } n > 0 \end{cases}$$

where q is a constant.

Then at the end of n months, you could unstack the calculation as follows:

$$\begin{aligned}
b(n,m) &= q \cdot b(n-1,m) - m \\
&= q\,(q \cdot b(n-2,m) - m) - m = q^2 \cdot b(n-2,m) - qm - m \\
&= q^2\,(q \cdot b(n-3,m) - m) - qm - m \\
&= q^3 \cdot b(n-3,m) - q^2 m - qm - m \\
&\;\;\vdots \\
&= q^n \cdot b(0,m) - q^{n-1} m - q^{n-2} m - \cdots - q^2 m - qm - m \\
&= 12000 \cdot q^n - m(q^{n-1} + q^{n-2} + \cdots + q^2 + q + 1).
\end{aligned}$$

The last series is geometric; summing it, we get

$$b(n,m) = 1200\,q^n - m\frac{q^n - 1}{q - 1}.$$

Setting $b(n,m)$ equal to 0 gives an explicit relationship between m and the cost of the car that explains the conjectured linearity that was noticed months earlier.

This description of the development is a much compressed version of the informal and gradual development that students in M^2 experience. And, in fact, the last part of the development wasn't attempted in several of the field test classes.

The monthly payment activity is typical of the kinds of activities we sought in developing M^2: Activities with easy entry, fundamental use of technology, and opportunities to develop and prove conjectures, make connections, and experience some basic mathematical methods.

Lessons learned

Developing M^2 has been a learning experience for all of us on the staff. Here are some reflections now that the development is finished.

- Reviews of early drafts of the chapters often contained comments like "high school kids could never do this kind of thing." Field tests showed otherwise. The materials had to be revised, and in many cases completely reworked, but in no case did we need to water down the level of mathematics for either students or teachers. I'm convinced that traditional curricula expect far too little from teachers and students.

- A colleague and I taught sections of the field test at two different local high schools. Observing classes or delivering occasional lessons is important, but the experience of teaching the course every day, in an authentic school setting, seeing first hand both students' fundamental difficulties and breathtaking insights, informs the development process in a way that nothing else can. Three-hour arguments at staff meetings about the sequencing of problems or the potential engagement of students are settled in 10 minutes in a classroom.

- High school students take delight in their own mathematical thinking. In addition to a teacher advisory board, we convened a student board to help us with questions of readability and level. At the first meeting, we asked students what they liked most about M^2. Now, the students had at this point been fitting functions to tables, proving things by mathematical induction, and working on other activities that are usually classified as pure mathematics. Four of the students immediately responded to our question, almost in unison, with something like "It's realistic." When prompted to explain, it became clear that they liked doing *real work*—work that allowed them to think for themselves and to exercise their creativity. Realistic contexts were nowhere near as important to them as a realistic style of work.

- Along these lines, students want to see how mathematics is used, but applications are just as engaging from inside as they are from outside mathematics. Indeed, by not separating or distinguishing these types of applications from each other, students enlarge their definitions of "real world" to include mathematical contexts, and in the end, this makes them much more able to apply mathematical thinking to all kinds of situations.

- Students at all levels can do this kind of work. Much of the field test of M^2 was done in senior topics classes; my students were the weakest students in their school taking a fourth year of mathematics. Poor performance in mathematics courses has many causes, but lack of ability to think in a characteristically mathematical way is, for the vast majority of students, not one of them.

References

1. National Council of Teachers of Mathematics, *Curriculum and Evaluation Standards for Teaching Mathematics*, NCTM, Reston, VA, 1989.

2. National Council of Teachers of Mathematics, *Principles and Standards for School Mathematics*, NCTM, Reston, VA, 2000.

3. Cohen, D. (ed.), *Crossroads in Mathematics: Standards for Introductory College Mathematics Before Calculus*, American Mathematical Association of Two-Year Colleges, Memphis, TN, 1995.

4. CBMS, *The Mathematical Preparation of Teachers*, AMS, Providence, RI, 2001.

5. National Research Council. Everybody Counts: *A Report to the Nation on the Future of Mathematics Education*, National Academy Press, Washington, DC, 1989.

6. Cuoco, A., Goldenberg, E.P., and Mark, J., "Habits of Mind: An Organizing Principle for Mathematics Curriculum," *Journal of Mathematical Behavior*, 15(4), 1996.

7. Cuoco, A., "Mathematics as a Way of Looking at Things," In *High School Mathematics at Work*, National Academy Press, Washington, DC, 1998.

8. Cuoco, A. and Levasseur, K., "Classical Mathematics in the Age of CAS," In *Computer Algebra Systems in Secondary School Mathematics*, James Fey et al (eds), NCTM, Reston, VA, pp. 97–116, 2003.

9. Cuoco, A. and Manes, M., "When Memory Fails: Putting Limitations to Good Use," *The Mathematics Teacher*, 94(6), 2001. Preprint available online at <http://www.edc.org/CME/>.

10. Allen, L., "Teaching Mathematical Induction: An Alternative Approach," *The Mathematics Teacher*, 94(6), 2001.

11. Education Development Center, Inc., *Mathematical Methods: Topics in Discrete and Precalculus Mathematics*, © EDC, 2000 (to appear as part of *The CME Project*, a four-year high school curriculum, Prentice Hall, 2007). Development supported by NSF Grant ESI 9617369.

12. Mathematical Sciences Education Board and National Research Council. *Reshaping School Mathematics: A Philosophy and Framework for Curriculum*, National Academy Press, Washington, DC, 1990.

13. Jackson, A., and Rossi, H., "Views on High School Mathematics Education," *Notices of the American Mathematical Society*, 43(8), pp 866–873, 1996.

14. Education Development Center, Inc., *Connected Geometry*, Everyday Learning Corporation, Chicago, 2000.

26

Alternatives to the One-Size-Fits-All Precalculus/College Algebra Course

Bonnie Gold
Monmouth University

Introduction

How do we want our future legislators, our future news reporters, our country's future parents to feel about mathematics? Do we want them to believe it is a collection of rituals, requiring special skills only achievable by a few and of no practical value? Or would we prefer that they see mathematics as a way of describing many aspects of the world, central to many issues that will affect their lives, and a subject in which they can achieve whatever level of proficiency they need?

If the last mathematics course students take is a traditional college algebra or precalculus course, the vast majority of the students leave feeling defeated in their ability to do mathematics and mystified as to the use or value of the subject. As someone at the conference, Rethinking the Preparation for Calculus, said, "There's something very wrong if the last course a student takes in a subject is named 'pre' anything." And yet, there is a wealth of mathematics, accessible to students at this level, that is being applied to a wide range of contemporary issues.

Can one size fit all?

A cornerstone of the American democracy is that all children should be given equal opportunity. Unlike many countries that start directing children to different academic tracks by age 12 or earlier, American education treats all children as potential national leaders. We try to give them a mathematical background that allows them to become top scientists. As a result, the standard school mathematics track leads to calculus. While this may be a reasonable policy at the school level, by the time students arrive at college, they have become unequal in many ways. Some have been stimulated by their school mathematics, while others have been crippled by their early mathematical experiences. Some have a clear interest in a mathematically-intensive discipline, while others are clearly focused on the humanities, business or social sciences and others are still undecided. One size no longer fits all (if it ever did), in college mathematics courses.

At the conference, many felt that the problem is that precalculus (or college algebra) is trying to do too many things for too many audiences. It's trying to prepare students to major in mathematics and the sciences, to be the terminal general education mathematics course, and everything in between, and is doing none of this well. The solution I developed at Monmouth was to break up our college algebra course into several separate courses each with a clear mission and clientele. At institutions where the standard course taken by everyone is precalculus, one could do the same at that level.

When I came to Monmouth University in 1998, college algebra was our largest course. It was taken as the prerequisite for precalculus for students who didn't do well on the placement examination but intended to go on to calculus, as the prerequisite for the business mathematics sequence (a semester of combined linear programming and precalculus, followed by a semester of applied calculus), and was required of biology and social science majors who would continue on to a statistics course. Since all these disciplines required college algebra, advisors used it as the default placement for almost all students. Even students who entered planning on a major not requiring college algebra would thus not have to take a second mathematics course should they change their major. Only students who were absolutely decided on a major in the humanities were placed in the course (Quantitative Reasoning and Problem Solving) the department had developed for general education purposes. By the time I arrived, even our future elementary teachers were often taking college algebra as their only college mathematics course.

In our college algebra course, we try to give students all the algebra they need for calculus. So the course includes linear, quadratic, polynomial, rational, exponential, and logarithmic functions, solutions of equations involving all of these kinds of functions, simplifying expressions involving these functions, and so on. For students continuing on to precalculus (which adds trigonometric functions and does more with functions in general—graphing, inverses, transformations, and so on), all of this material is necessary. We use all of this material in our standard calculus sequence. But students in biology don't need to know how to solve $\sqrt{2x - 3} = 5 - \sqrt{x + 7}$. They do need to understand linear and exponential growth, and be able to recognize the distinction between them. They must understand the idea of a rate of change, have a bit of appreciation of effects of scaling, and know enough trigonometry to handle vectors in physics. Students in the social sciences need to be able to correctly use formulas in statistics involving sigmas, have an understanding of rates of change, including what units are involved in a particular context, and be able to use and interpret graphs and tabular data. Our business mathematics sequence uses linear, quadratic, exponential and logarithmic functions, but with much less symbolic manipulation than the standard calculus course. Because of the amount of time devoted to algebraic manipulation in the college algebra course, the students in these disciplines were not learning the things they actually needed for their majors and, since they mostly got Cs and Ds on the second or third try, they were not learning symbolic manipulation either.

An alternative: Splitting the course

The best way to break up college algebra or precalculus to serve client disciplines depends on the particular institution's programs and student body. At Monmouth, the college algebra audience was sent in four directions.

- For elementary education majors (who NEVER should have been in the course) we developed a new course, Foundations of Elementary Mathematics that, as the Mathematical Education of Teachers document (Chapter 1 in [1]) strongly recommends, gives these students a deeper understanding of the mathematics they will be teaching.

- For biology majors we developed two courses, Introduction to Mathematical Modeling in the Biological Sciences and Calculus for Biologists, only the first of which is required. We hope the better students in the first course will be inspired to continue to the second (and some do).

- For social science majors, we developed Mathematical Modeling in the Social Sciences, which also is an acceptable prerequisite for the business math sequence.

- For students who eventually go on to a standard calculus course, we retained College Algebra.

In addition, we retained Quantitative Reasoning and Problem Solving as our general education quantitative literacy course for students in majors without a specific mathematics requirement.

The two modeling courses are at the college algebra level. We study primarily linear and exponential models, with some time spent on quadratic models in Mathematical Modeling in the Social Sciences and on power function models in Mathematical Modeling in the Biological Sciences. These are topics from the college algebra course. However, we replace time spent learning symbolic manipulation by time spent looking at a range of applications related to the disciplines whose majors take the course, including how to interpret answers in terms of the original problem. We use many data-driven projects taken from the media or texts in their fields. We also make extensive use of the computer; each course involves at least seven computer lab projects done by students working in pairs. (This would work equally well with graphing calculators.) We use Excel as the computational tool, since it, or a similar spreadsheet, will be available to most of these students once they graduate and start working. (In addition, all students at Monmouth are required to take an introduction to information technology, which includes some work in Excel.) We assign for homework primarily the problems from the texts that are relevant to students' fields and use examples from those fields on exams as well as in labs.

Before developing these courses, I spoke with the chair of the biology department and then to the whole department at one of its meetings. I also spoke to the chairs of the social science departments, whose students had been taking college algebra as a prerequisite for statistics, and to the faculty member responsible for their statistics course, who is in the psychology department. I wanted to find out, from the perspective of the faculty in these fields, what they needed students to get from college algebra.

For the social sciences, the main skill they wanted was correct understanding of order of operations. They also felt that students in our general education offering didn't develop the necessary level mathematical sophistication. (The students in our quantitative reasoning course are, on average, even more math phobic than our social science majors. Since this is their last mathematics course, and the purpose is, in part, to make them less math phobic, we cover topics rather gently, which doesn't get them to the level of sophistication the statisticians in the social science departments wanted.) The biologists wanted a bit more—their students need some knowledge of exponential and logarithmic functions for chemistry and a bit of trigonometry for physics. I looked at what was available among current textbooks and designed a course for social sciences based on the Kime-Clark text [2], and one for biology based on the Crauder-Evans-Noell text [3]. Once I had a tentative syllabus and text, I again discussed the courses with the client departments to ensure the new courses would meet their needs. The departments were delighted to be consulted by the mathematics department about what they wanted for their students. They were very cooperative and helpful and promptly changed their requirements to make these courses required of their majors.

Students are much more responsive in these courses than they were in college algebra. They can see the use of the mathematics they are learning and are not overwhelmed by attempts to learn too many techniques in too short a time. There are some problems that don't go away. Most students take these courses in their first semester or two, and many haven't yet adjusted to college and the need to take responsibility for their own learning. But the old complaint - "Where will I ever use this?" - has gone away, and students view the courses as important to their future work. Because we're not trying to cram so much into the courses, there is time to make sure those who are working (which seems to be a higher proportion than in the old college algebra course) actually understand what we're trying to teach. Faculty seem to find these new courses less painful to teach because it's easier to interest students in the material.

Advising issues

The head of our advising center was very worried when we added these new courses, because many students come in undeclared or change their major in their first year or two. We've worked hard with the advising center to minimize situations in which students need to take an additional mathematics course if they change majors. If students come in genuinely undeclared (rather than simply vacillating between two majors), we suggest that they wait a semester before taking mathematics (unless they place into calculus, since all majors except elementary education will accept calculus as a substitute for their requirement). The education school (with the mathematics department's support) won't accept anything else as a replacement for the course we developed for future elementary teachers. We require students to be sophomores for that course, since by then they have started to think of themselves as teachers as well as students and take it more seriously. Thus, those undeclared majors who think they might teach at the elementary level are encouraged to wait until their sophomore year.

We give special help to the small number of students who start out majoring in the social sciences or biology and then change their major to a subject requiring calculus (mathematics, chemistry, computer science, software engineering). The two modeling courses are acceptable as replacements for each other. Because students get a course oriented toward their current interests, they take it more seriously than they would a course unrelated to their proposed major. Yet if they later decide to change majors, their graduation isn't delayed by the mathematics course they took.

My department feels strongly that a course emphasizing symbol manipulation, rather than concepts, is an inappropriate terminal mathematics course. Therefore, we removed the standard college algebra from the list of courses satisfying the mathematics component of the university's general education requirements. It still yields three credits toward the total number needed for graduation, but it doesn't satisfy the mathematics requirement. This got the attention of advisors fairly quickly and made implementing these changes relatively simple. In addition, the department secretary screens the list of students registered for college algebra and telephones all those whose major doesn't require calculus to warn them that college algebra does not meet the general education requirements.

Scheduling issues

One potential problem with having a range of entry-level courses is that it's much easier to fill students' schedules if there are many sections of a given course. There are a variety of ways one can partition the students flowing through college algebra or precalculus, depending on institutional enrollment patterns. Of the various college algebra alternatives we offer, only the course for biology majors has fairly few sections—three per year. To ensure that these fill and are offered at times biology students can take them, we schedule them in consultation with the chair of biology. For all other courses we have sufficiently many students to fill at least three sections per semester, and scheduling them has not been a problem.

Articulation issues

After our Undergraduate Studies Committee approved these new courses, I contacted our three main two-year colleges that feed into Monmouth University. I asked the chairs of their mathematics departments if I could visit them to discuss the changes we were making and explore what could be done in terms of their students who planned to transfer here. Of the three, the closest one was initially rather hostile to the new courses, since we are only one of three major client schools for their graduates. However, within a year they had developed a version of our social science course that meets our requirements and also meets the requirements of the main state university (Rutgers) where many of their students continue. One of the other two feeder schools already had a course similar to our social science version, and the last one was interested in at least trying one out. Two of the three schools were also interested in what we were doing with the elementary school teachers. They were not happy about our plan not to give general education credit for college algebra. On further investigation, it turned out that the lock-in articulation agreement to which Monmouth University has agreed promises that, if their students complete an associate's degree at their school, then their college algebra course will meet Monmouth's general education requirements. However, our students can't simply take a course at another school to get around our requirement—the agreement only applies to students who complete a full associate's degree.

In retrospect, it would have been better if I had contacted our neighboring two-year schools earlier in the process and involved them in the course development. I was new to an institution with a large transfer population, and this is one mistake I won't make again! But the chairs of the other schools were appreciative of my taking the time to visit them to discuss what we were doing.

Another way to cut the pie

Portland Community College has split college algebra into three versions: College Algebra for Liberal Arts, College Algebra for Business, Management, Life and Social Science, and College Algebra for Math, Science and Engineering, which have the same course number, but a terminal letter, A, B or C, to distinguish among them. All three courses cover linear functions investigated graphically, numerically, symbolically, and verbally as well as logarithmic, exponential, polynomial and rational functions. They differ primarily in the applications considered. The business and science versions also include solving systems of equations. College Algebra for Liberal Arts is considered a terminal course, but the other two are accepted interchangeably as the prerequisite for Elementary Functions. This latter leads to the standard calculus sequence, to Calculus for Management, Life and Social Science, as well as to Discrete Math I. Presumably most students planning to major in business or the social sciences start in College Algebra for Business, Management, Life and Social Science, while those going into the sciences start in College Algebra for Math, Science and Engineering. If they change their major, they can still continue without retaking a course. All three versions were designed to be transferable and are accepted by the Oregon universities. The state has guidelines that say what must be taught in a course in order to give it a specific title, for example College Algebra. These guidelines are fairly broad—thus allowing for the variants Portland Community College has developed—and consequently courses vary between institutions, although they all include the same core material. College Algebra for Liberal Arts fulfills the college-level mathematics requirement at the other universities, but may transfer in with a different title. The other two versions are accepted as college algebra.

Summary

Splitting college algebra into several courses more relevant to students' majors requires an understanding of the needs of the particular institution. It also requires a few days of effort talking with client disciplines. However, the time thus spent does wonders for the image of the mathematics department at the institution and results in much less grief for faculty teaching these courses, as students are more motivated to learn the mathematics involved. And, in the long run, it should also result in a less math-phobic and antipathetic public.

Acknowledgement: My thanks to Jerry Kissick at Portland Community College for information on their variants of college algebra.

References

1. *The Mathematical Education of Teachers*, v. 11 of CBMS "Issues in Mathematics Education", CBMS, Washington, DC, 2001; or online at http://www.cbmsweb.org/MET_Document/index.htm.

2. Kime, Linda, Judith Clark, and Beverly Michaels, *Explorations in College Algebra*, 3rd edition, John Wiley & Sons, Inc., New York, 2005.

3. Crauder, Bruce, Benny Evans, and Alan Noell, *Functions and Change: A Modeling Alternative to College Algebra*, 2nd edition, Houghton Mifflin Company, Boston, 2003.

Theme 6. Influencing the Mathematics Community

Changes are taking place. New materials are being developed. New pedagogies are being implemented. Data are being collected at schools here and there. Conferences are being held where colleagues meet to discuss the issues. Tools to assess student learning are being developed and new programs are being evaluated. Conversations are taking place between members of the mathematics community and colleagues in partner disciplines and between mathematicians who teach at the collegiate level and at the high school level. However, the biggest challenge confronting refocusing the courses below calculus is to launch a national initiative. This is a complex and immense undertaking. In this section, Bernard Madison stresses, once again, the importance of drawing on what was learned during the calculus reform efforts and building on our successes. Recalling that the sound-bite for the calculus movement was "a pump, not a filter," Bernard asks: "What is the headline for precalculus reform? Is there a nugget or snippet that will bring to mind the goals of this effort?" Naomi Fisher and Bonnie Saunders describe a national program to help mathematics and mathematical science departments strengthen their undergraduate programs. Sheldon Gordon summarizes what has been done-to-date to refocus the courses below calculus, and he outlines an ambitious, comprehensive, collaborative plan for implementing systemic change.

One of the reasons given to support the need for precalculus reform is the reform of calculus. Some argue that calculus has changed little, and some argue that what changes there have been have not helped. Others have quite different views. However one views calculus reform, it did serve to galvanize the community we address on curricular and pedagogical issues. This and significant societal changes, technology advances for one, make the challenges of precalculus reform very different from those of its ancestor, calculus reform. Nonetheless, our effort should draw strength and ideas from our past experiences while avoiding pitfalls and building on past successes. . . .

Proposals for change must be clear about subjects and goals. They must take into account the other priorities of the mathematics community and the forces that constrain actions. Proposals must be honestly presented and patiently tested. Most of all, in considering calculus and the preparation for calculus, we must not forget that most US citizens will never study or use calculus. Whatever is proposed must respect the general education of all students.

Launching a Precalculus Reform Movement: Influencing the Mathematics Community
Bernard L. Madison

The population of students falling into the category of those who need to take a mathematics course, but do not intend to advance to a calculus sequence is a widely varying group, which is difficult to classify. Even knowing who they are is difficult. If college algebra is the only option for them, they are mixed in with other students who intend to take calculus. Putting a name to the group is difficult, but for simplicity this paper refers to these students as rest-of-us students. . . .

In many math departments, the students enrolled in courses below calculus who do not intend to take calculus account for the majority of enrollment. The numbers alone make it critical that these students' needs be addressed. Beyond whatever mathematics knowledge and skills the students learn in these courses, they are forming attitudes, or reinforcing earlier attitudes, about mathematics that are likely to be fixed at this time.

Mathematics Programs for the "Rest of Us"
Naomi D. Fisher and Bonnie Saunders

Any initiative to refocus the courses below calculus must be a collaborative effort among the MAA, AMATYC, and NCTM, since each has a significant interest in these courses. However, a collaboration of this scale and complexity is unprecedented.... Perhaps the most important activity needed to launch a national initiative is to conduct a reasonably large-scale data collection project to collect and analyze information on the student population that is actually taking these courses.... Unlike the calculus reform initiative a wide variety of projects have already developed effective and innovative materials that bring new visions to the courses below calculus, including precalculus, college algebra, quantitative reasoning, and developmental algebra. What is needed is a cohesive plan to identify and publicize model programs that have adapted and implemented these projects.... What is needed is a comprehensive effort designed to influence members of the community to get them to try new versions of the courses themselves or to allow and encourage other members of their departments to do so.... To be successful, this initiative must reach out beyond the mathematics community for support and assistance.

Where Do We Go From Here? Creating a National Initiative to Refocus the Courses below Calculus
Sheldon P. Gordon

27

Launching a Precalculus Reform Movement: Influencing the Mathematics Community

Bernard L. Madison
University of Arkansas

Introduction

In 1982 I was chair of a mathematical sciences department that included statistics and computer science and a PhD program in mathematics. Resources other than booming student enrollments were scarce. In an act of frustration I wrote a letter to the presidents of the American Mathematical Society (AMS) and the Mathematical Association of America (MAA) asking for help in addressing the problems facing my department and many other departments. That letter opened the door to my involvement, over the next decade, in several interconnected efforts where influencing the mathematics community was a central and critical focus, including the Mathematical Sciences in the Year 2000[1] project, *Calculus for a New Century*,[2] *A Challenge of Numbers* [1], and *Moving Beyond Myths* [2].

The above and other initiatives in the 1980s and 1990s were successful in influencing the mathematics community to revitalize undergraduate mathematics and to reform calculus. The effects of these successes and the lessons learned bear heavily on efforts to influence that community again and improve the preparation for calculus. One of the reasons given to support the need for precalculus reform is the reform of calculus. Some argue that calculus has changed little, and some argue that what changes there have been have not helped. Others have quite different views. However one views calculus reform, it did serve to galvanize the community we address on curricular and pedagogical issues. This and significant societal changes, technology advances for one, make the challenges of precalculus reform very different from those of its ancestor, calculus reform. Nonetheless, our effort should draw strength and ideas from our past experiences while avoiding pitfalls and building on past successes.

The community

I mentioned community. The community I address consists largely of academic mathematicians—faculty at universities, colleges (four-year and two-year) and schools. This is a large and diverse community, with a variety of priorities. The 2000–2001 Combined Membership List (AMS, AMATYC, AWM, MAA,

---

[1]Mathematical Sciences in the Year 2000 (MS2000) was a joint project of the Board on Mathematical Sciences and the Mathematical Sciences Education Board at the National Research Council. The aim of MS2000 was assessment and renewal of US undergraduate mathematical sciences.

[2]*Calculus for a New Century* was the 1987 national colloquium that brought national attention to calculus reform.

and SIAM)[3] lists 54,030 individuals, with approximately 50,000 of these being members of AMS or MAA. The National Council of Teachers of Mathematics (NCTM) has approximately 80,000 individual members, including many elementary and middle school teachers. The community we need to address, or try to influence, in order to launch a precalculus reform movement, is mostly contained within this community of approximately 135,000. From one view, since preparation for calculus is contained in school mathematics, this issue is one for the 80,000 NCTM members. From another view, since much of the preparation for calculus does occur in colleges, and because colleges' curricula and courses are more susceptible to piloting reform efforts, the target audience could be considered within the 50,000 AMS and MAA members. It is this latter view that will dominate most of the analysis in this paper.

The professional activities within this community range from full-time attention to creating new mathematics in the research-intensive environments to full-time teaching of basic topics in arithmetic and algebra. Priorities for professional lives and measures of success are likewise spread over a large landscape.

There are pressure points for influencing this large and varied community, and I outline a few from my own experience—in national reform efforts, as department chair for ten years, as dean for ten years, and in various positions with the AP[4] Calculus program of the College Board for twenty years. In much of what I say I draw parallels with the calculus reform movement, which is part of the rationale for this effort and addressed essentially the same audience.

Need for change

There must be need for the proposed changes. Since the need for mathematics education is rarely only local, need for change is national. Since preparation for calculus implies a single purpose—getting ready for calculus—the argument for need has to hinge on the needs of calculus. So, should this be re-titled improving calculus? If rethinking the preparation for calculus has benefits other than those realized through calculus, then the argument needs recasting to include these other benefits. Not all students in precalculus courses go on to calculus. How will the reform courses serve them?

The arguments for the importance of calculus have been made far and wide. Some believe that these arguments have been made too well. They view calculus as a magnet with too much influence in school and college mathematics curricula. Arguments for improved calculus education rather than more calculus education mute much of this debate. Arguments for changes in precalculus that increase success in calculus are on the edge of the debate. Arguments for a larger share of the curriculum will probably fail; a smaller more efficient share would have more appeal.

Are there differences in the preparation for calculus for different student populations? College calculus attracts a variety of students with wide-ranging interests and education goals, from art majors to business and engineering students. Some students are in calculus for general education; others are there to learn the techniques for applications in other courses or in their chosen profession; still others are there because calculus is seen as a good testing ground for study in their selected majors.

The preparation for calculus has been mostly algebra and trigonometry, with considerable time spent on algebraic manipulation. Calculus courses have also emphasized the techniques of differentiation and integration. The need for these techniques has been reduced by technology, so some calculus courses have shifted to emphases on the concepts of calculus. Many of the students mentioned above will never use

---

[3] AMATYC = American Mathematical Association of Two-Year Colleges; AWM = Association for Women in Mathematics; SIAM = Society for Industrial and Applied Mathematics.

[4] Advanced Placement (AP) Calculus is a program of the College Board. Two AP Calculus courses, AB and BC, are taught in high schools following the AP Calculus course descriptions. Each year examinations are given and graded on a 5-4-3-2-1 scale with these grades being roughly equivalent to A-B-C-D-F. Many colleges and universities award college credit for one semester of college calculus for grades of 3 or better on the AB examination and two semesters of college calculus for grades of 3 or better on the BC examination.

the techniques of calculus, but the ideas of calculus—for examples, rate of change, approximation, and optimization—have relevance in many intellectual endeavors. So, perhaps the need for preparation for calculus is much less now, or much different.

Whatever the proposal, need must be demonstrated, clearly and cleanly, taking into account the goals of preparation for calculus. What are the consequences of not acting, and what will we get for our work?

Compatible with other priorities

Probably the most important consideration when seeking to influence the mathematics community is considering how what is being proposed fits with other priorities of this community. Some in the community have as their highest priority production of original research. Maintaining enrollment levels are critical at some institutions. Transferability of credit is critical, especially at two-year institutions. Teaching by graduate assistants is a big part of the precalculus instructional programs of many institutions. The needs of client disciplines are critical. How will improving the preparation for calculus interact with these and various other priorities? Will it take time away from research? Can the reformed courses be taught by graduate assistants and adjunct faculty?

For many students, the preparation for calculus occurs in school mathematics, and, in the view of many college faculty, all the preparation for calculus should be in school mathematics. Consequently, improving the preparation for calculus means changes in school mathematics, and several other forces come into play. Are teachers prepared to teach the proposed courses? (This was a major issue in changing AP calculus and requiring graphing calculators.) Exit testing from high school is in place in several states and being considered in others. College entrance and placement examinations must be considered.

Implementing any significant reform of precalculus courses will need to be linked with other efforts and take pressures and priorities into account. If these are not taken into account, they may become tethers to moving ahead with the reform efforts.

Empirical evidence

Improved education is difficult to document. Nonetheless, if the mathematics community is to be convinced to take a different path to calculus, that path has to be lighted by a few candles of successful innovations. These innovations have to pass some tests. Can they be multiplied and transferred to other student populations with other teachers? Will they require more resources, both fiscal and human? Are these resources available? Is the evidence of improved education strong enough or significant enough to warrant the risks of change? How will surrounding educational efforts be affected?

Although others know much more than I about the success of precalculus innovations, different approaches have different appeals, which is important when attempting to influence others. Among the approaches I know are:

- Just-in-time prerequisites
- Front-loaded prerequisites
- Independent value courses

Just-in-time calculus courses insert prerequisite material from algebra, geometry and trigonometry as needed. Front-loaded calculus courses cover prerequisite material in the first portion of the course. This model has been used in AP calculus courses in high schools. In fact, prior to 1998, the AB calculus course description included the 'A' material, which was prerequisite material. The 1998 course description dropped the 'A' material from the course description but kept the AB rubric. Both just-in-time and front-loaded models appeal because they focus on success in calculus, and strand fewer students with just the precalculus experience.

The third model consists of courses that have educational value beyond preparation for calculus. These include problem-based courses in algebra and trigonometry, where the precalculus material is learned in real-world contexts. The effectiveness of these courses versus the effectiveness of a streamlined route to calculus needs study. The route that has independent value for general education is much easier to sell to both students and policymakers.

Need a headline

What is the headline for precalculus reform? Is there a nugget or snippet that will bring to mind the goals of this effort? All of us have dismissed and decried these sound bite approaches to US politics and advertising, but we have used them big time in various efforts. In calculus reform, there were two, "A Pump, not a Filter," and "Lean and Lively." In current teacher education, Liping Ma's PUFM [3] acronym (which stands for Profound Understanding of Fundamental Mathematics) implies broad changes in what teachers need to know and how they need to learn. In the early days of the efforts to stimulate reform in collegiate mathematics, Ken Hoffman[5] had his own version of "It's the economy, stupid"[6] that demonstrates the need for a more focused headline. Hoffman's headline was "x is the problem." Some[7] would argue that this is indeed an appropriate headline for a precalculus effort. Hoffman claimed that everyone had his or her candidate for x and that x varied across a large domain that included teaching by graduate assistants, bad textbooks, and the public schools. As long as the discussion has this kind of dispersion, focused efforts will be deterred. Focus is necessary and the headline needs to capture that focus.

At the risk of stepping outside my charge to write about influencing the mathematics community, let me suggest that our current precalculus suffers from two ills: too many students and too much material. Many students take courses that have as their central purpose preparation for calculus, but these students never take calculus. Much of the material in courses prior to calculus is not necessary for calculus. Both these ills are rooted in the dual purposes of these courses—preparing for calculus and general education. Neither is served well. So, perhaps our headline is "less is more."

What are the issues?

Calculus reform was partially driven by changes in technology, by the introduction and rapid development of widely available computers and calculators. Is technology a major issue in the preparation for calculus? To what extent do we know how computing power, graphing technology, and computer algebra systems change preparation for calculus?

Another impetus for calculus reform was to slim down the topics, to reduce the size of textbooks— toward a leaner course. Are we trying to cover too much in precalculus? The second half of lean and lively is lively. Calculus courses were seen as needing more life, more discovery and less narration. Surely, lean and lively are issues for improving the preparation for calculus.

More contextual problems was a major issue in calculus reform. That may not be an issue here if the aim is the quickest route to calculus where the contextual problems can be attacked with both precalculus notions and calculus notions.

---

[5]Ken Hoffman was a faculty member at Massachusetts Institute of Technology but spent many years in Washington working with the Joint Policy Board for Mathematics, and the Board on Mathematical Sciences and the Mathematical Sciences Education Board at the National Research Council.

[6]In the 1992 Presidential Campaign, James Carville, a Clinton strategist, had this slogan posted above his desk at Clinton Headquarters in Little Rock. It became the headline of the Clinton strategy.

[7]In "What Mathematics Should Everyone Know and Be Able to Do?" (draft, 2001) Arnold Packer argues that one should not use x and y as variables names until the junior year of college.

Improved communication skills was an issue in calculus reform. Students needed to learn to read, write, and explain mathematics. Surely, improved communication is an issue in preparation for calculus.

Are there issues in improving preparation for calculus that are not issues in improving education in calculus? The articulation issue is different for precalculus since precalculus is available in all high schools. The calculator and computer issue is different, especially in precalculus courses with the single purpose of success in calculus.

Stamp of approval

Mathematics faculties and departments have many demands on their time and resources. Very often priorities are set by what is viewed as important by agencies or organizations they respect or give support to their work. Therefore any effort to influence this community needs the stamp of approval from some of these agencies or organizations.

The calculus reform movement received several stamps of approval in 1987. Two of the more significant ones were the Calculus for a New Century (CNC) colloquium (with Sloan Foundation support) in the bully pulpit of the National Academy of Sciences and the announcement by the National Science Foundation of a program of grants for calculus reform projects. The proceedings of the CNC colloquium were published by the MAA and soon thereafter calculus reform projects were underway at prestigious institutions (for examples, Harvard University and Duke University). The movement had prestige. Prestige combined with projects that had a range of institutions involved—universities, liberal arts colleges, and two-year colleges— paved the way for many more efforts at many different institutions.

The NSF support of the conference on "Rethinking the Preparation for Calculus" was a beginning, and the publication of this volume by MAA adds more luster. AMS is partnering with Mathematicians and Education Reform (MER) in efforts aimed at precalculus and other collegiate courses. These stamps will help. Now, real support is needed.

Real support

Calculus reform was an approved activity after the CNC colloquium in 1987, but without the impetus of the NSF program of grants for reform projects, the reform would have moved more slowly. Part of the reason for the colloquium was to generate proposals for the NSF program, and that effort was successful. One should not judge the effectiveness of the NSF program of grants solely on the reports of the NSF-funded projects. Many projects were carried out with support from educational institutions and other agencies or organizations. And one major change, the rewriting of the AP calculus course descriptions, was not NSF-funded, but no doubt was strongly influenced by NSF-funded projects.

Improving the preparation for calculus will require real support—money and energy—from a variety of sources. It is likely that much of this will need to come from educational institutions themselves since there is considerable competition for agency and foundation funds for reform educational efforts.

Top-down and bottom-up

Efforts to influence change in mathematics programs in colleges and universities are much more likely to succeed if they grow from within the mathematics faculty and have support and encouragement from the university administration. Calculus reform had strong support from university disciplines whose students study calculus, notably the physical sciences and engineering. Considerable work was directed at these client disciplines to secure their support. It was the president of the National Academy of Engineering who spoke the phrase "pump, not a filter" in his keynote to the CNC colloquium. This support translated to support from university administrations.

There is considerable interest among university administrators for improved courses in mathematics, especially precalculus mathematics. This interest centers on improved student learning and, as a consequence, better student retention. Leaner and livelier courses would probably be popular, and if they lead to more student success, they will receive strong support from many university administrators.

Bottom-up initiated changes are difficult in some situations. School mathematics teachers have far less freedom to change and to experiment with innovations than do college faculty; and two-year college faculty may be restricted from innovations because of transfer agreements.

AP calculus and technology: the levers of calculus reform

There were two major levers that affected the reform of calculus courses and significantly broadened its effect. These were: (1) the development of affordable graphing and calculators that support a computer algebra system (CAS) and (2) the revision of the AP calculus course descriptions to include many of the features of reform calculus courses.

The calculus reform movement began with two small conferences, one at Williamstown College in 1983 and another at Tulane University in 1986. The large national colloquium CNC was held in October 1987. About this time hand-held graphing calculators were appearing and soon became generally available at affordable prices. In 1989, the AP Calculus Development Committee (responsible for test construction, course descriptions and policy recommendations) began discussions about the use of calculators on the AP calculus examinations. Subsequently, scientific calculators were required in 1993 and 1994, and graphing calculators were required from 1995 onward. By 1999, calculators with CAS were permitted.

Because AP calculus was taught in high schools across the country to approximately a quarter-million students, the AP calculus policies have considerable immediate effect. The preparation for the implementation of the use of calculators in AP calculus was extensive and was centered on a program of TICAP[8] workshops held following the 1993–1996 AP calculus readings at Clemson University. This effort prepared hundreds of AP calculus teachers to teach thousands of students using graphing calculators.

Around 1994, the AP calculus Development Committee began discussions of rewriting the AP course descriptions both to reflect the use of technology and to incorporate some of the ideas of reform calculus courses in the AP calculus courses. A conference of reformers and traditionalists from the mathematics community was held to get advice, and consensus exceeded most expectations. A draft was produced, made public and discussed at several national meetings. In 1996 the revised course description was adopted for implementation in 1998, giving time for teachers to prepare. Some had considered the AP calculus program to be an impediment to reform since AP calculus is modeled after a typical college course. Quite the opposite was the case; AP calculus provided one of the first national recognitions of the reality of calculus reform.

Fifteen years separated the Williamstown Conference in 1983 and the rewritten AP calculus course descriptions in 1998. In between these two dates, there was considerable discussion within the mathematics community about reformed calculus ideas. The discussions were healthy and should serve to inform and inspire the reform of preparation for calculus.

What about strategy?

Advertising campaigns, political campaigns, or fund-raising campaigns usually have intricate and expensive strategies mapped out. Should a campaign to influence the mathematics community have a strategy? Probably, and the conference for which this paper is written should give the general structure of a campaign.

---

[8]TICAP= Technology Intensive Calculus for AP.

The MS2000 project was aimed at the assessment and revitalization of undergraduate mathematics. When MS2000 began in 1985–86, the calculus reform effort was still reasonably small. Revitalization of calculus, not necessarily reform, was a part of MS2000, but there were other surrounding issues: remediation, resources for departments, faculty supply and qualifications, precalculus mathematics, general education mathematics, and teacher education, to name a few. But in 1986 it became clear that the calculus reform movement was gathering steam, so the MS2000 project joined the calculus reform movement to sponsor the CNC colloquium. Some argued that this focused too much effort on one part of undergraduate mathematics. Others argued that if we could reform and revitalize calculus, then the effect would spread to other parts of undergraduate mathematics. The latter view prevailed, the saddled horse taken. History will see if that was correct.

The above describes a strategy of the MS2000 project that is essentially opportunistic in nature. The precalculus movement needs to be alert for such opportunities. The technology lever is still there; the AP lever is less relevant than it was for calculus. Better general education is a possible powerful argument, if, in fact, better preparation for calculus means better general education.

Whatever the opportunities, there needs to be public discussion of the issues. The conference, "Rethinking the Preparation for Calculus," is a part of that. The 1993 conference/workshop, "Preparing for a New Calculus," and the subsequent proceedings [4] have not received much attention. This effort was probably too early in the calculus reform effort; attention was still focused on what the new calculus would be.

Conclusion

In order to influence the mathematics community, proposals need to be well reasoned and important. Mathematicians have high standards for proofs and preponderance of evidence. They are busy with their research and teaching, and usually hold strong views on educational issues. Their experiences in learning mathematics are not typical. Most of them had little trouble with precalculus or calculus topics, and consequently they see limited reasons to make significant changes in the system that worked so well for them. On the one hand, the mathematics community is a tough audience; on the other, it is a rich source of questions and ideas that sharpen our arguments.

As I rethink the preparation for calculus, several questions surface. What is preparation for calculus? Is it twelve years of school mathematics or is it one school or college course? Can we afford the attention that calculus now receives? How can we make the system more efficient? How can we give more independent value to our courses? Why do we always seem to be going somewhere? How does precalculus take shape with full use of graphing and CAS devices? What's the best preparation for understanding the ideas of calculus?

Proposals for change must be clear about subjects and goals. They must take into account the other priorities of the mathematics community and the forces that constrain actions. Proposals must be honestly presented and patiently tested. Most of all, in considering calculus and the preparation for calculus, we must not forget that most US citizens will never study or use calculus. Whatever is proposed must respect the general education of all students.

Acknowledgements: The author is indebted to Tom Rishel who read a draft of this paper and made suggestions that improved the final version; however, all the mistakes are the author's.

References

1. Madison, Bernard L. and Hart, Therese A., *A Challenge of Numbers: People in the Mathematical Sciences*, National Academy Press, Washington, DC, 1990.

2. Committee on Mathematical Sciences in the Year 2000, National Research Council, *Moving Beyond Myths: Revitalizing Undergraduate Mathematics*, National Academy Press, Washington, DC, 1991.

3. Ma, Liping, *Knowing and Teaching Elementary Mathematics*, Lawrence Erlbaum Associates, Mahwah, New Jersey, 1999.

4. Solow, Anita (Editor), *Preparing for a New Calculus*, MAA Notes #36, Mathematical Association of America, Washington, DC, 1994.

28

Mathematics Programs for the Rest-of-Us

Naomi D. Fisher   and   **Bonnie Saunders**
University of Illinois at Chicago   *University of Illinois at Chicago*

Introduction

Beginning in 2001, the American Mathematical Society (AMS) and the Mathematicians and Education Reform (MER) Forum began a three-year project, funded by the National Science Foundation, to help mathematics/mathematical science departments strengthen their undergraduate programs. Entitled *Excellence in Undergraduate Mathematics: Confronting Diverse Student Interests*, the project focuses on the different groups of students in mathematics and their particular mathematical needs. The goal is to develop curricula and instruction that is valuable to these different student populations. The project seeks to build a network of departments that are committed to (1) reviewing and assessing how well their undergraduate program is working for their students, and (2) revising existing courses and developing or adapting new courses to afford all their students a meaningful experience in learning mathematics.

The heart of the project is six integrated 4-day workshops, two in each academic year, each of which will bring together faculty teams of 2 to 4 members from some 20 or more colleges and universities to discuss curricular, instructional and implementation issues relating to the mathematics offerings for different groups of students. The first workshop, *Excellence in Undergraduate Mathematics: Mathematics for the "Rest of Us,"* was held on December 6–9, 2001 at Arizona State University. It focused on students who fulfill their mathematics requirement with a course or courses below the calculus level. This student population, which accounts for the majority of instruction in most mathematics/mathematics sciences departments, represents a variety of academic interests.

Prior to the workshop, the departments were asked to complete a comprehensive departmental survey. The purpose of the survey was to obtain detailed profiles including:

- information about the student population of the institution
- information about the students enrolled in mathematics courses
- information about the mathematics faculty
- courses for students in the focus population of the workshop
- descriptions of three successful educational initiatives
- descriptions of three challenges faced by the department

This information was summarized and distributed to the participating departments as part of the follow-up of the workshop.

Note on "rest-of-us" terminology

The population of students falling into the category of those who need to take a mathematics course, but do not intend to advance to a calculus sequence is a widely varying group, which is difficult to classify. Even knowing who they are is difficult. If college algebra is the only option for them, they are mixed in with other students who intend to take calculus. Putting a name to the group is difficult, but for simplicity this paper refers to these students as rest-of-us students. The term refers to those students who need or want a mathematics course but do not intend to take calculus. The term rest-of-us course refers to any course that may be taken by rest-of-us students whether or not the course is intended to be on the calculus track. We will distinguish between rest-of-us students and other students in rest-of-us courses whenever possible.

Profile of participating departments

Thirty-three departments participated in the December workshop, which included departments from research universities (11), other comprehensive universities (11), four-year colleges (5), and two-year colleges (6). These departments represented seventeen different states from across the country as shown on the map.

Students

The total number of students represented by the participating institutions is over 200,000. These students represent a cross section of America as represented in the nature of the institutions and their geographical locations, and the cultural background of the students.

Departments do not often have available detailed information on the students enrolled in their classes, so it can be difficult to represent trends among students taking mathematics courses. It is particularly difficult to find information about rest-of-us students. Many of these students are mixed in the lower-level courses with students who *do* intend to complete a calculus sequence, and many of these students are marginal to the mathematics world. However, it appears that 42% of the students enrolled in a math course at the participating institutions are enrolled in a course that does not require calculus. In contrast, the number of mathematics majors is a slim slice of the pie at 2.6% of all students studying mathematics, but if we include only the pure mathematics major, the percentage reduces further to 1.4%.

Faculty

The number of faculty in the participating departments ranged from 11 at a small liberal arts college to 100 at one of the large state universities—the average was 31. Frequently, adjunct faculty do the majority of the mathematics teaching at this level. Among the reporting departments from four-year institutions, 33% of the faculty are adjunct employees. This figure varied from 0% at a small state university to 79%

at a private urban university. Looking at just the 2-year schools in the study, the percent of adjunct faculty jumps to almost 60%. All of the 2-year schools in the survey employ at least 30% adjuncts and one department reported that 83% of the faculty are adjunct employees.

Courses for rest-of-us students

In the initial survey of the departments, the departments were asked to describe the courses offered for students, who need to meet some mathematics requirement, but do not intend to take calculus. The following chart summarizes this information by type of course: developmental, college algebra, trigonometry and precalculus, statistics, finite math, elementary education, math for liberal arts, and modeling.

Course Type	# of schools reporting	Average class size	Comments
Developmental	16	22	Six departments reported more than one course of this type.
College algebra	14	29	Most of the departments felt that this traditional and remedial course was not working.
Trigonometry	12	39	Four departments listed separate trigonometry courses. One listed a college algebra with trig course. The others are precalculus courses. Most departments felt that these courses do not work for students who do not intend to take calculus.
Elementary education	9	29	Most of the departments reported hands-on learning, use of manipulatives, and problem solving as part of these courses. Five departments reported having a 2-course sequence.
Finite mathematics	9	30	Business majors usually take this course but there are courses designed for general education that includes education majors. Many students will continue to a calculus course, most likely a calculus designed for business majors.
Statistics	17	29	Many of the departments are developing statistics courses. Some are designed for general education requirements; some for other majors like business, biology, political science, geography, psychology, forestry, or environmental science. One department mentioned two courses.
Math for liberal arts	17	30	Two departments mentioned courses that are designed for future teachers. One course was interdisciplinary. Most mentioned general education requirements.
Modeling	6	21	Modeling was also often mentioned as part of the liberal arts courses.

The second column in the chart indicates the number of schools that reported at least one course of that type. It should be noted that, although all of the 25 departments that completed the survey reported at least one course, a department did not necessarily report all the courses offered at this level.

The third column reports the average class size. For large lectures this number represents section size. Class size varied from 10 or 12 at small schools to 45 or 60 at some of the larger universities.

Successes and challenges

Participating departments came to the workshop anticipating discussion of issues that were already of great concern to them. In advance of the workshop, each department was asked to identify and describe three successful educational initiatives in the department/institution and three initiatives that the department would like to develop to meet the needs of rest-of-us students.

We can identify four major areas in which the participating departments are attempting change to improve the quality of education offered to those students who will not be continuing in mathematics.

Technology

Fourteen departments listed improvements in some sort of technology: web-based learning, computer labs, graphing calculators, use of specialty software in courses.

Development of new courses

Nineteen schools listed development of specialty courses including quantitative literacy, modeling or other application-driven courses, and math for elementary education majors, technology and business. Nine schools mentioned new statistics courses at this level.

Changes in Pedagogy

Changes mentioned by the departments included new placement procedures, peer mentoring or resource centers, smaller class size, self-paced or group work alternative assessments, using real-life applications, and teaching problem-solving.

Collaboration with other institutions

Types of collaboration reported included long term education reform efforts with other colleges; articulation between colleges and community colleges, as well as interdisciplinary collaboration on courses and/or requirements within the institution.

Those areas where departments see future challenges often overlap with areas where others have already seen success. For example, some departments report success with their courses for elementary education majors while other departments are seeking to develop these courses. We summarize the challenges in several categories:

- **Adjunct faculty.** A large number of adjunct faculty are employed to teach rest-of-us courses. This provides many challenges for departments. Two departments mentioned better coordination of courses, for example, by developing strong coordinators, as a means of improving education with a large adjunct faculty.

- **Collaboration.** One department mentioned a university-wide goal of a new curriculum mandate to "help students learn to use and value the lenses of different disciplines, and seek the connections between them." Other forms of collaboration cited were the development of interdisciplinary courses or links between math courses and other disciplines, increasing the understanding of mathematics courses in other disciplines, understanding what is being done in the rest of the country, and merging statistics courses which are sometimes offered by each client department separately.

- **Courses.** The types of courses departments would like to see developed or improved include quantitative reasoning, alternative and less rigorous business courses, teacher preparation courses, and, more generally, reform oriented-courses.

- **Pedagogy.** Departments are looking for improved teaching methods including updated technology, project-based activities, more problem-solving in the classroom, and skill modules.

- **Concern for students.** Departments have concerns about inadequately prepared students, placing students into the right math course and high attrition rates. They are seeking techniques to motivate students, attract students to reformed courses, improve student attitude toward mathematics, reduce math phobia, and increase student participation in class.

Workshop program overview

The workshop program is aimed at helping departments better understand the issues underlying the creation and implementation of successful courses for the targeted student groups. In turn, this will help the faculty make informed decisions and plan changes thoughtfully. The program looked at issues and exemplary courses in several ways, including presenting department case studies, focusing on particular student-client groups, and discussing courses in depth. Subgroups of departments in peer institutions discussed what problems they face and what improvements they would like to make.

Thinking about the problem

In the opening session the participants were challenged to think about the interaction between the courses below calculus and the students who take one or more of these courses to complete the mathematics requirement of their majors. Participants were asked to consider a 2 by 2 matrix in which the columns indicate whether or not the course curriculum and instruction are satisfactory, and the rows indicate whether or not the students in the course are satisfactory.

Course

		S	U
Students	S	(1)	(2)
	U	(3)	(4)

This gives four possibilities: 1) both the course and the students in the course are satisfactory; 2) the students are satisfactory, but the course is not; 3) the course is satisfactory, but the students are not; and 4) neither the course nor the students are satisfactory. This breakdown suggests different ways to analyze the problem. For example, for the third possibility that the course is satisfactory, but the students are not, there are two options for adjustments, either change the course or change the students. The second option is not as outlandish as it appears. What it suggests is that the mechanisms for allowing students to take the course need adjustment. Making changes in advising procedures or placement exam, or both, would be ways to improve the situation. The most challenging category is the one in which the course is judged unsatisfactory and the students do not succeed. Some points that come up in considering these different scenarios are:

- developing the goals for a course involves a negotiation between the mathematical curriculum and recognized academic needs of the targeted students—keep in mind how these students will use mathematics in their subsequent studies.

- planning instruction so that the targeted students are likely to succeed.

Department case studies

Benny Evans, Oklahoma State University, Amy Cohen, Rutgers University, and Eric Kostelich, Arizona State University, presented department case studies of their respective departments.

Evans discussed how a new course, *Functions and Change*, was developed in response to the department's problems with college algebra. The department identified some eight different major areas represented among the students in college algebra, and then set out to talk with faculty in these client disciplines about the mathematical capabilities they would like their students to have. These conversations revealed that the college algebra course, even if students succeeded, was not preparing students for using mathematics in client disciplines. Some of the ways in which the new course is more relevant and useful to the students in the client disciplines interviewed are using linear, exponential and power functions, including the qualitative idea of rates of change throughout the course, and relying on the use of graphing calculators or spreadsheets.

Cohen explained how some of the political savvy and instructional expertise that the department had developed in running workshops as part of an intensive calculus course were channeled at the precalculus level. One example is improving the course on *Topics for the Liberal Arts*. The course coordinator, Charlie Sims, regularly updates the syllabus, sets the exercises, and briefs the instructors on expectations. As an instructional innovation, he has hired undergraduate graders, selected from graduates of the course who have earned A or B+; this works well for the faculty and enrolled students, as well as providing jobs for a non-standard group of mathematics undergraduates.

Kostelich spoke about the different viewpoints of administration and the mathematics faculty with regard to introductory courses. The administration wants students to be able to pass required mathematics courses in politically acceptable numbers; the math faculty wants courses with substantive content at a university level with broad applicability. The department developed a First Year Mathematics (FYM) program that is taught by a specially hired cadre of full-time instructors. The courses in FYM incorporate modern pedagogy and technology, and emphasize conceptual understanding as well as manipulation. The TAs assisting in the courses have a two-week training program to prepare them for assisting in these courses.

From the department case studies, it's clear that departments face many common problems including high failure rates, student dissatisfaction, and faculty dissatisfaction in teaching the targeted courses. A variety of strategies may be used to address the problems such as developing an alternative course for a targeted student population, dramatically changing the instructional format for the course, and incorporating successful innovations from other courses. Each of the departments has reaped benefits from investing in changes: greater student and faculty satisfaction with the course, tangible administrative recognition, and an improved reputation for the mathematics department. Among the lessons learned that other departments can heed is the importance of communicating with faculty in client disciplines, enlisting faculty support and avoiding faculty opposition, and developing good working relationships with the community colleges whose graduates enroll in the university.

Courses for targeted groups of students

Another aspect of the workshop program was highlighting two groups of student clients among those taking the courses below calculus to fulfill their mathematics requirements: students majoring in elementary education and liberal arts majors whose major does not require calculus. Cathy Kessel, a mathematics education consultant, and Rosamond Welchman, CUNY Brooklyn College, discussed the importance of keeping in mind *how* prospective elementary teachers learn mathematics as well as the mathematical content they study. Thus, courses for prospective teachers simultaneously should aim to deepen the students' understanding of mathematics and enable them to implement effective teaching strategies in school classrooms.

Michael Starbird, University of Texas at Austin, advocated a course aimed at liberal arts students that makes "great ideas of mathematics" accessible to these students. When having to make the choice between mathematical elegance of the presentation and the students' comprehension and appreciation, the students' positive experience should be the favored choice. In teaching about powerful ways of thinking mathematically to liberal arts students, the instruction should present the material to allow for the investigation of the ideas without depending on a lot of notation, for example.

Two fundamental issues and discussion sessions

The program also included discussions of preparing graduate students for their teaching responsibilities using case studies, and participation of under-represented groups in mathematics, as exemplified by the SUMS Project at Arizona State University.

Solomon Freidberg, Boston College, posed the question, "How can we stimulate mathematics instructors to understand and successfully embrace the challenge of teaching, especially non-mathsci students?" He advocated an experience-based approach based on group discussions and analyses of case studies that are designed to present a variety of pedagogical and communication issues. In support of this approach, he cited some results of the use of case studies from The Boston College Case Studies Project (BCCase), which Friedberg directs, at 19 colleges and universities. Graduate students, including foreign students, can discuss the case studies effectively. The case studies provide a window into American university culture for the foreign students, and promote thoughtful dialogue about teaching for all the participants. The case studies provoke spirited discussion among experienced faculty and help them continue to analyze their teaching.

Kate Sisulak, Arizona State University, talked about the Institute for Strengthening Understanding of Mathematics and Science (SUMS) an intense math and science program for high school students in which they are enrolled in a university level mathematics course for university credit. This nationally recognized project, directed by Joaquin Bustoz, has had a great impact in the participation of Native American students in mathematics, and on the number of mathematics majors at Arizona State University. Among other lessons that can be learned from this project is the critical role of counseling and mentoring students.

Breakout sessions gave the participants the opportunity to discuss particular courses in greater detail, and also to learn about the situation with respect to the lower level courses on other campuses represented at the workshop.

Participant assessment of the workshop

Participants tended to describe their expectations in coming to the workshop in terms of interest in ideas for a particular course or courses, and these courses varied from department to department. In contrast, their assessment of the workshop experience showed a greater appreciation of the issues under consideration, and interest in expanding the course offerings in their department. At the highest level, the workshop inspired and energized some participants. In these cases, participants recognized the importance of reaching students in the targeted courses, and that teaching these students can be exciting and positive.

The networking experience is particularly appreciated and worthwhile. It was valuable to meet with other departments that are concerned about the targeted courses. This affords a sense of perspective about the efforts of one's own department in comparison to other departments, an opportunity to revisit and expand ideas, and an opportunity to collect new resources.

Actions plans and follow-up to the workshop

At the closing session of the workshop, the members of each department team met to begin outlining plans to effect change in their departments. Many of the participants were very excited about the possibilities of making positive changes in how their department teaches rest-of-us students. The action plans were to be discussed within their department and/or institutions and implemented in the near future. Six months after the workshop many of the participating departments reported on their progress. A sampling of these initiatives by participating departments include:

- University of Portland, a private institution in Portland, Oregon, has implemented improvements in three courses: elementary statistics, a workshop precalculus especially for prospective middle school teachers, and a calculus course that includes timely precalculus material.

- Northern Arizona University has made improvements in their placement system and is field testing an "Interactive Notetaking System."

- Trinity College, a small liberal arts school in Washington, DC, is implementing changes in their curriculum that are designed to provide students with more choices in an environment where only a few courses can be offered.

- Miami-Dade Community College is implementing the course "Geometry for Educators."

- Roosevelt University, a private comprehensive university in Chicago, held a forum with all science and mathematics professors to discuss curriculum issues in the elementary mathematics sequence at Roosevelt.

- Rutgers University in New Jersey is introducing a course that will count for general education credit in the liberal arts colleges and that also is particularly valuable for prospective elementary school teachers.

- Southern Utah, a teaching university of 6,000 undergraduate students, has begun a Masters in Education with Emphasis in Mathematics where they will take teachers from where they are to a level IV (BS in Math Education). They are also trying to work with the school districts in their rural area to raise the level of mathematics.

Summing up

In many math departments, the students enrolled in courses below calculus who do not intend to take calculus, account for the majority of enrollment. The numbers alone make it critical that these students' needs be addressed. Beyond whatever mathematics knowledge and skills the students learn in these courses, they are forming attitudes, or reinforcing earlier attitudes, about mathematics that are likely to become fixed at this time.

One reason the numbers are so large is that a great variety of disciplines require one or more mathematics courses. But appropriate mathematical content and mathematical skills differ from one major to another, although being able to think conceptually, for example, should transcend all the courses. Mathematics departments should consider offering several courses at this level with each designed for one or more of the targeted student populations.

How the courses are taught is of equal importance as to what is taught. The method of presentation should take into account what purpose the mathematics serves for the students. Some instructional considerations are using technology, such as graphing calculators, to allow students access to problems that are beyond their skills in algebraic manipulations; using hands-on manipulatives suitable for elementary classrooms in teaching prospective teachers; and presenting and discussing ideas without unduly using specialized mathematical notation.

Recognizing that these students learn appropriate mathematics well should be a goal of mathematics departments. In addition, it would be well if these students completed their formal study of mathematics with a positive view of mathematics and a regard for its usefulness in their major field.

More details on this project can be found on the MER webpage, www.math.uic.edu/mer

Acknowledgements: The authors acknowledge the help of Tad Tyler for all his work in helping us prepare the departmental survey and analyze the results.

This project is based upon work supported by the National Science Foundation under Grant Number 0088794. Any opinions, findings and conclusions or recommendations expressed in this material are those of the authors and do not necessarily reflect the views of the National Science Foundation.

Appendix: Participating departments

AMS-MER Workshop
Excellence in Undergraduate Mathematics: Mathematics for the "Rest of Us"
Arizona State University, Tempe, AZ
Thursday, December 6–Sunday, December 9, 2001

Arizona State University, Tempe, AZ
Boston College, Boston, MA
Bowie State University, Bowie, MD
Portland State University, Portland, OR
Brooklyn College, Brooklyn, NY
Butler County Community College, Butler, PA
Eastern Michigan University, Ypsilanti, MI
Edinboro University of Pennsylvania, Edinboro, PA
Gallaudet University, Washington, DC
Grand Rapids Community College, Grand Rapids, MI
Loras College, Dubuque, IA
Miami-Dade Community College, Miami, FL
Mississippi State University, Mississippi State, MS
Northeast Mississippi Community College, Booneville, MS
Northern Arizona University, Flagstaff, AZ
Occidental College, Los Angeles, CA
Oklahoma State University, Stillwater, OK
Roosevelt University, Chicago, IL
Rutgers University, Piscataway, NJ
Maricopa Community College District, Scottsdale, AZ
Southern Oregon University, Ashland, OR
Southern Utah University, City Cedar City, UT
Texas Tech University, Lubbock, TX
Trinity College, Washington, DC
University of Arkansas at Little Rock, Little Rock, AR
University of California at Santa Barbara, Santa Barbara, CA
University of Illinois at Chicago, Chicago, IL
University of New Haven, West Haven, CT
University of North Carolina at Pembroke, Pembroke, NC
University of Portland, Portland, OR
University of Texas at Austin, Austin, TX
Voorhees College, Washington, DC
Wake Technical Community College, Raleigh, NC

29

Where Do We Go From Here? Creating a National Initiative to Refocus the Courses below Calculus

Sheldon P. Gordon
Farmingdale State University of New York

Background on subsequent activities

In the four-month period following the *Rethinking the Preparation for Calculus* conference in October 2001, there were three other invited conferences regarding the undergraduate mathematics curriculum.

1. CRAFTY's *Curriculum Foundations Summary Workshop* (November 2001), organized by Bill Barker and Susan Ganter and supported by the NSF and the Calculus Consortium for Higher Education [1]. CRAFTY (the MAA committee on Curriculum Renewal Across the First Two Years) had previously organized a series of 11 workshops in which leading educators from 17 different quantitative disciplines came together to discuss and inform the mathematics community of the mathematical needs of their students today. The summary workshop was held to unify the suggestions from the individual workshops.

2. The *Forum on Quantitative Literacy* (December 2001), organized by Bernard Madison, sponsored by the Woodrow Wilson Foundation and funded by the Pew Charitable Trusts [2].

3. *Reforming College Algebra* (February 2002), organized by Don Small on behalf of the MAA Task Force on the First College Level Mathematics Course and supported by the Consortium of Historically Black Colleges and Universities [3].

Each of these conferences focused on the mathematical needs of students in courses below calculus. Although the CRAFTY *Curriculum Foundations Workshop* did not look at these courses specifically, the recommendations from most of the quantitative disciplines were directed at courses such as college algebra and precalculus, because these are the courses that provide the mathematical foundation for students in most other disciplines. The final reports [4] were mailed to the chairs of all mathematics departments in the country.

It soon became clear that there was a need to bring together the principals from each of the four conferences to see to what extent there was a common philosophy among the four groups. The goal was to see if it was possible to channel the momentum from the four groups into a unified, national initiative that would refocus this portion of the curriculum that affects several million students each year [5]. With support from the NSF and the Calculus Consortium for Higher Education, a working meeting was held at the MAA headquarters in April 2002. The intent of this meeting was to:

1. Identify the common elements from among the four groups;
2. Carefully delineate the differences between them;

3. Prepare a formal report to the MAA Committee on the Undergraduate Program in Mathematics (CUPM) on the group's thinking about the courses below calculus. CUPM's 2004 Curriculum Guide [6] recommendations for the mathematics curriculum based much of its discussion on the courses below calculus on this report.

4. Prepare a comparable document informing the writing team that is in the process of revising the AMATYC *Crossroads Standards*; and

5. Plan toward a national initiative that would influence the mathematics community and other related constituencies to refocus the courses below calculus.

Subsequent working meetings of the group were held at all succeeding MAA national conferences.

Either in attendance at the working meeting or participating in the discussions prior to and following the meeting were principals from each of the four conferences. There were also official representatives of various MAA committees, including the Task Force on the First College Level Mathematics Course, CUPM, CRAFTY, the Committee on Quantitative Literacy (CQL), the Committee on Service Courses (CSC), the Committee on Two Year Colleges (CTYC), the Committee on Articulation and Placement (CAP), and the Committee on Professional Development. In addition, the presidents of AMATYC and NCTM, the head of the writing team for the revisions of the AMATYC *Crossroads* Standards, and the director of the Mathematical Sciences Education Board (MSEB) were also involved. In total over 40 individuals are presently involved in the working group for this initiative.

The common elements

At the April 2002 meeting, the participants focused entirely on the courses below calculus, most notably college algebra and precalculus and the relationship between these courses and quantitative literacy. They intentionally did not address remedial level algebra courses, leaving them explicitly for the AMATYC writing team, nor did they discuss any of the other mathematics courses at this level, such as statistics, finite mathematics, and survey of mathematics. A brief outline of these discussions is presented here; the complete report produced for CUPM is available from the author.

There was an amazing degree of convergence of thought and philosophy regarding all of these courses. This can be seen in the articles [1–3] describing each of the other three conferences as well as the other articles in this volume on the *Rethinking the Preparation for Calculus* conference. Perhaps the most impressive aspect is the fact that the identical themes and recommendations came from almost all of the quantitative disciplines represented in the *Curriculum Foundations* workshops [4]; moreover, most of those individuals were not at all aware of the efforts to revitalize calculus in the mathematics community.

For instance, the main points made by the physicists in their original report to CRAFTY were:

Conceptual understanding of basic mathematical principles is very important for success in introductory physics. It is more important than esoteric computational skill. However, *basic* computational skill is crucial. Development of problem solving skills is a critical aspect of a mathematics education. Courses should cover fewer topics and place increased emphasis on increasing the confidence and competence that students have with the most fundamental topics. The learning of physics depends less directly than one might think on previous learning in mathematics. We just want students who can think. The ability to actively think is the most important thing students need to get from mathematics education. Students need conceptual understanding *first*, and some comfort in using basic skills; then a deeper approach and more sophisticated skills become meaningful. Computational skill without theoretical understanding is shallow.

The engineers emphasized:

One basic function of undergraduate electrical engineering education is to provide students with the conceptual skills to formulate, develop, solve, evaluate and validate physical systems. Mathematics is indispensable in this regard. The mathematics required to enable students to achieve these skills should emphasize concepts and problem-solving skills more than emphasizing the repetitive mechanics of solving routine problems. Students must learn the basic mechanics of mathematics, but care must be taken that these mechanics do not become the focus of any mathematics course. We wish our students to understand various problem-solving techniques and to know appropriate techniques to apply given a wide assortment of problems.

The business faculty recommended that:

Mathematics is an integral component of the business school curriculum. Mathematics departments can help by stressing conceptual understanding of quantitative reasoning and enhancing critical thinking skills. Business students must be able not only to apply appropriate abstract models to specific problems, but also to become familiar and comfortable with the language of and the application of mathematical reasoning. Business students need to understand that many quantitative problems are more likely to deal with ambiguities than with certainty. In the spirit that less is more, coverage is less critical than comprehension and application. Courses should stress problem-solving, with the incumbent recognition of ambiguities. Courses should stress conceptual understanding (motivating the math with the 'whys'—not just the 'hows'). Courses should stress critical thinking.

Very similar sentiments were voiced by the representatives from business, industry, and government at the QL Forum regarding the mathematical preparation of students for today's increasingly quantitative workplace. These views are enunciated very forcefully in articles in the volume, *Quantitative Literacy: Why Numeracy Matters for Schools and Colleges*, edited by Lynn Steen and Bernard Madison [7].

In summary, there was total agreement among the working group members that college algebra and precalculus courses, as presently constituted with a primary emphasis on the development of algebraic skills, are not working for a variety of reasons, including:

1. At most schools, these courses have unacceptably high DFW rates.

2. These courses do not motivate large numbers of students to go on to further mathematics courses.

3. These courses do not adequately prepare most of the relatively few students who do go on to subsequent mathematics courses.

4. These courses do not serve the present-day mathematical needs of most other quantitative disciplines, where a deep level of conceptual understanding of mathematics is deemed more valuable than a very high level of facility in manipulating symbols.

5. These courses do not provide the students with the type of intellectual skills and understanding that are needed in the workplace or that would enable them to be effective citizens.

Yet, according to the data such as the most recent CBMS study [5], some two million students take these courses each year. The overwhelming majority of them take these courses only to fulfill some general education requirements that are imposed (and often prescribed in extreme detail) by people and groups outside the mathematics department. The working group believes that:

• These courses should have a solid algebraic spine, but algebraic techniques should not be the focus of the courses.

• These courses should have a strong emphasis on conceptual understanding and be deep intellectual experiences for the students.

• It is at least as important to prepare students conceptually for succeeding mathematics courses as it is to prepare them algebraically.

• These courses should focus heavily on mathematical modeling and realistic problem-solving, and that interpretation of results should be a vital component of an applied problem.

- Data analysis should be an integral part of all of these courses and should be used to connect the mathematics to its use in most other quantitative disciplines.
- Technology has an important and meaningful role to play in both the teaching and learning of mathematics.
- The development of writing and communication skills should be an important and significant aspect of these courses.
- The quantitative literacy theme should permeate all of these courses.

Developing a national initiative

The meeting in Washington (and the follow-up meetings) also addressed some longer-term strategies to bring about a climate in which change in the courses below calculus could take place. The working group developed a three-year plan to launch a movement to rethink and refocus these courses. To accomplish this, we hope to influence either NSF or the Department of Education or both to develop a funding program that is at least as large as the NSF's calculus initiative. Such an initiative would spur the development and implementation of new approaches to these courses, as well as adaptation of previously developed innovative approaches.

However, there are some significant differences between the initiative being organized to rethink the courses below calculus and the efforts to revitalize calculus during the 1990s. First, and perhaps most importantly, the changes proposed for calculus did not significantly change the content of the course—they did introduce some new topics, such as differential equations via slope fields; they changed the focus in the course to achieve a better balance between graphical, numerical, and symbolic approaches; and they introduced the use of technology to support both the teaching and learning of mathematics. But a reform calculus course was clearly recognizable as a calculus course by anyone in the mathematics community.

Some of the proposed changes to the courses below calculus go substantially further in terms of changing the very nature of the courses. Perhaps the greatest challenge to be faced is changing some very deep-rooted beliefs, both within and without the mathematics community. People who think of college algebra as consisting primarily of a collection of algebraic techniques to be practiced and mastered may not recognize some of the alternative courses as being *algebra* courses. NCTM has been working for years at the school level to broaden the definition of algebra to encompass all types of algebraic reasoning and algebraic representations, as well as just symbolic operations. The same kind of effort will be needed at the college and university level.

A second deep-rooted belief held by many in the mathematics community is that college algebra and precalculus courses exist primarily to prepare students for calculus and, more indirectly, to produce the next generation of mathematicians. Underlying this outlook is the strong belief among many mathematicians that it is necessary for *all* students to replicate as much as possible of their own mathematical training. But when significant changes are made to the courses below calculus, that replication may no longer occur. The data shows that the reality is quite different from this perception, given that so few students taking these courses ever go on to take what traditionally was considered freshman mathematics. But changing the attitudes will represent a true change in the culture of mathematics education for many. One way to do this is to identify some successful models by which students can move toward and into higher mathematics without having a strong algebraic theme in the preparatory courses. Certainly, relatively few professional mathematicians ever use the full array of algebraic techniques they learned outside the introductory mathematics classroom.

Another major difference between the calculus revitalization movement and the proposed initiative to refocus the courses below calculus is that the former was basically an academic effort—the key was to convince other mathematicians and some people in a handful of allied disciplines of the need to change some aspects of calculus. However, the proposed initiative necessarily extends well beyond the academic arena. In some states, general education requirements, including the specific course content and allowable textbooks, are specified by state education departments and even state legislatures. In many university systems, the courses are specified by academic senates or other external bodies. At many institutions, particularly two-year colleges, transfer and articulation agreements limit the changes that can be made in courses. There are many individuals who are subjected to such external requirements, who express tremendous frustration at not being able to change the courses they give to better serve the needs of their students. One major challenge we face is to convince these external bodies to change some of the requirements that they have laid down.

Still another difference between the calculus revitalization efforts and a movement to refocus the courses below calculus is that the former basically started at ground zero. The initial stages of the calculus movement involved the development and testing of new materials that implemented the ideas being discussed. As one unanticipated outgrowth of the effort at revitalizing calculus, various individuals have already developed a variety of alternatives to the traditional courses below calculus; descriptions of many of those projects and the materials they have developed are included in this volume.

To develop a national initiative to refocus the courses below calculus, the following activities, which fall into several distinct categories, have been planned.

A. Developing an organizational structure

Any initiative to refocus these courses must be a collaborative effort among the MAA, AMATYC, and NCTM, since each has a significant interest in these courses. However, a collaboration of this scale and complexity is unprecedented. Thus, it is necessary to develop an appropriate structure that will satisfy the needs of the three organizations, as well as to identify the special roles that each organization will play. One suggestion that has been made is for the three society presidents to appoint a joint presidential commission to assume responsibility for the initiative. Within MAA, the overall responsibility for the initiative has been given to CRAFTY. With a functioning organizational structure in place, the working group has an official mandate from the three societies that will enable the group to approach various groups (governmental and others) outside of the mathematics community that need to be persuaded to encourage a new focus for these courses. The official mandate will also help in obtaining the large scale funding needed for this initiative through the organizations, rather than through an ad hoc group.

As part of the organizational issues, we also have to take the following steps.

- The 40 member working group already involves the official participation of a large number of MAA committees, including the Task Force on the First College Level Mathematics Course, CUPM, CRAFTY, CAP, CTYC, CSC, CQL, and the Committee on Professional Development. We need to forge links to other MAA committees, such as the Committee on the Mathematics Education of Teachers (COMET), the Committee on Sections, the Committee on Mathematics Across the Disciplines, and the Committee on Industrial and Government Mathematicians. We also need to forge links to comparable groups in AMATYC and NCTM.

- Increase the links with the writing team that is currently revising the AMATYC *Crossroads Standards*.

- Develop greater connections with MSEB that go beyond the present involvement of its past-director, Carole LaCampagne, in the working group.

- Develop a network of people who are interested in this initiative. One special emphasis will be to involve the Project NEXT fellows, as well as the participants in the joint AMATYC/MAA NEXT project for two-year college faculty. Many of the Project NEXTers have been teaching for as long as 10 years now and have reached a stage in their careers where they may welcome the opportunity and challenge of becoming involved in a national initiative.

- Develop a regional structure to coordinate efforts at the local section and affiliate level.

B. Collecting data

Perhaps the most important activity needed to launch a national initiative is to conduct a reasonably large-scale data collection project to collect and analyze information on the student population that is actually taking these courses. Such a project should determine:

1. Who are the students who take these courses and why do they take the courses? What are their majors? What requirements do the courses fulfill?
2. Where do the students come from? Do they come from the prerequisite courses given at the same institution, from prerequisite courses at a feeder college, or from high school? How many are actually repeating the course?
3. How do the students do in these courses?
4. What subsequent math courses do these students take and how do they do in the follow-up courses?
5. How do the students do in subsequent courses in other quantitative disciplines?

Such data have been collected at several institutions. For instance, Steve Dunbar has been conducting such a study at the University of Nebraska for more than 12 years, tracking each of the approximately 120,000 students who have taken mathematics. He has found that, of the students who successfully complete their college algebra course, only about 10% ever go on to start calculus I; less than 1% go on to start calculus II; and virtually none has started calculus III. Mercedes McGowen has found very similar results at William Rainey Harper College, a large two-year institution. Both studies are reported on in detail earlier in this volume [8, 9]. Very comparable results (though unpublished) have been found at other schools. Yet, the underlying philosophy on most campuses is that college algebra courses are intended to prepare students for mainstream calculus.

If appropriate funding can be obtained, we would like to provide small grants to many schools to assist them in obtaining the appropriate data and preparing reports and possible articles.

In addition, it would be very helpful to conduct a separate project to determine just what mathematical ideas and techniques are needed in other disciplines and in the workplace. We need to know precisely what the consumers of our students really need and want. Much of that information regarding the needs of the other disciplines has begun to emerge from the *Curriculum Foundations* project; some of the comparable information on the needs of today's business, industry, and government was reported on at the *Forum on Quantitative Literacy*.

C. Identifying and publicizing model projects and programs

Unlike the calculus reform initiative, as mentioned before, a wide variety of projects have already developed effective and innovative materials that bring new visions to the courses below calculus, including precalculus, college algebra, quantitative reasoning, and developmental algebra. Many are described elsewhere in this volume. Each project has its own unique vision, but each vision is clearly in the spirit of what the working group believes is appropriate for the students who take these courses. These projects are also consistent with the visions that underlie the new CUPM Curriculum Guide, the AMATYC *Crossroads* Standards and the NCTM Standards for courses at these levels. Each project has the flexibility to be easily adapted to many different settings and to the needs of many different groups of students and related disciplines.

What is needed is a cohesive plan to identify and publicize model programs that have adapted and implemented these projects. In particular, we need to:

- Identify a variety of institutions that have refocused their college algebra and related courses to reflect the kinds of changes that need to be brought about. These institutions would serve as model programs to prove the existence of successful efforts. They would also provide specific information on student performance in such courses and in succeeding courses in mathematics and other fields, as well as information on implementation issues.

- Stimulate the adaptation and implementation of existing projects at new sites.
- Stimulate the expansion of model programs at institutions that have already implemented new versions of courses in a few sections of one or more courses.
- Stimulate the development and implementation of several new model projects through a series of grants that could be awarded as part of the funding for planning and development phases of this initiative, as discussed below.
- Track student performance and attitudes throughout these model projects and programs over several years to develop some specific data to help in on-going efforts to develop the national initiative.
- Encourage individuals at each model institution to write articles describing their experiences and results, to give presentations on their efforts, to give workshops, and to disseminate any curricular materials they may have developed to help in their implementation efforts.

At MAA, the Committee on Service Courses has accepted the responsibility to spearhead this effort with the assistance of CRAFTY, CUPM, and CTYC.

D. Influencing the mathematics community

Perhaps the greatest obstacle to widespread change in the courses below calculus is the attitude that many in the mathematics community hold toward these courses—that they exist to prepare students for mainstream calculus. What is therefore needed is a comprehensive effort designed to influence members of the community to get them to try new versions of the courses themselves or to allow and encourage other members of their departments to do so. These efforts should include:

- Organize panel sessions at national meetings of MAA, AMATYC, NCTM, and ICTCM.
- Organize contributed paper sessions and poster sessions at national meetings to provide the opportunity for individuals to report on the results of their innovative efforts.
- Produce volumes in the MAA Notes series and similar series. Some recent and forthcoming volumes that address these issues are:
  1. The current volume of the proceedings of the *Rethinking the Preparation for Calculus* conference
  2. The final results of the *Curriculum Foundations* project, *Voices of the Partner Disciplines*, edited by Susan Ganter and Bill Barker
  3. A series of volumes coming out of the *Forum on Quantitative Literacy* edited by Lynn Steen and Bernard Madison
  4. The CUPM curriculum recommendations
  5. A volume that we would like to see produced is an MAA Notes volume consisting of reports from, or on, each of the model projects and programs mentioned earlier.
- Articles and reports in a variety of journals. For instance, The AMATYC Review recently devoted a special issue to changes in college algebra.
- Panels at regional meetings of MAA sections, AMATYC affiliates, and NCTM affiliates.
- Develop strategies for faculty development for the part-time faculty who teach these courses at most two- and four-year institutions.
- Develop strategies for faculty development for the TAs who teach these courses at most universities.

At MAA, responsibility for the faculty development efforts has been assumed by the Committee on Professional Development's PREP program. At AMATYC, the challenges associated with faculty development to support refocusing these courses is one of the most important goals of the organization's leadership.

E. Forging connections

To be successful, this initiative must reach out beyond the mathematics community for support and assistance. In particular, we see the need to:

- Develop links to other disciplines to gain their support for the initiative. This has already begun through the *Curriculum Foundations* workshops and the *Forum on Quantitative Literacy*. The external support, most likely from the social and life sciences, will be as critical to this initiative as the support of the engineering disciplines was to the calculus effort. A second round of Curriculum Foundation workshops involving many of these disciplines is being undertaken by the MAA's Committee on Mathematics Across the Disciplines.

- Develop links to business/industry/government to seek their support from the point of view of preparing students for today's workplace and for effective citizenship; this will also be critical in building a case for change.

- Develop links to other potentially concerned groups, such as AMS, ASA, SIAM, MER (Mathematicians and Education Reform), and the Consortium of Historically Black Colleges and Universities.

At MAA, responsibility for these efforts has been assumed by the Committee on Mathematics Across the Disciplines and the Committee on Quantitative Literacy, with assistance from CRAFTY and CTYC.

F. Planned outcomes

The scope and complexity of the challenges of refocusing the courses below calculus are immense. There are many different aspects that need to be addressed that go well beyond convincing the mathematics community to rethink these courses. Some of these challenges include the need to:

- Develop links to the testing industry to stimulate the development of a new generation of placement and related tests that better reflect the content of NCTM *Standards*-based curricula in the high schools and refocused versions of college mathematics offerings.

- Develop plans to rethink the mathematical training of prospective teachers, who are often required to take college algebra as their final mathematics experience. Such efforts need to be made in conjunction with COMET, the PMET project, NCTM, and AMATYC.

- Develop materials that could be used to influence college and university administrators to encourage and support, both academically and financially, implementation of projects that refocus the emphasis in college algebra and precalculus courses. This would include some official statements from the professional organizations, a position statement on the need to change the courses, and supporting evidence about the success and benefits of some of the model projects.

- Develop a generic PowerPoint presentation that could be used, or adapted, by many people who will be giving talks and presentations on the goals of the initiative.

- Develop materials that can be used by individuals in a region to influence state agencies that are responsible for statewide or university system-wide mandates about which courses students must take or what the content of such courses must be.

- Affect transfer and articulation agreements between individual schools and across university-wide systems.

- Develop a series of faculty development workshops to prepare faculty to teach these new versions of the courses. These workshops would be held in conjunction with national and regional meetings of MAA, AMATYC, and NCTM. They could also be offered directly to departments and collections of faculty from neighboring schools to assist them in completely changing their programs. The latter will be an expansion of AMATYC's program of travelling workshops on technology and implementation of the *Crossroads* Standards, of the MAA's PREP program offered by the Committee on Faculty Development, and of a series of comparable workshops currently being offered by the MER network.

- Influence the publishing industry to publish-and-nurture innovative textbooks instead of the publish-and-allow-to-perish policy that occurred with most of the innovative calculus texts that emerged.

- Create a website for the exchange of information and ideas regarding the initiative.

- Influence either the NSF or the Department of Education to develop a large scale funding initiative to fund the development and implementation of new approaches to the courses below calculus. The influence will be based on the data being collected, the model programs and projects, calls for change from within the mathematics community in general and the three professional organizations in particular, calls for change from the other disciplines, and calls for change from business, industry, and government.

Clearly, all of these activities will require the investment of much time, effort, and considerable money. One of the first efforts by our group will be to seek a large grant to fund the planned activities. The NSF has already provided a relatively small grant to help get the initiative organized and off-and-running.

Acknowledgment: The work described in this article was supported by the Division of Undergraduate Education of the National Science Foundation under grants DUE-0089400 and DUE-0310123. However, the views expressed are not necessarily those of the Foundation or the projects.

References

1. Barker, William and Susan Ganter, "Fundamental Mathematics: Voices of the Partner Disciplines," in this volume.

2. Madison, Bernard, "Preparing for Calculus and Preparing for Life," in this volume.

3. Small, Don, "College Algebra: A Course in Crisis," in this volume.

4. Ganter, Susan and William Barker, editors, *Curriculum Foundations Project: Voices of the Partner Disciplines,* MAA Report, Mathematical Association of America, Washington, DC, 2004.

5. Lutzer, D., Maxwell, J. and Rodi, S., *Statistical Abstract of Undergraduate Programs in the Mathematical Sciences in the United States: Fall 2000 CBMS Survey,* American Mathematical Society, Washington, DC, 2002.

6. Pollatsek, Harriet, et al, *Undergraduate Programs and Courses in the Mathematical Sciences: CUPM Guideline 2004*, Mathematical Association of America, Washington, DC, 2004.

7. Steen, Lynn and Bernard Madison, editors, *Quantitative Literacy: Why Numeracy Matters for Schools and Colleges*, National Council on Education and the Disciplines, Princeton, NJ, 2003.

8. Dunbar, Steven, "Enrollment Flow to and from Courses below Calculus," in this volume.

9. McGowen, Mercedes, "Who are the Students Who Take Precalculus?," in this volume.

Ideas and Projects that Work: Part I

Just as our students say, "Show me an example," participants at the conference, *Rethinking the Preparation for Calculus*, asked for examples of ideas and projects that work. The authors of the five papers in this section discuss some of the larger issues connected with curricular materials that they have developed. The next section, "Ideas and Projects: Part II," contains fifteen short, descriptive papers for particular projects.

In this section, Doris Schattschneider describes an alternative one-year college calculus course that integrates a review of precalculus concepts to help students, who have taken precalculus but are not prepared for calculus, succeed in calculus. Bill Fox reiterates a common theme in this volume—the need to work with our colleagues in other disciplines—as he describes a new college algebra course that is based on modeling and applications. Dan Kalman describes a course in elementary mathematical models for the general education student that is a hybrid of two approaches, college algebra and liberal arts math. Brigitte Lahme, Jerry Morris and Elias Toubassi present a case for integrating laboratories into a precalculus course. Gary Simundza describes the "Fifth Rule" for helping students understand mathematical concepts: the direct experience of mathematical processes.

A course that integrates precalculus review with calculus takes careful planning, but does not require fundamental changes in calculus content or teaching style. An integrated approach can be used in calculus courses anywhere in the spectrum from traditional to reform, and in special calculus courses such as those designed for majors in business or the biological sciences. It can be used in a technology-dependent course or without any technology at all.... There is now a small core of institutions (perhaps 40 or more) offering a version of this course using materials we developed. It is a quiet revolution, but certainly has made a difference where it has been tried.

College Precalculus Can Be a Barrier to Calculus: Integration of Precalculus with Calculus Can Achieve
Success
Doris Schattschneider

The most important challenge facing our graduates in the future is data overload. Data and statistics are found everywhere: newspapers, magazines, television, and radio. There is no escape.... As a result, we proposed a different sequence of courses based upon mathematical modeling and applications using college algebra. It seemed that the goal of the college algebra of the past was to prepare students for calculus, but most would never take calculus.... We felt this new course sequence would better serve our students by preparing them to solve problems that they may face in their own disciplines and in real life.

College Algebra Reform through Interdisciplinary Applications
William P. Fox

It would be difficult to overstate the importance of algebra in mathematics. If mathematics is the language of science, then algebra provides the alphabet, vocabulary, and syntax. In particular, the traditional college algebra course covers the elementary functions of analysis: linear and quadratic functions; polynomial and rational functions; roots, exponentials, and logs. These functions are inescapable in the most elementary applications of mathematics to other subjects. In a word, they are ubiquitous all over the place....

Liberal arts mathematics courses offer an alternative exposure to mathematics for general education students. These courses emphasize lofty educational goals, hoping to communicate something of the beauty, power and fascination of our discipline. The development of specific manipulative skills is given much less emphasis. This essay concerns a hybrid of these two approaches to general education mathematics.

Elementary Math Models: College Algebra Topics and a Liberal Arts Approach
Dan Kalman

The purpose of the labs is to have students investigate real-life situations that can be addressed with the tools learned in precalculus. The labs provide students with an opportunity to step back from the exercises that are in the text and allow them to consider problems that require more from them in terms of critical thinking skills. While many of the exercises in the text are one-step calculations asking for a single answer, a typical lab project asks questions in an open-ended way that requires students to first identify the correct tools to use and then to use them in an appropriate way. In addition, these labs give students a chance to discuss mathematics in small groups, debate strategies, convince fellow students of the validity of their arguments, and then put it all together into a polished, written report detailing their findings and conclusions. The labs also allow students to explore mathematical ideas from different points of view—graphically, verbally, numerically, and symbolically—which helps students solidify their understanding of the concepts.

The Case for Labs in Precalculus
Brigitte Lahme, Jerry Morris and Elias Toubassi

Since the beginning of the calculus reform movement, the Rule of Three has become an essential feature of mathematics education.... Shortly after the Rule of Three construct was widely adopted and became part of the mathematics educator's lexicon, it evolved into the Rule of Four. If analytical, graphical, and numerical representations are all different lenses through which mathematical relationships can be viewed, they are, in many cases, language-independent substitutes for connections or processes that can be described in words.... There is reason to suggest that the Rule of Four should be broadened to a "Rule of Five": the fifth component of mathematics understanding is direct experience of mathematical processes. That is, in addition to providing opportunities for obtaining usable data through active experimentation, the act of experimenting itself can promote mathematical concept acquisition.

The Fifth Rule: Direct Experience of Mathematics
Gary Simundza

30

College Precalculus Can Be a Barrier to Calculus: Integration of Precalculus with Calculus Can Achieve Success

Doris Schattschneider
Moravian College

Introduction

The articles collected in this volume make one fact very clear: precalculus means many things to many people, and serves many different functions. One of the primary functions of a precalculus course (whether in high school or in a college or university setting) is to serve as a preparation for the first course in calculus. It is this "preparing for calculus" course, specifically in the college or university setting, that we assert is failing in its stated purpose. The dismal percentage of those who complete precalculus and continue on to complete the calculus I course makes it clear that something is very wrong. Tinkering with content, technology, and pedagogy may improve the precalculus course, but these repairs do not address several fundamental problems. We discuss here an alternative one-year college course that integrates the review of precalculus topics on a just-in time basis; it has shown documented success in helping a larger number of students succeed in completing calculus I.

The problem and a response

In recent years, as mathematical techniques have gained increasing importance in a wide variety of fields, the requirement of at least one term of calculus has become standard not only for the physical and mathematical sciences, but for many other college majors. This has created a serious problem for those who teach calculus: roughly half the students with this requirement arrive at college inadequately prepared to take calculus. The most standard response to this lack of preparation is to require the completion of a college precalculus course before enrolling in calculus I. This cohort of students is large, and the standard response, rather than providing access to calculus, most often serves a gate-keeping function that typically filters out 60% or more of the students from completing a term of calculus. A typical student in this situation has done poorly in mathematics in high school or has been away from high school mathematics for several years, and so comes not only with insufficient or rusty math skills, but an attitude of apprehension about mathematics, low self-esteem with regard to the ability to do mathematics, and a view that mathematics is a jumble of formulas and tricks without context.

The assignment to a precalculus course, which is to prepare students for calculus, often does little more than reinforce these views. The students are in an essentially remedial course (not great for self-esteem) in which important skills and ideas are retaught, but only with the promise that "you will need this next term in calculus." In this course, calculus remains a mysterious subject, providing no context to the topics covered. For those (typically 50% or fewer) who complete the college precalculus course, calculus is a great challenge, introducing new symbols and abstract concepts, expecting students to remember algebra skills and fundamental ideas covered months ago in a setting that used none of those symbols and words. In the calculus course, many students will insist that they have never seen or used certain techniques simply because the context is so different. (For example, solving simple linear equations certainly is covered in precalculus, but linear equations were never solved for something called dy/dx or y'.) Inevitably, teachers need to review still again the non-calculus skills that are essential to solve calculus problems. There is often a high degree of frustration on both the part of the students and of the teachers; in the precalculus course and (for the survivors) in the calculus course that follows it, there is low morale in both camps.

In 1988, the math faculty at Moravian College reached the decision that something different needed to be tried—almost anything could not be worse than perpetuating the typical situation described above. Since our precalculus course existed only to prepare students for calculus and was not a prerequisite for any other course, the result was to throw out the precalculus course and replace it with a two-semester course entitled Calculus I with Review. The course would serve the same student population, and in one year's time cover all the material in our one-semester calculus I course, but at a slower pace, integrating the review of precalculus concepts and skills as they were needed, in the context of presenting calculus concepts and solving calculus problems. The goal was to improve the numbers completing the first course in calculus, as well as to try to change attitudes towards learning mathematics. Although the idea of an integrated course hardly seemed radical, we found only a handful of other schools that used this approach; Amherst College was one.

Finding materials to use with a calculus text was difficult: there were review manuals on algebra, trigonometry, and scores of precalculus texts, but there were no books that had precalculus review material presented in a calculus context, or presented in a way that could easily supplement a calculus text. After trying to use an algebra review manual for the first year, we concluded that we needed to develop materials ourselves and made a formal proposal to FIPSE (the Fund for the Improvement of Post-Secondary Education, Department of Education) to produce materials, class-test them, and assess student performance in the integrated course [4]. The proposal was funded, and a 2-year project ensued that produced the first draft of *A Companion to Calculus*, later published by Brooks/Cole [6]. It also produced an extensive assessment that made clear this approach was far more successful than the traditional precalculus-followed-by-calculus. The Calculus I with Review course used the same calculus text as that used in the "regular" one-semester calculus I, and at the end of the year, students had covered the same calculus syllabus as calculus I, so were prepared to continue to calculus II if they so desired. (The regular calculus I course is the first of a three-semester calculus sequence and has as a prerequisite a solid background in algebra and functions (precalculus) from high school or another institution.)

In 1993, at the end of the two-year grant period, a dissemination conference at Moravian College was well attended by other institutions that were interested in trying the integrated approach. The project outcomes were also noticed by FIPSE, and we were invited to become mentors in a two-year dissemination project (1995–1997) to six other institutions as they developed their own versions of the integrated course and tried it out on their campuses [5], [7]. The mentored institutions in this project were DePauw University, The George Washington University, Hudson Valley Community College, University of Puerto Rico – Mayaguez, University of Puerto Rico – Rio Piedras, and the United States Military Academy at West Point. FIPSE wanted to find out if the model was transportable to various sizes and types of institutions. Some of the results of that project are summarized in Appendix I. Every one of the institutions showed improvement in the percentage of students who successfully completed the equivalent of precalculus/calculus I, and ultimately four of the six institutionalized the course. During the course of this dissemination project and afterward, several other institutions established an integrated calculus-with-precalculus course on their campuses. There is now a small core of institutions (perhaps 40 or more) offering a version of this course. It is a quiet revolution, but certainly has made a difference where it has been tried.

Some details on the course

A course that integrates precalculus review with calculus takes careful planning, but does not require fundamental changes in calculus content or teaching style. An integrated approach can be used in calculus courses anywhere in the spectrum from traditional to reform, and in special calculus courses such as those designed for majors in business or the biological sciences. It can be used in a technology-dependent course or without any technology at all. If the goal is to have students arrive at the same place as those who complete the regular one-semester calculus course, then it should use the same text and technology as used in that course, and the integrated review needs to be planned with that calculus text and technology in mind. In order for the integrated review to be successful, those planning the course need to be aware of the most common "precalculus" difficulties that students encounter in the calculus course, and plan to address these problems with deliberate intervention at the appropriate times. Some algebraic techniques need to be reviewed more than once, as the need for them arises in different contexts.

When we at Moravian developed our review materials for the integrated course, we spent a lot of time discussing each topic covered in the calculus I course and identifying the main weaknesses of students that could be addressed. For example, to understand how (and when) to use the chain rule of differentiation, students need to understand how to recognize composite functions and how to decompose them. Extra examples and practice problems to help students to understand this process are needed. When finding minima and maxima, they need to solve equations that often have variables with negative or fractional exponents, or have quotients of functions. They lack the algebra skills to solve these equations, and that is where the just-in-time review comes in. Needing to solve these calculus problems provides strong motivation to learn the skills necessary to accomplish the task. Success in solving calculus problems using the refreshed skills also provides an immediate reward. For each calculus topic, we produced a companion chapter in our text that reviewed basic ideas and techniques that were needed in order to understand and solve problems related to the calculus topic. Since the review material was written to directly support a calculus topic, we used standard notation for calculus terms, and discussed algebraic techniques to solve typical calculus problems [2].

The syllabus of the two-semester integrated course consisted of an interleaving of our review material with the calculus I syllabus. We made a deliberate decision to introduce and study only algebraic functions in the first semester of the Calculus I with Review course, rather than cover the trigonometric, exponential, and logarithmic functions as well. The rationale for this decision was that the transcendental functions were more difficult and needed substantial time to be covered; we wanted students to learn the essentials of limit and derivative in the first semester, not just have a course on functions (i.e., a mostly precalculus course). The syllabus for our course became an outline in the front pages of our *Companion* to illustrate how to interleave its chapters with topics in a calculus text (this outline is given in Appendix II). Others using the *Companion* or other review materials might choose to omit, add, or rearrange the order of some review topics, but the essential point is that thoughtful planning must go into deciding what, when, and how to give the just-in-time review in support of the calculus topics. In teaching the course, usually readings and assignments are from one book, either the *Companion* or the calculus text, but in some instances, both books may be used for a single topic. Typically, some sections of the *Companion* are covered first to give background and algebraic review, and then some sections in the calculus text are covered. Tests include questions on the precalculus topics as well as the calculus topics. Pacing is particularly important in this course. In the first term, when there is a great deal of review as well as the introduction of the difficult new concepts of limit and derivative, the pace is slower than in the typical calculus course. In the second term, the pace can be gradually accelerated—this is especially helpful to prepare students who will continue to the calculus II course. Throughout the course, topics are treated in four modes: verbal, symbolic, numerical, and visual.

Student outcomes

The goals for our integrated course were clear: to provide better access to calculus for those entering college inadequately prepared and to increase the number of students completing the first course in calculus. In the 1991–1993 FIPSE project at Moravian, we undertook a four-part assessment to measure how well the integrated course met those goals. Detailed results of the assessment are in our final report to FIPSE [3]; here we summarize the main points. Past data was gathered on enrollment, attrition, and completion of our precalculus-calculus I sequence, and then the same data was tracked for the integrated course. Later, all institutions in our FIPSE dissemination project gathered this same data for their courses. In addition, during the dissemination project period, two institutions offered both the integrated course and the precalculus-calculus I sequence, and tracked this data for both groups. The cumulative data appears in Appendix I. The data made clear that there was far less attrition between completing the first semester and continuing the second semester in the integrated course than in the precalculus-calculus sequence. Typically, 80%–95% who completed the first semester in the integrated course continued. Also, the percentage of students who completed the full year of the integrated course was substantially higher (from 50% to 100% higher) than the percentage of students who completed the precalculus-calculus sequence.

A second measure of student achievement was to compare performance on common exam questions. At Moravian, some precalculus questions from a placement exam (taken by students before entering the integrated course) were repeated on the final exam; individual answers to these common questions on each of the two tests were compared for each student. At institutions that offered both precalculus and the integrated course, common questions appeared on both final exams. The results showed overwhelmingly that students in the integrated course had mastered precalculus concepts and, when compared with those in precalculus, significantly outperformed them. At both Moravian and other institutions, another set of questions on the final exam in the integrated course also appeared on the final exam in the regular calculus I course; this allowed comparison of performance of students in the two courses. Those in the integrated course performed almost as well as those in the calculus I, a good achievement, considering the fact that the average SAT score of those in the integrated course was about 100 points lower than those in the regular calculus I, and the additional fact that many students in calculus I had previously taken a high school calculus course. It should be pointed out that when the performance of adult students in the integrated course was compared with all students in the regular course, they outperformed this group!

The last two assessments were of student attitudes and teacher attitudes. Both groups were assessed at the end of each semester of the integrated course. Most students showed a positive change in attitude towards learning mathematics and an increase in self-confidence in doing mathematics. Teachers reported some problems in teaching the course for the first time, in using two sets of materials and in carrying out the integration of material, but were very positive about the course and the improvement of morale among students.

Teaching outcomes

There are some interesting outcomes of teaching an integrated course; some are predictable, and others unforeseen. A predictable outcome for teaching is that there is more time for teaching both the precalculus review and the calculus than in the separate precalculus and calculus I courses. This happens for at least two reasons. First, in the integrated course, since the precalculus review is directly in support of the calculus topics taught, some material often covered in a precalculus course is not included. Second, in the integrated course, many precalculus topics only need to be reviewed once, as needed, rather than twice (once in the precalculus course and again the next semester when forgotten, but needed, in the calculus course). This luxury of extra time to absorb, understand, and practice is a great help to students who have often been left behind in math courses in the past or who need to get back up to speed after a long absence from math. An unforeseen outcome is that those teaching the regular calculus I course were happy to have more homogeneous classes of better-prepared students; those students who in the past would have struggled to keep up in calculus I were now in the integrated course. Another benefit of having the integrated course is that when sections of the integrated course are offered at the same time as sections of the regular calculus I course, students in the latter course who find it difficult to keep up can have the opportunity to change into the integrated course rather than drop out of calculus I altogether. Several institutions that offer the integrated course allow students to make this course change during the first four weeks of the course. On the opposite end of the scale, some schools identify early in the integrated course those who plan to continue to calculus II and then encourage those students to enroll in the regular calculus I after taking just the first semester of the integrated course. This model uses the review and the slower pace of the integrated course as a warm up for the regular course, and then when those students enroll in the faster-paced calculus I, they have a head start and can adjust more easily to that faster pace, which will be continued in calculus II.

In the study conducted by Moravian of the institutions in its FIPSE project (and in informal reports from other institutions), teachers uniformly reported that they enjoyed teaching the integrated course far more than the traditional two-semester precalculus-calculus sequence. A frequent comment was "I would never go back to what we had before." While teacher satisfaction may not be a primary goal of the course, it clearly influences the classroom atmosphere and even the process of teaching.

Challenges in establishing the course

Perhaps the greatest challenge to establishing an integrated calculus/precalculus course is to convince colleagues both within and without the department that this is worth the investment of time (it takes little money, if any). The precalculus course is entrenched on most campuses, and to change what has always been done takes effort. If the focus is on helping those students who need to complete some calculus to achieve their goal in a positive manner, then "selling" the course is easier. Data and testimony from other similar institutions that have established the course can help convince departmental colleagues and deans. One of the primary tasks that we performed as mentors was to sit down with departments and deans to discuss how the course worked and to show the data from our own assessment. Another convincing factor was to note that there is an economic problem with the precalculus-followed-by-calculus sequence. Many students spend much time in remedial courses and often (more often than not) end up not enrolling in or completing calculus. This is costly to them and also wasteful of faculty resources.

Placement is a crucial issue that must be addressed. Institutions that serve a student population with wide-ranging preparation have an especially important task in identifying those who lack the most basic algebra skills and need to complete a remedial course such as college algebra before being ready to take the integrated course. The integrated course provides review of necessary ideas and skills for calculus, but assumes that its students are at a level where review can be effective. Simply relying on an SAT score is not sufficient for placement; an evaluation of high-school records and a placement instrument that tests specific skills needed for calculus is a much better approach. An interesting fact that emerged in our own assessment and in assessments at other institutions in our dissemination project was that there was no correlation at all between SAT scores and grades in the integrated course. (In fact, at one institution, some students were mistakenly placed in the integrated course rather than in a lower remedial course, and their grade distribution, from A to F was roughly the same as those who had been placed in the course using the SAT as a placement tool.)

At institutions where the precalculus course serves as a prerequisite to courses other than calculus, the integrated course does not replace the precalculus course, but serves the population who intend to take calculus. The identification of such students is an important part of the advising process. Two-year colleges (such as Hudson Valley Community College and Northampton Community College) and some large state institutions have had success in instituting the integrated course while retaining the precalculus course for a different population.

As with any change, you need to have a core of teachers willing to try the new approach. They (or the whole department) need to plan what material to use for the precalculus review and how to integrate it with the calculus text. In preparation for this article, I contacted about thirty institutions using the integrated approach and asked what materials they were using for the precalculus review. Many were using the *Companion* or another supplemental text, *Just in Time Algebra* [10]; others were using a precalculus text, and still others depended on their own materials, using handouts; one institution set up a tutorial center where review materials could be found.

A natural question arises concerning the amount of course credit for the two-term integrated course. Most institutions (including Moravian) give the same credit as that for two one-term courses. It is easy to justify this—in the integrated course, in the first term, a student will see the fundamental ideas of limit and derivative, not just precalculus topics. At a few institutions where the precalculus course was a non-credit (remedial) course, the two-term integrated course receives the same credit as a one-term calculus I course.

Other models for the integration of precalculus review in the calculus I course

There are some additional models for an integrated precalculus/calculus I course that have been developed by others. A notable model is the workshop, or laboratory approach, in which the use of technology and guided investigation are central to the course. One of the leading reform calculus projects, *Project Calc*, centered at Duke University, featured this approach. *Workshop Calculus*, developed at Dickinson College is most similar in its goals to the course that we have described earlier [8], [9]. The *Workshop* course also is one year in length, and integrates review of precalculus concepts and skills with the introduction of calculus ideas and problems. The teacher's role is primarily that of a guide, as students work through investigations and exercises in a laboratory setting, using technology. Another model followed by some institutions is to augment their regular one-semester calculus I course with an extra one or two class meetings a week in which the review of precalculus ideas is covered on a just-in-time basis. While this approach can certainly help those who need some extra review, it does not alleviate the pressure experienced by students who simply need extra time to absorb the new concepts as well as receive integrated precalculus review.

Some final comments

At the conference Rethinking the Preparation for Calculus (October 2001), there wasn't a single participant who felt that all was well with the course called precalculus. It is the nature of faculty to want to improve content and pedagogy of courses, and we spend much of our teaching lives doing just that. My point in this article has been to describe a relatively simple, yet to some, radical alternative to trying to improve the college precalculus course intended to prepare students for calculus. Let's offer students who need extra time and review in order to succeed in calculus just that opportunity, but not in a calculus-free setting. Instead, let's introduce them to the interesting ideas of calculus, and in this setting, and as needed, review the necessary background and techniques for them to succeed in calculus. Many institutions both large and small, private and public, have tried this approach and report gratifying results. I will be happy to supply more information on the course, institutions of which I am aware that offer such a course, and answer any other questions. There is additional information on our website [1] and in the references that follow.

References

1. Web site: http://www.calculus-with-precalc.org

2. Sevilla, Alicia and Kay Somers, "Integrating Precalculus Review with the First Course in Calculus," *PRIMUS*, Vol. III, No. 1, 1993, pp. 35–41.

3. Schattschneider, Doris, Final Report, FIPSE Project "Development of course materials to integrate precalculus review with the first course in calculus." Moravian College, 1993.

4. Schattschneider, Doris, Kay Somers, and Alicia Sevilla, "A report on a FIPSE-funded project to integrate precalculus with the first course in calculus," in *Preparing for a New Calculus*, Anita Solow, ed., Mathematical Association of America, Washinton, DC, 1994, pp. 121–124, 180.

5. Schattschneider, Doris, Alicia Sevilla, and Kay Somers, "Instituting Calculus-with-Precalculus on Different Campuses," *MER Newsletter*, Vol. 10, No. 2, spring 1998.

6. Ebersole, Dennis, Doris Schattschneider, Alicia Sevilla, and Kay Somers, *A Companion to Calculus*, Brooks/Cole, 1995.

7. "Moravian College Cluster: Integration of Precalculus Topics into the First Course in Calculus," *Lessons Learned from FIPSE Projects IV*, Fund for the Improvement of Post-Secondary Education, U.S. Dept. of Education, 2000, pp. 259–261.

8. Hastings, Nancy Baxter, *Workshop Calculus: Guided Exploration with Review* (computer-based), Volumes I and II, Springer Verlag, 1996, 1998; currently distributed by Key College Publishing.

9. Hastings, Nancy Baxter, *Workshop Calculus with Graphing Calculators: Guided Exploration with Review*, Volume I (with Barbara Reynolds) and Volume II, Key College Publishing, 1999.

10. Mueller, Guntram and Ronald I. Brent, *Just in Time Algebra and Trigonometry*, Addison Wesley, 2000.

Appendix I
Comparative attrition and completion data from six institutions

In the traditional model of a precalculus course followed by calculus I, a significant percentage of the students who complete the precalculus course choose not to continue to calculus I. That, coupled with a high withdrawal and failure rate in calculus I translates into a low completion rate of the two-course sequence of precalculus followed by calculus I. Here is completion data from the FIPSE dissemination project institutions, prior to the adaption of the integrated course, and in two instances, from a control group for the year the integrated course was introduced on a pilot basis

Completion of Precalculus Followed by Calculus I

Institution	Data period	% of those enrolling in Precalculus who completed Calculus I
DePauw U.	1988–1995	33%
Geo. Wash. U.	1992–1995	41%
	1996–1997 control group	47%
Moravian C.	1986–1988	40%
Hudson Val. CC	1996–1997 control group	17%
UPR Mayaguez	prior to 1995	less than 40%
UPR Rio Piedras	1996–1997 control group	less than 40%

The completion data on the integrated one-year course shows a dramatic difference. The data given in the chart is for the year 1996–1997, the second year of the project, when all of the adapting institutions were teaching the integrated course. Where the course has been institutionalized, these completion rates have persisted. In the following chart, the completion rate is the percentage of those who enrolled in the year-long integrated course who successfully completed the course.

Integrated Calculus with Precalculus (CWP I & II) Enrollment and Attrition 1996–1997

	Enrolled in CWP I	completed CWP I	continued to CWP II	completed CWP II	Completion Rate
DePauw U.	85	83	64	61	72%
Geo.Wash. U.	89	86	66	65	73%
Moravian C.	85	66	59	47	55%
Hudson Val. CC	43	22	20	15	35%
UPR Mayaguez	129	94	93	90	70%
UPR Rio Piedras	68	48	45	40	59%
West Point	45	42	41	38	84%

Appendix II
Topics outline: Calculus I with Review

Outline of Integration of *Companion* chapters with chapters in a Calculus text

Section in Companion	Section in Calculus text
Introduction	
Symbols and Notation	
Modes of Communication	
Cartesian Coordinates	
The Cartesian Coordinate Plane	
Graphs	*Coordinate Geometry and Lines*
Lines and Their Equations	
Parallel and Intersecting Lines	
Distance between Two Points	
The Circle	
Functions	
Function Notation	
Domain and Range of a Function	
Different Ways to Represent Functions	*Functions and Graphs*
The Graph of a Function	
Special Classes of Functions	
Transformations of Graphs	
Limits	*Tangent and Velocity Problems*
Algebraic Combinations of Functions	*Limit of a Function*
Algebraic Simplification of Functions	*Calculating Limits Using Limit Laws*
Inequalities	*Precise Definition of Limit*
If-then Statements	

Continuous Functions	*Continuity*
Polynomials	
Zeros of a Polynomial Function	
Composition of Functions	
Domains of Functions	
The Role of Infinity	
Graphical Interpretation:	
Horizontal Asymptotes, Vertical Asymptotes	*Limits at Infinity, Horizontal Asymptotes*
Algebraic Manipulations: Finding Asymptotes	*Infinite Limits, Vertical Asymptotes*
Rates of Change	
Problem Solving	
Applications	*Tangents, Velocity and Other Rates of Change*
Secant and Tangent Lines	
Rules of Differentiation	
Negative and Rational Exponents	*Derivatives*
Decomposition of Functions	*Differentiation Formulas; Chain Rule*
Simplifying Derivatives	
Angle measures	
Definition and Evaluation of the Trigonometric Functions	
Properties of the Trigonometric Functions	*Review of Trigonometry*
Domain, Range, and Graphs of the Trigonometric Functions	
Combining Functions with the Trigonometric Functions	*Derivatives of the Trigonometric Functions*
Implicitly Defined Functions	
Solving Equations That Contain dy/dx	*Implicit Differentiation*
Iteration	*Higher Derivatives*
Rate of Change of Rate of Change	
Setting up Equations for Related Rates Problems	*Related Rates*
Problem-Solving Strategies for Related Rates Problems	
Tangent Line Approximation	*The Differential and Tangent*
The Differential	*Approximation*
Rules of Exponents	*Exponential Functions*
The Natural Exponential Function	*Derivatives of Exponential Functions*
One-to-One Functions	
Inverse of a Function: Domain, Range, Graph	*Inverse Functions*
Finding the Inverse	
Definition and Properties of Logarithmic Functions	*Logarithmic Functions*
Graphs of Logarithmic Functions	*Derivatives of Logarithmic Functions*
Solving Equations with Logarithmic and Exponential Functions	*Exponential Growth and Decay*
Extreme Values of a Function	*Maximum and Minimum Values*
Solving Equations to Find Critical Values	
Setting Up Functions to Solve Extreme Value Problems	*Applied Maximum and Minimum Problems*
Solving Inequalities	*The First Derivative Test*
Graphical Interpretation	*Concavity and Points of Inflection*
Putting It All Together	*Curve Sketching*
Antidifferentiation as the Inverse of Differentiation	
Finding Antiderivatives	*Antiderivatives*
Substitution for Antiderivatives	

31

College Algebra Reform
through Interdisciplinary Applications

William P. Fox
Francis Marion University

Traditional college algebra has been taught at Francis Marion University since before the school's establishment as a four-year institution in 1970. The majority of students performed poorly in these courses. We initially tried an experiment using applications and projects to motivate the college algebra. We integrated real-world problems in the form of projects, applications, and activities to motivate students to better understand the principles of algebra. Performance improved and feedback from most of the students was positive. Based on the overall positive experience, two new freshmen algebra courses were added using modeling and problem-solving as their framework.

In this paper, we discuss courses, our salesmanship, a few illustrative examples of the applications/projects used, and student comments.

Introduction

Typical of most liberal arts curricula, the college algebra course at Francis Marion University was designed to achieve mastery in (intermediate) algebra. Currently, the traditional course is offered as either a self-paced course or as a lecture course. In these courses, the mastery of manipulation skills is the focus. These skills generally include as a minimum: operations of numbers, linear equations, inequalities on a number line, polynomials and their operations, factoring, simplification of expressions, exponents and their simplification, radicals, rational functions, exponential functions, and logarithms. We asked ourselves if all these skills are important and if so, for what reason? Mathematics professors know that these skills are very perishable. Use them often or lose them! Our most common exclamation about our own majors is, "if this student could only do the algebra."

While I was at the United States Military Academy from 1992–1998, we gave an examination in July to all our entering freshmen. The freshman class usually has a math SAT average of about 650. Additionally, we mailed out to all students that were accepted the list of high school mathematics skills that they should know. The testing results showed a significant decrease over time in the performance on this test indicating a decrease in high school mathematics algebraic skills. All students were required to pass this test with a score of 80 or above by the end of the first year. However, most students did not retain these skills for subsequent courses requiring mathematics. This was a considered a "Pass and Go" requirement (see [6]).

Reform of college algebra now questions this mastery principle in intermediate algebra. Students find college algebra to be a repeat of their high school experience. Other than fulfilling a course requirement, these algebra skills serve little real purpose in their college education. Stop for a minute and ask yourself what mathematics does a college graduate really need. If you are really honest, it will not be the college algebra of yesterday. The most important challenge facing our graduates in the future is data overload. Data and statistics are found everywhere: newspapers, magazines, television, and radio. There is no escape. For any mathematicians who have ever done consulting: "Has anyone ever handed you an equation, let alone ever asked you to simplify it?" Allow me to answer this—"No."

As a result, we proposed a different sequence of courses based upon mathematical modeling and applications using college algebra. It seemed that the goal of the college algebra of the past was to prepare students for calculus, but most would never take calculus. Also, achieving mathematics maturity through acquiring manipulation skills is not the best preparation for the calculus. Many of the skills in the algebra will not be revisited in calculus and thus will be lost. We felt this new course sequence would better serve our students by preparing them to solve problems that they may face in their own disciplines and in real life. As you will see, enough of the algebraic structure and manipulation is retained to recognize the course as algebra, but the focus is now changed toward mathematical modeling and problem-solving.

It is important for mathematicians to recall the reform calculus movement followed by confronting the core curriculum or the 7-into-4 problem for engineering students (see [1] and [3]). The 7-into-4 problem concerned the seven topics that engineering students need to cover in the four semesters of core mathematics that were allowed. In developing possible solutions to the 7-into-4 problem, members of the Mathematics Department at the USMA listened to their colleagues in client disciplines and tried to restructure the mathematics courses (calculus and other topics) in response to the specified needs. We believed this same approach should be applied to college algebra. We went out and talked to our colleagues who require and use college algebra in their coursework: biology, chemistry, physics, psychology, and business to a name a few. It was amazing to see what they thought was in our algebra courses and what abilities their students should have. Low on their list was manipulation skills, but high on their list were the concepts and connections of the key ideas. These included, but were not limited to the following: slope, proportionality, linear relations, lines, systems of equations, and linear regression. It was amazing that the theme of linearity, which we take for granted and which our students have long forgotten, ranked so high on their list.

Mathematics departments need to continue to listen. Today, we should not be telling our client departments what mathematics their majors need. We might not even recognize those disciplines today, as compared to when we took them in college. We need to listen to their needs. We need to share with them our offerings. We need to allow them to choose what their majors take and we need to be prepared to create new courses, if applicable.

Our new courses

Our two-course sequence seeks to interconnect five mathematical topics of linear models, non-linear models, functions, applications, and data. Together these topics can empower students to solve real-world problems. The mathematical concepts of statistics, data modeling (linear and nonlinear), graphical interpretation, and functions (linear and nonlinear) are taught so that they can be applied to problems from a real-world context. These concepts and topics are further connected to other disciplines through the interdisciplinary nature of the projects and applications that we use. College algebra can be used to prepare students for calculus and to prepare them to solve problems that they will see in real life. We believe that the important concepts of college algebra and the connections to other disciplines are critical to student retention of the material. For once, the student sees a purpose to the mathematics being taught. Students, especially new freshmen, need to see mathematics in a new light. They see this as an exciting endeavor that they can use in their everyday lives.

Let us briefly outline the topics in our new two-course modeling and problem-solving sequence. The first course includes basic statistics and displays of data, linear equations, linear inequalities in one and two variables, absolute value, linear functions, systems of linear equations, graphical linear programming, and linear regression. The second course includes statistics and data displays, linear functions, nonlinear functions, inequalities and linear programming, polynomial functions, exponential and logarithmic functions, and conic sections. Note the intentional overlap of some key material for statistics and applications of linear functions. Based on placement, students might start in either course. All students at Francis Marion University are required to complete six hours of mathematics for their general education requirement, and these courses are designed to fulfill this requirement for many students.

Interdisciplinary applications and projects are used to show the diversity of applications of mathematics and to facilitate our educational (mathematical) threads. One of the ways in which we weave together the content to attain student growth in problem-solving is through the use of mathematical threads. The threads that we chose for these courses were scientific computing (technology), mathematical modeling, and communicating in mathematics (writing and connectivity). Technology removes the tediousness of doing some calculations and obtaining graphs. It allows for a deeper understanding as well as increased empowerment, discovery, and experimentation. All students are required to have and use a graphing calculator for these courses. Modeling provides the motivation for the students to want to learn to solve problems. Interdisciplinary models show the students how mathematics is used in other disciplines and in other courses they might take. Communication is used to show interpretation of results and relevance of mathematics across disciplines both in the sciences and humanities. These three threads are addressed in a significant way throughout the two-course sequence. Students are expected to model elements within each project, use technology to assist in solving their mathematical model when applicable, and then communicate their solution.

Motivation through applications

Application projects are integrated throughout the new courses to motivate student learning and reinforce critical concepts (see [2]). Projects are used after a topic is taught to provide the student with the opportunity to synthesize concepts into a real-life context. In our course, the topics chosen to motivate student growth through projects are statistics, linear functions, linear programming, linear regression, non-linear functions, exponential and logarithmic models, and conic sections (parabolic motion). These projects are written to draw on interdisciplinary topics in social sciences, business/economics, and the physical and life sciences. Each project is a group project where groups consist of one, two, or three students. Students pick their own groups, because many of our students are commuters. Each project has a writing component, because we feel writing (communications) is essential to understanding, critical thinking and communicating about mathematics. Mathematics is much more than just the numerical answers that the student wrote down in high school.

Each project, as a minimum, includes the following:

- A title page showing the title, the names of the students in the group, date.

- The solution to the project either typed or neatly printed. Included is a problem statement in the student's own words, definitions of all variables, sketches (if applicable), formulation of all equations, solutions, including appropriate graphs, answers to all specific questions, and a brief discussion of the strengths and weaknesses of their solutions.

- References.

The objective or goal of each project is to provide connectivity through an example for the use of the mathematical topic and to allow for a deeper understanding or interpretation of the material. The following are examples of some of these interdisciplinary projects.

Example 1. Linear functions: Stretching a spring   (Physics and Mathematics)

The dimensions of an object tend to change when forces are applied to the object. For example, when opposite forces are applied to both ends of a spring, the spring is either stretched or compressed. Unless the spring is damaged, it will return to its original dimension when the forces are removed. Objects that return to their original dimensions after the applied forces are removed are called elastic objects. Their study has led to the study of interesting relationships. Many objects react this way: baseballs when hit, golf balls when struck, and so on. Those objects are much harder to gather data for analysis than a spring so we will stick with a simple spring.

The amount of deformation of an elastic object is proportional to the forces applied to deform the object. This statement was first published by Robert Hooke (1635-1703) and is referred to as Hooke's Law. This law has many applications, including archery bows, shock absorbers, and automobile bumpers.

In this lab, you will analyze Hooke's law by using the spring, some weights, and a measurement apparatus. For your spring, measure the length of the spring without any forces being applied. Record this length. Next choose at least ten weights (by using the gram weights or combinations of weights). For each force (the weight) applied to the end of the spring record the additional length (stretch) of the spring.

Requirements:

1. Make a table of your data using the stretch of the spring versus the weight (force) applied.

2. Make a scatterplot of your data. Let force be y and the stretch be x. Describe in words what your data look like.

3. Representing Hooke's law mathematically as $F = kx$, verbally describe the graph of this equation. What does k represent in this equation?

4. From your scatterplot, estimate the slope and the intercept of your relationship. Write your model in equation form.

5. What is the physical meaning of the slope of your line?

6. What does the y intercept tell you about the physical system?

7. Use your model to estimate the values of weights for each of the input values of x. Plot both your data and the equation of your model together on the same axis. Is your model a reasonable fit? Why or why not?

8. Is the value of k unique? Do you think it can be used to calculate the stretch with other springs? Compare your value of k with two other group's value of k. What do you think now?

9. Exceeding the elastic limits: Is it possible to violate the linear relationship? Use your model to determine the stretch of your spring using your own body weight to stretch the spring (estimate your weight and ensure it is dimensionally correct in units). Is this answer reasonable?

Example 2. Linear programming: Producing tires   (Business and Mathematics)

Firestone Tires headquartered in Akron, Ohio, has a plant in Florence, SC, which manufactures two type of tires: SUV 225 radials and SUV 205 radials. Demand is high because of the recent recall of tires. Each 100 SUV 225 radials require 100 gallons of synthetic plastic and 5 lbs. of rubber. Each 100 SUV 205 radials require 60 gallons synthetic plastic and 2 1/2 lbs. of rubber. Labor costs are $1 per tire for each type tire. The manufacturer has weekly quantities available of 660 gallons of synthetic plastic, $750 in capital, and 300 lbs. of rubber. The company estimates a profit of $3 on each SUV 225 radial and $2 on each SUV 205 radial. How many of each type tire should the company manufacture in order to maximize their profits?

Requirements:

1. List the decision variables and define them.

2. List the objective function.

3. List the resources that constrain this problem.

4. Graph the feasible region.

5. Label all intersection points of the feasible region.

6. Plot the objective function in a different color (highlight the objective function line, if necessary) and label it the ISO-Profit line.

7. Clearly indicate on the graph the point that is the optimal solution.

8. List the coordinates of the optimal solution and the value of the objective function.

9. Assume now that the manufacturer has the opportunity to sign a nice contract with a tire outlet store to deliver at least 500 SUV 225 radial tires and at least 300 SUV 205 radial tires. Use graphical methods to help recommend a decision to the manufacturer. Support your recommendation.

10. If the manufacturer can obtain an additional 1,000 gallons of synthetic plastic for a total cost of $50, is it worth it to obtain this amount? Determine the new optimal solution caused by adding this level of resource.

11. If the manufacturer can obtain an additional 20 lbs. of rubber for $50, should they obtain the rubber? Determine the new solution caused by adding this amount.

12. Write a one-page letter to your boss of the company that summarizes the results that you found.

Example 3. Exponential and logarithmic functions: The terror bird   (Biology and Mathematics)

During an archaeology dig in Florida, a prehistoric femur bone was found. As the dig continued, many more bone fragments were found that allowed the paleontologists to recognize the prehistoric bones of the bird as a Titanus Walleri. The Titanus Walleri lived over 2 million years ago on the oak and grass savanahs of what is now Florida. Scientists believe that the Titanus Walleri was a bizarre predatory bird, and it has been called the "terror bird" (see Figure 1 for an artist's rendering).

Your team has been asked to assist in determining the size of this terror bird from its fossils. Data have been collected on numerous large dinosaurs that roamed the earth millions of years prior to the terror bird as well as birds that currently inhabit the area near Florida.

Dinosaur fingers and wrists fused into a feather-supporting shaft. Terror birds turned wing bones into weaponry.

Figure 1.

Terror Bird Data

Femur circumference (cm)	Weight (kg)
.7943	.0832
.7079	.0912
1.0000	.1413
1.1220	.1479
1.6982	.2455
1.2023	.2818
1.9953	.7943
2.2387	2.5119
2.5119	1.4125
2.5119	.8913
3.1623	1.9953
3.5481	4.2658
4.4668	6.3096
5.8884	11.2202
6.7608	19.9500
15.1360	141.2500
15.8500	158.4893

Requirements:

1. Plot the data of weight (as the dependent variable) versus the circumference of the femur (as the independent variable). Very briefly describe the shape. Is it linear or a curve?

2. Using the concept of natural logarithms, calculate the natural logarithms of each data point for weight and femur circumference. Now, plot the natural logs of the data (natural log weight versus natural log femur circumference). What does the plot look like now? Does it look like a straight line?

3. Using any method, find the slope and the y-intercept of this line. Write down this equation using LN W (for natural log of weight) and LN FC (for the natural log of the femur circumference) in the equation.

4. Using the properties of logarithms, transform the equation back into its original form (probably in the form $y = ax + b$).

5. Using this equation, predict the weight of the terror bird that has a femur circumference of 21 cm. That is, evaluate the equation for weight when the femur circumference is 21 cm. Does your value seem reasonable?

6. Write a short paragraph to the scientists, explaining how you used mathematics to answer their scientific question.

Example 4. System of equations: Balancing chemical equations   (Chemistry and Mathematics)

In your chemistry class, you find you are working on an experiment. You find that chromium compounds exhibit a variety of bright colors. When solid ammonium dichromate, $(N_4H_2)Cr_4O_7$, a vivid orange compound, is ignited, a spectacular reaction occurs. Although the reaction is actually somewhat complex, let's assume here that the products are solid chromium (III) oxide, Nitrogen gas (consisting of N_2 molecules), and water vapor. Solid compounds are annotated with an (s) and gas compounds are annotated with a (g) after the chemical compound within the equation. The unbalanced equation is:

$$(NH)_4 2Cr_2O_7(s) \rightarrow Cr_2O_3(s) + N_2(g) + H_2O(g)$$

Requirements:

Using the four chemicals, nitrogen (N), hydrogen (H), chromium (Cr), oxygen (O) and multipliers $\{a, b, c, d\}$, consider the equation:

$$a(NH)_42Cr_2O_7(s) \rightarrow bCr_2O_3(s) + cN_2(g) + dH_2O(g)$$

1. Write this as a system of equations in matrix form.
2. Solve the values of $\{a, b, c, d\}$ that balance the chemical equation.

Other projects   The table below lists some of the other projects that we have used in these courses. The authors will provide the project scenarios and solutions, if requested. We create eight to nine new projects each semester for these courses.

Name of the Project	Topic Area
Mark McGwire's homerun pace	Statistics
Calories and fat content	Statistics
Relations of GPA and SAT Scores	Statistics
Braking distance versus reaction time	Linear functions
Cost of US postage stamp	Systems of linear equations
Automobile repair in SC	Linear programming
Manufacture and sale of tennis shoes in SC	Linear programming
Flight of a cannon ball	Parabolic motion
Flight of a baseball	Parabolic motion

Student growth through projects

While solving these projects, our students were surprised that there were many correct answers. Most students entered this course believing that mathematics problems had only one correct answer that there was only one way to obtain the answer. For example, as we discussed the linear function project results, more students appreciated the idea that their selection process for the two points used to derive a linear equation determined all other results. Also, students struggled, at first, with factors that could affect the use of their model in the future. After hearing many other student responses, it became apparent that the students were beginning to grasp the connections between mathematics and external factors.

Our linear regression project introduced the students to the idea of best fit. We discussed the idea of a best-fit line during an earlier linear function project, and subsequently the students were going to try to use the concept. Graphical examination of the linear relationship as well as numerical tables of relative error did a lot to illustrate this concept.

The "terror bird" project was very interesting to the students. They were interested in this fossil and determining the weight of this prehistoric bird. Transforming the data into a log-log data set (using natural logarithms) was very doable. Each group was able to perform the transformation, plot the data, and obtain an estimate for the slope and intercept. Yet, very few groups could transform the model back from the log-log form into exponential form, $y = ax^b$.

To follow up on our projects, we include a question or two similar to parts of the project on each major test. So the final course grade includes the group submission of the projects and their individual performance on tests. Students realize that the understanding of the work in the project is important to their success on the test and in the course.

Student growth is also achieved within our threads: technology, modeling, and communications. Within each course, we see growth in the ability to use the graphing calculator to assist in solving problems, to model basic mathematical course related problems and to communicate their solution (see [1], [4] and [5]).

Student responses

Students were asked to write reflective summaries of their project experiences in the form of portfolios. Student responses were not all positive. Some students were very apprehensive about projects, the work involved, and the value of projects.

Students who stated that they did not like the projects said that because the projects were too much work and were not like the problems in the book. One student claimed that projects were not good, because he or she was used to being taught mathematics through many examples. These reasons were the only non-positive comments from a survey of over a hundred students. These comments, in our opinion, are really positive comments. Projects are supposed to be different, challenging, and make students think critically about the mathematics that they are learning. Students who learn only through repetitive examples may not truly understand the concept and may be merely memorizing a procedure.

Many students provided more positive feedback. As one student wrote, "...projects provided more help than an endless series of tests would have. Either would have covered the material but projects are something we can refer to time and time again." Another student wrote, "...this way, you realize how you may need this information one day." One student said that after doing these projects, he now looks at newspaper articles differently, in that, he sees whether or not he can build the mathematical model to get or check their results. Most of the students directly stated positive comments about the experience and that they felt it would make them better students in the future.

Faculty growth and development

We have established an open line of communication to our other departments that transcend these courses. Some of the members of the mathematics department and members of other discipline departments have joined together to create these projects and joint articles. Currently the following interdisciplinary projects are being developed that fit many other mathematics courses.

Title	Discipline	Mathematics Topics
Capture-recapture	Biology	Statistics and algebra
Population survivability	Biology	Algebra
Genetic transfer across borders	Biology	Markov chains and dynamical systems
Tracking turtles	Biology	Modeling
Bridge swinging	Physics	Algebra and trigonometry
Planetary motion	Physics	Algebra and conic sections
Moons of planets	Physics	Algebra
Radon	Physics	Statistics
Chemical balancing	Chemistry	Algebra and matrices
Diffusion	Chemistry	Differential equations
Saving $ for college	Economics	Algebra

The development of joint projects has enabled faculty to work together and get to know each other. These broaden our own perspectives and are an important component of our professional development.

Conclusions

Based upon the student comments, anecdotal evidence from faculty teaching the new courses, and the success of our students in follow-on mathematics courses, we claim that the two new courses in our freshman sequence, Math Modeling and Problem Solving I & II, are successful. They use a subset of skills from the college algebra sequence and concentrate on the student's ability to truly master these skills through motivation, modeling, problem-solving, projects, and writing with mathematics. We also measure success by the student growth in our mathematical threads as measured through the student portfolios. The portfolios illustrate growth in each area of our threads during the semester. The effort, sophistication, use of technology, and communications of the mathematical results into understandable language improve throughout the semester. Additional departments have begun to request and allow their students to take these new courses. Finally, our own mathematics department has unanimously endorsed a plan to move towards one four-credit-hour course, joining the best ingredients of the traditional and reform algebra courses.

References

1. Fox, William P., Chris Arney, Kelley Mohrmann, Joe Myers, and Rich West, "Core Mathematics at the United States Military Academy: Leading into the 21st Century," *PRIMUS*, Vol. V(4), 1995, pages 343–367.

2. Fox, William P. and R.D. West, "College Algebra-Drills or Applications," *PRIMUS*, Vol. XI (1), March 2001, pages 89–99.

3. Dossey, John, Editor, *Confronting the Core Curriculum*, MAA Notes #45, Mathematical Association of America, Washington, DC, 1998, pages 17–32.

4. West, Richard D., "Evaluating the Effects of Changing a Core Mathematics Curriculum," *Proceedings of the Seventh Annual International Conference on Technology in Collegiate Mathematics*, Addison-Wesley, 1996, pages 530–534.

5. West, Richard D., "Evaluating the Effects of Reform," MAA Notes #49, Mathematical Association of America, Washington, DC, 1999, pages 219–223.

6. West, Richard D., "Skills vs. Concepts," in this volume.

32

Elementary Math Models:
College Algebra Topics and a Liberal Arts Approach

Dan Kalman
American University

Introduction

It would be difficult to overstate the importance of algebra in mathematics. If mathematics is the language of science, then algebra provides the alphabet, vocabulary, and syntax. In particular, the traditional college algebra course covers the elementary functions of analysis: linear and quadratic functions; polynomials and rational functions; roots, exponentials, and logs. These functions are inescapable in the most elementary applications of mathematics to other subjects. In a word, they are ubiquitous all over the place.

It is tempting, therefore, to prescribe algebra as the minimal quantitative component of a higher education. Unfortunately, for many students who study algebra in college, their mathematical education goes no further. For these students, the significance of algebra is largely lost. It is as if they studied an alphabet, vocabulary, and syntax, but never got to read any literature.

Liberal arts math courses offer an alternative exposure to mathematics for general education students. These courses emphasize lofty educational goals, hoping to communicate something of the beauty, power, and fascination of our discipline. Developing specific manipulative skills is given much less emphasis.

This essay concerns a hybrid of these two approaches to general education mathematics. Elementary Mathematical Models (EMM) was developed in the mid 1990s at American University, a selective medium size university in Washington, DC. It is offered at the lowest level of the mathematics curriculum and fulfills the general education mathematics requirement. Like a traditional college algebra course, EMM seeks to make students familiar with the elementary functions. But like a liberal arts math course, it strives for an intrinsic educational significance independent of utility for any other courses.

Classroom materials for EMM evolved into a textbook, now available from the MAA [1]. Additional resources are available on the web [2]. The course has been used successfully for about five years at American University and a few other institutions. In the discussion below, I will provide an overview of the course and its clientele as I observe them at American University, concluding with a summary of student reactions and performance.

Content and organization

The core of the EMM course is a sequence of progressively more complicated growth models: arithmetic growth, quadratic growth, geometric growth, mixed models, and logistic growth. The first three of these give rise to important families of elementary functions, linear, quadratic, and exponentials, respectively. Logs are introduced as part of the material on exponential functions. Mixed models, which are a combination of arithmetic and geometric growth, give rise to shifted exponentials of the form $Ae^{bt} + c$. This core material can be expanded in a variety of ways, covering closely related units on polynomials and rational functions, additional properties of logs, linear regression, or chaos.

In EMM, mathematical topics are always introduced in the context of some realistic modeling problem, and the mathematical exposition never strays far from the applied context. Algebra is presented when and where it is needed and in conjunction with numerical and graphical methods. I made a commitment in developing the course to omit any topic or skill that could not be immediately justified in the context of application. That is, I only included topics that students could see they needed. It should not be surprising that relatively little of the traditional material on elementary functions had to be discarded in this way. Afterall, those topics are in the traditional curriculum because they *are* useful. EMM simply exploits this fact by developing each topic within the context where it is needed. EMM students never ask, *Why is this topic useful?*

By design, EMM incorporates themes that appear repeatedly throughout the course. Acquiring a deep understanding of these themes and retaining that understanding beyond the end of the course are among the liberal arts goals of the course. Some of the themes are methodological and procedural, others are more philosophical. For example, the modeling framework as a broad methodology for applying mathematics constitutes a philosophical theme. I hope students will, by the end of the semester, understand that models involve simplifying assumptions; that there is a spiral aspect to model development, evaluation, and refinement; that mathematical analysis provides a powerful tool for tracking the consequences of simplifying assumptions; and that mathematical models have both strengths and limitations.

A more procedural theme concerns the use of discrete models with simple recursive patterns. The general framework is a sequence of data values or predictions, and the recursive pattern specifies how each successive term is obtained from its predecessor. In arithmetic growth, the recursive pattern is addition of a constant; in geometric growth it is multiplication by a constant. These patterns are expressed as difference equations. For example, $a_{n+1} = a_n + 2$ describes an arithmetic growth pattern where each term is found by adding 2 to the preceding term. With the exception of logistic growth models, all the simple difference equations have corresponding solutions, which express the terms of a sequence, a_n, as a function of n.

As part of this theme, there is a repeated pattern of development for each family of growth models. In each topic, we look at several examples of different phenomena (e.g. metabolization of drugs, pollution in a body of water, repeated loan payments) and find a simple recursive pattern that is common to all of them. The pattern is expressed as a specific kind of difference equation. Systematic exploration of these difference equations leads to solutions comprising a family of elementary functions. Graphical, numerical, and analytic properties of these functions then reveal information about the evolution of the corresponding models.

EMM makes no attempt to survey modeling in a broad way. In fact, EMM is not a modeling course at all. Its primary goal is not to teach students all about mathematical modeling. Rather, models of a very particular kind, discrete recursive models, are used as a vehicle for motivating the study of elementary functions. In the process, I hope students will acquire a realistic sense of how mathematics is truly applied and experience some of the creativity and judgement that goes into applying mathematics to real problems.

Students

The students of EMM have as a minimum the traditional college preparatory mathematics coursework at the secondary level. This typically means something like three years of high school math, often algebra 1, algebra 2, and geometry. However, the specific algebraic skills of my students can be quite low. Presumably, this reflects both imperfect mastery of material in the earlier courses, as well as attenuation over time of whatever skills were originally learned. In fact, students who master and retain three years of high school mathematics would probably not be placed in EMM at American University.

Part of the design of EMM was a reaction to working with students for whom abstract symbol manipulation is nearly meaningless. I see these students often. Typically they are willing to work diligently in the course. They also often have moderate to good number sense and can function quite well in the concrete context of a particular problem. Somehow, though, the language of algebra just does not work for them.

This point is worth elaborating. One of the advantages of algebraic notation is its ability to capture numerical relationships or patterns in a very general, yet very succinct way. Take for example, the difference equation $a_{n+1} = 3a_n$. To anyone with a command of algebra, this says in a very compact way, each term is three times the preceding term. I regularly see students to whom it says nothing of the kind. These students easily apprehend and work with the pattern of successive tripling of terms in a sequence. When that sequence is part of a model, they can handily use numerical methods to answer questions about the model. In short, they understand everything about the model that I want them to understand. But, whereas to me the equation captures most or all of that understanding, for these students, the equation does not. They cannot quite see what the point of the equation is and can only translate between the abstract notation of the equation and their own clear understanding of the pattern with great effort and concentration.

For these students, the context of a particular problem or family of problems is critical. They cannot attach any meaning to the symbolic notation otherwise. EMM tries to assist these students by keeping the mathematics firmly in a meaningful context. I hope students will use their concrete number sense, general reasoning ability, and verbal skills to help them endow the symbolic notations with meaning.

I do not want to leave the impression that EMM students *all* struggle with algebra. On the contrary, the audience is quite diverse, and frequently one or two of the students in EMM have even completed a semester of calculus. For the strongest students, the contextually thick buildup to each symbolic abstraction can be a bit annoying. But overall, these students do not find the course boring or trivial. The realistic models make the material interesting, and the constant connections between mathematical procedures and application issues is something that is unfamiliar to just about all of the students.

Classroom practice

The EMM course does not presuppose any particular style of presentation. When I teach the course, I use a mixture of lecture-discussion, small group activity, and computer exercises. Some of my colleagues have used mainly traditional lecture style presentation. The course is intended to be language rich, and reading and writing are heavily emphasized.

The application contexts and recursive models lend themselves nicely to empirical investigation. This is easily supported by technology in a number of ways. I have successfully used graphing calculators and special purpose computer activities designed for the course. These were implemented in an authoring package called Mathwright [4] and are freely available over the internet [2]. They have the interaction style of a webpage, allowing students to enter data and equations in a natural and intuitive way and explore graphical and numerical properties of models. It is also possible to provide similar kinds of computer activities using a spreadsheet program, such as Excel. I imagine that notebooks in Mathematica or Maple could also be created, but I have not done so.

Sample lesson

To illustrate the evolution of ideas in the course, here is an outline for one topic, mixed models. This material would be covered in about one and one-half weeks. For additional samples, see [3]. Mixed models are characterized as a combination of geometric growth and arithmetic growth and by a difference equation of the form:

$$a_{n+1} = r a_n + d$$

with r and d fixed constants. These models arise in a variety of contexts, including:

- pollution in a body of water
- metabolization of medicine with repeated doses
- amortized loan payments
- sums of geometric growth models

Given the formula $1 + r + r^2 + \cdots + r^{n-1} = (r^n - 1)/(r - 1)$ for the sum of a geometric series, a simple pattern analysis leads naturally to the solution of a mixed model difference equation. This is presented in terms of a numerical example of the following sort. Consider the sequence a_n, where $a_0 = 200$, and each successive term is 3/4 of the preceding term, plus 100. We systematically generate several terms of the sequence, without actually carrying out any of the arithmetic:

$$a_0 = 200$$
$$a_1 = 200(.75) + 100$$
$$a_2 = 200(.75^2) + 100(.75) + 100$$
$$a_3 = 200(.75^3) + 100(.75^2) + 100(.75) + 100$$

Here, the right-hand side of each equation is obtained by multiplying every term on the preceding line by .75, and then appending an additional increment of 100. Students quickly recognize the pattern and are soon led to discover

$$a_n = 200(.75^n) + 100 \frac{1 - .75^n}{1 - .75}.$$

It is a small step from this result to the generalization:

$$\text{if } a_{n+1} = r a_n + d, \quad \text{then } a_n = a_0(r^n) + d(1 - r^n)/(1 - r).$$

Students study the graphs and the numerical tables for these sequences and observe that for $r < 1$, the values of a_n level off to an equilibrium value. In one lab period, they explore a model for repeated drug doses numerically and graphically, discovering that the equilibrium value does not depend on the initial dose of medication, but is proportional to the size of the repeated dose and inversely proportional to the percentage of the drug which is eliminated from the body between doses. They use these observations to determine the size of the repeated dose required to achieve a predetermined equilibrium level of medication retained in the body.

The algebraic lessons associated with this unit focus on algebraic rearrangement and solving equations. One of the goals of the course is to make it clear to students why algebraic rearrangement is necessary and useful. In this unit, they see that the natural form for a solution to a difference equation, for example:

$$a_n = 200(.75^n) + 100 \frac{1 - .75^n}{1 - .75},$$

can be expressed more compactly in the form:

$$a_n = 400 - 200(.75^n).$$

Each form has value in some context. The first form shows clearly the significance of the parameters a_0, r, and d. The second is more convenient for computation and shows at once that the graph is a vertically shifted exponential, with equilibrium value 400. In this way, the abstract practice of algebraic rearrangement is observed to have significance in a very concrete way.

A second repeated appearance of algebra occurs in connection with inverting the functions which arise as solutions to difference equations. In the context of the example discussed above, we might ask when the amount of drug will reach a value of 350. Using the simplified form of the equation for a_n, that leads to:

$$350 = 400 - 200(.75^n).$$

Here, algebraic rearrangement is again used, this time to reduce the equation to one which the students can solve with logs:

$$.25 = .75^n.$$

In this unit, there is also a discussion of fixed points and their role in determining equilibrium values. If x is fixed by the recursive operation *multiply by .75 and add 100*, then it follows that:

$$.75x + 100 = x.$$

This provides an alternative route to the equilibrium value of 400.

Assignments for this unit include applications of mixed models that arise naturally in all of the areas cited above. Students can formulate models for these application areas and use numerical, graphical, and symbolic methods to predict future behavior of the models. They also work with the idea there are several different justifications for using a mixed model (or any other kind of model). In some of the applications, simple assumptions about mechanisms at work in the model (like elimination and repeated ingestion of medications) lead to a mixed model. In other cases, the data are simply observed to fit closely to a mixed model pattern. In yet other situations, a previously defined geometric model is summed to obtain a mixed model. For example, a geometric growth model for annual oil consumption leads directly to a mixed model for world petroleum reserves.

Student reactions and performance

Students have been very supportive of the goals and framework of the EMM course. Between 80% and 90% describe the course as interesting and worthwhile. Each semester, there are a few students (perhaps 3 or 4 in a class of 30) who make comments like these:

> *Best math course I ever took; I am usually awful at math but I really understood this course; first time I actually enjoyed a math course; I was dreading math but this turned out to be one of my favorite courses.*

There also are always a few students who object to the emphasis on writing and *thinking*. It is rare for a student to find the material so easy that it offers no intellectual challenge. On the other hand, the stronger students would find that a traditional college algebra course can be completed in a much more mechanical way and with much less in the way of conceptual demands. Perhaps these are the same students who complain: *Too much writing. I never had to write essays in a math class before.*

In my classes, almost every student who makes a reasonable effort over the course of the semester completes the course successfully, ending with a grade of C or better. Roughly half to two thirds of my students receive final grades of B or above.

I have not gathered any data on the long term instructional goals that provided a primary motivation for creating this course. I do not know how well EMM students retain ideas about mathematical models in general, about how and why math is applied, about the specifics of recursive patterns and functions. To that extent, I cannot really substantiate how effective the course is.

Conclusion

As a teacher, I find the EMM course very satisfying. It has a coherent story line and an evolving conceptual thread that stretches from the simplest models to current concepts in chaos. There are well-defined long-term instructional goals, and the repeated emphasis of aspects of these goals enable students to make progress through the entire semester. I am confident that the course is laying a foundation for the quantitative demands of other general education courses in the natural and social sciences. But at the same time, I feel that EMM students have an opportunity to learn something significant about how math is used, and why math is important. In addition, positive student reactions reinforce the philosophical convictions that inspired the course to begin with.

References

1. Kalman, Dan, *Elementary Mathematical Models: Order Aplenty and a Glimpse of Chaos*, Mathematical Association of America, Washington, DC, 1997.

2. ——, *EMM Resources for Teachers*. http://www.dankalman.net/emm.

3. ——, *Entry Level College Mathematics: Algebra or Modeling?*, AMATYC Review, Vol. 24, number 2, Spring, 2003, pp 65–76.

4. *Mathwright Software*, http://www.mathwright.com.

33

The Case for Labs in Precalculus

Brigitte Lahme
Sonoma State University

Jerry Morris
Sonoma State University

Elias Toubassi
University of Arizona

Background

The goal of this position paper is to make the case for the use of labs in a precalculus course. The observations made in this paper are based on our experience with the precalculus program at the University of Arizona, where supplementary lab assignments are integrated into the precalculus curriculum. While these lab assignments frequently involve the use of technology, they are not computer labs in the traditional sense; rather, they are multi-step, real-life problems that students explore in a group setting.

First, we set the context for the precalculus course at the University of Arizona. All students intending to take beginning mathematics courses, including precalculus, must take a readiness test. The tests currently in use were developed in California by the Mathematics Diagnostic Testing Project. The tests primarily cover topics from algebra, geometry, and trigonometry. Most of the students who end up in the precalculus course hoped to start in first semester calculus. Most have had four years of high school mathematics ending in a course equivalent to precalculus and in some cases they have had calculus.

The precalculus course at Arizona is a four-credit course that meets in three 50-minute sessions and one two-hour lab session. The text for the precalculus course is *Functions Modeling Change: A Preparation for Calculus* by Connally, Hughes Hallett, et al [2]. It is followed by three semesters of the reform calculus text by the Consortium schools written by Hughes Hallett, et al [5]. The precalculus course adopts a balanced approach, focusing on both concepts and procedures. As with all the Consortium material, it stresses understanding and multiple ways of representing mathematical ideas. It aims to build a solid foundation for calculus by focusing on key classes of functions. The exercises in the text are varied, with some challenging problems and some using real data. The material encourages students to think symbolically, numerically, graphically, and verbally. Finally, technology is integrated into the course with the form and degree of use varying by instructor and student preparation. All precalculus classes use TI graphing calculators, which are also part of the calculus sequence.

Why labs?

The purpose of the labs is to have students investigate real-life problems that can be addressed with the tools learned in precalculus. The labs provide students with an opportunity to step back from the exercises that are in the text and allow them to consider problems that require more from them in terms of

310

critical thinking skills. While many of the exercises in the text are one-step calculations asking for a single answer, a typical lab project asks questions in an open-ended way that requires students to first identify the correct tools to use and then to use them in an appropriate way. In addition, these labs give students a chance to discuss mathematics in small groups, debate strategies, convince fellow students of the validity of their arguments, and then put it all together into a polished, written report detailing their findings and conclusions. The labs also allow students to explore mathematical ideas from different points of view— graphically, verbally, numerically, and symbolically—which helps students solidify their understanding of the concepts.

The labs are very different from problems students may see in the text. In particular, the labs are not short answer questions requiring only a few calculations; in some cases there is no right answer, only a model that approximates a natural phenomenon. The labs require written explanations supported with data, numerical calculations, graphs or spreadsheets, which ultimately lead to some conclusions. The lab report is meant to stand alone and be understandable by anyone with a basic knowledge of mathematics. In addition to helping students gain deeper insight into the concepts of precalculus, the labs are one of several ways that the instructor can use to gauge how well students understand these concepts.

Format of the labs

The lab assignment is handed out one week in advance with a brief discussion of the overall idea behind it. During the intervening week, students read the assignment and may have to collect data for use during the lab. Students work in groups of 3-4 that can be formed in different ways. Some instructors let the students select their own partners, while others select groups at random or just assign groups they think will work well together. During the lab, students analyze data, discuss mathematical models, find answers to specific questions, look for patterns, and come up with some conclusions. Students continue to discuss the lab after class and write a report that is due the following week. Some instructors require one report per group with the lead author rotating among the group while others require a report from each student.

Some advantages of the labs

The labs are intended to enhance the writing/explanation part of the course. In the reformed calculus text, many of the problems require answers that are explanatory in nature, i.e., students have to express answers in verbal form, not just numerical format. This is often difficult and frustrating for students, especially for those whose view of mathematics is limited to numerical calculations. Here are some of the main advantages of labs:

- The labs give the students experience communicating mathematics through writing. Most students have never been required to do this before but will be expected to do it in subsequent mathematics courses.

- The lab environment allows students to work together on math problems where they have to explain their thinking to fellow students. Thus, students are forced to communicate mathematics to each other in order to finish the lab assignment.

- The labs are an opportunity to reach the students in the course who put very little effort into homework problems involving written explanations. The fact that labs are done in class guarantees that students will use this time to write and communicate mathematics.

- The labs help students learn to explain the ideas in precalculus. This skill is necessary to succeed in a reformed calculus course. Thus, doing the labs gives students who go through the reformed version of precalculus an advantage over those who do not. Some calculus instructors have noticed that these students seem to experience less frustration doing calculus than those who have not had the precalculus course.

- The lab assignments present practical problems that help change the students' flawed or narrow views of what mathematics is. Many students, especially in this course, come into it thinking that a problem requiring a written explanation is not really a mathematics problem. Therefore, they think that such problems should not be included in a mathematics course. By giving them examples of practical problems like those in lab reports, they become accustomed to accepting such problems as a part of what they view to be mathematics.

- The lab reports give the instructor a window into the way students think about concepts. It alerts the instructor as to which concepts are understood and which ones need additional attention.

- The labs provide students with the opportunity to use multiple strategies to solve problems including the use of formulas, graphs, tables, and verbal descriptions. The labs also provide a good opportunity to use technology such as graphing calculators, Excel, and the web. The ability to use different strategies and tools makes students more competent problem solvers.

- The labs add an important dimension to the course which otherwise may appear to some students as merely a catalog of functions. They provide students with an opportunity to discover practical interpretations of mathematical objects.

- The labs provide students with lots of practice organizing, displaying and analyzing data in real-world contexts. These are important mathematical skills that relate directly to other areas in life.

- The labs are designed to promote teamwork, and improve writing and problem-solving skills. All of these are skills that future employers look for in college graduates.

Some challenges associated with the labs

Focusing too much time on lab reports takes away from the time needed to improve algebra skills. Since a lack of algebra skills is usually the reason students end up in precalculus, a careful balance must be achieved. To solve this problem, some instructors choose to do labs only every two weeks and use the lab days in between to concentrate on algebra intensive worksheets. It is also possible to include labs where algebra skills are necessary to solve the problems posed in the labs. (For instance, please see the Revenue Lab, which is the fourth sample lab given below.)

Asking students to work in groups always brings up issues involving fairness. This is particularly the case when each group submits one report and all students get the same grade for unequal amounts of work. Good students are sometimes resentful of having to work in groups, since they often have higher standards than their partners. It is a good idea to remind students that being a productive member of a team will be expected of them in their careers. To ensure that it is not possible for some students to get a free ride, it is a good idea to require every student to be the lead author for at least one lab and to remind students to omit an inactive group member's name from the group report. Sometimes it is also necessary for the instructor to reassign groups, allow students to switch groups or to include attendance requirements during lab days.

Many of the labs involve calculations done by graphing calculators, especially those involving scatter plots and linear regression. Since students do not have the same or roughly the same, kind of calculator, this makes such a lab assignment difficult to manage. Since students often become anxious and frustrated when calculator instructions do not apply to their calculator, it is a good idea for instructors to announce early in the course the type of calculator they will use for in-class demonstrations. For students who use different calculators, it helps to collect or design calculator instructions for as many different calculators as possible. Putting students with the same type of calculator together in a group for the first lab and having additional instructors (teaching assistants, other precalculus instructors) available to help with calculator problems is also helpful.

It is sometimes hard to motivate students for some labs since they are overwhelmed by the technical details of the lab and fail to see the big picture. For these difficult labs, a good introduction to the lab by the instructor can prevent a lot of frustration.

Below are examples of four labs.

Body Parts

The following is an excerpt from Jonathan Swift's, *Gulliver's Travels.*

> The seamstresses took my measure as I lay on the ground, one standing at my neck and the other at my midleg, with a strong cord extended, that each held by one end, while the third measured the cord with a rule an inch long. Then they measured my right thumb, and desired no more; for by a mathematical computation, that twice round the thumb is once round the wrist, and on to the neck and waist and by the help of my old shirt, which I displayed on the ground before them for a pattern, they fitted me exactly.

This project will focus on finding mathematical relationships between various body parts. The first relationship we want to investigate is the length of the ulna (the bone extending from the elbow to the wrist, on the side away from the thumb) and the length of a person's shoe.

- Without taking any measurements, can you make a conjecture regarding the relationship between the length of one's ulna and the length of his/her shoe?

- Begin by measuring the length of each person's ulna and the length of his/her shoe, both in inches. Be careful to be very precise when making these measurements. Keep the male data separate from the female data. Make one table which represents each female in your group's ulna and shoe lengths and another table which represents each male in your group's ulna and shoe lengths. Include in these tables results from a few of the other groups.

- Make two scatterplots—one which corresponds to the male data and the other the female data. Make sure your plots contain a well-labeled set of axes. Based on your scatterplot, can you guess a relationship between the length of one's ulna and the length of one's shoe?

- Think of the length of your shoe as a function of the length of your ulna. What is the appropriate domain and range? Calculate the best-fit linear models for the male and female data.

- Based on your model, suggest a rule of thumb describing the mathematical relationship between the length of one's shoe and the length of his/her ulna.

- Suppose someone has an ulna that is 10 inches. Predict the length of her shoe.

Can you think about any other body parts that might be related? How about arm span and height? Describe and conduct an experiment that finds the mathematical relationship of a person's height and his/her arm span. (Please note a similar activity is presented at: http://mathforum.org/alejandro/frisbie/math/leonardo.html.)

Goals: The lab gives students the chance to collect and organize data by hand and to use their graphing calculators. In addition to being introduced to the idea of fitting a function to data points and using the function to make predictions, they learn how to use linear regression on their calculators.

Advantages: The "Body Parts" lab is a good lab to do early in the semester. The students begin by taking measurements of each other. They move around and share their data with different groups, and this creates a nice atmosphere where students get to know each other. The lab involves everyone since measuring, recording, and entering data into the graphing calculator can each be done by a different group member.

Challenges: The most common problem associated with this lab is the nature of the measurements themselves. Most of their numbers are very close together, and there are not that many data points to consider.

Many students are therefore confused by the scatter plot that they see and do not see how a line of best fit is relevant to the problem. The instructor can help with leading questions and encouraging the students to discuss their concerns about the model as part of their lab reports.

Modeling the Hours of Daylight

The website http://aa.usno.navy.mil/data/docs/RS_OneYear.html provides a way for you to obtain a table of the times of sunrise/sunset for one year at different locations worldwide. To start this lab, please go to the website and obtain sunrise and sunset times for Fairbanks, Alaska, Seattle, Washington, Honolulu, Hawaii, and Barrow, Alaska. It is the goal of this lab to use the given information to find functions that model the amount of daylight as a function of time (in days) in Fairbanks, Seattle, and Honolulu. Your assignment is to write a report that explains how you come up with a mathematical model and how you can use your model to answer different questions about the amount of daylight in these cities. The following questions will help you to explore the topic.

1. Given the data for 1999, can you make predictions about the hours of daylight in the future and also in the past?
2. What kind of function do you expect to model the behavior of the amount of daylight?

In fact, there are infinitely many functions that model this kind of behavior. Suppose for each city, we choose the function $f(t) = A\cos(Bt) + C$ to model the amount of daylight, where t is given in days. You can choose a date for $t = 0$, that is, $t = 0$ could represent January 1 or any other day of the year. Remember that a trigonometric function is characterized by three things: its period, amplitude, and midline. In our function $f(t)$, these three quantities are related to A, B, and C (not neccesarily in the same order).

1. What is the practical interpretation of the period?
2. What is the practical interpretation of the amplitude?
3. What is the practical interpretation of the midline?
4. What is the practical interpretation of A?
5. What is the practical interpretation of B?
6. What is the practical interpretation of C?

What kind of information do you need to compute the amplitude, period, and midline of a trigonometric function? On which day do we have the most hours of daylight? On which day do we have the fewest hours of daylight? (In the northern hemisphere!) For each of the three cities do the following:

1. Find A, B, and C.
2. Plot the function $f(t)$. Plot $y = \cos(t)$.
3. Explain how you have to transform the graph of the normal cosine function to get the graph of $f(t)$.

Having found the three functions that model the given situations, do the following:

1. Explain in geographical terms why A and C differ for the three cities, but B stays the same.
2. Use your functions to predict the hours of daylight in each city on October 8, 2010.
3. Use your graphs to find out the days on which there will be 12 hours of daylight. In what seasons does this happen? Is there a name for those dates?

We haven't looked at the data given for Barrow, Alaska yet. You may wonder about the following points:

1. What is the meaning of all the dashes and stars in the table?

2. What would the function of daylight look like? Does it still look like a trigonometric function?

3. Do you have a geographical explanation for the dashes and stars?

4. Can you think of any other towns that will exhibit a similar kind of behavior? Where are they (latitude)?

Please don't just write this lab as a list of answers to the posted questions! The questions are simply there to help you explore the topic. The lab report, however, should be a flowing text explaining how you come up with the mathematical model of a given situation and how you can use this model to point out interesting features of the situation. Don't forget the introduction and conclusion!

Goals: Many students have experience with the sine and cosine function coming into a precalculus class, but few have seen them applied to model periodic phenomena. This lab provides students with the opportunity to understand the concepts of period, amplitude, and midline in terms of a real-word problem.

Advantages: For students who have only understood sinusoidal functions in terms of the unit circle or triangles, this lab develops a more complete conceptual understanding of the sine and cosine functions. By doing this lab, students are forced to interpret input and output values of the function and to see the physical interpretation of midline, period, amplitude, and horizontal shift in the context of this problem.

Challenges: Students generally find this to be one of the most challenging labs that they do over the course of the semester. Many are overwhelmed by all the numbers in the tables of data that they are given and incorrectly think that the point of the lab is to enter all of the data into their calculator and graph it. It is very helpful to have a thorough introduction to the lab before the students begin. During this introduction, the instructor can prompt students for the times of year when we have the most and the least amount of daylight and guide students toward a method of determining the critical features of the function to be calculated. The instructor can also point out details for students to look out for when doing the lab, like conversion from hours and minutes to decimal representations of hours.

The Richter Scale

The following description of the Richter Magnitude Scale can be found on the webpage of the U.S. Geological Survey (USGS) at http://neic.usgs.gov/neis/general/richter.html. It was abridged from *The Severity of an Earthquake* [8], a U.S. Geological Survey General Interest Publication.

Seismic waves are the vibrations from earthquakes that travel through the Earth; they are recorded on instruments called seismographs. Seismographs record a zig-zag trace that shows the varying amplitude of ground oscillations beneath the instrument. Sensitive seismographs, which greatly magnify these ground motions, can detect strong earthquakes from sources anywhere in the world. The time, locations, and magnitude of an earthquake can be determined from the data recorded by seismograph stations.

The Richter magnitude scale was developed in 1935 by Charles F. Richter of the California Institute of Technology as a mathematical device to compare the size of earthquakes. The magnitude of an earthquake is determined from the logarithm of the amplitude of waves recorded by seismographs. Adjustments are included for the variation in the distance between the various seismographs and the epicenter of the earthquakes. On the Richter Scale, a magnitude is expressed in whole numbers and decimal fractions. For example, a magnitude of 5.3 might be computed for a moderate earthquake, and a strong earthquake might be rated as magnitude 6.3. Because of the logarithmic basis of the scale, each whole number increase in magnitude represents a tenfold increase in measured amplitude; as an estimate of energy, each whole number step in the magnitude scale corresponds to the release of about 31 times more energy than the amount associated with the preceding whole number value.

At first, the Richter Scale could be applied only to the records from instruments of identical manufacture. Now, instruments are carefully calibrated with respect to each other. Thus, magnitude can be computed from the record of any calibrated seismograph. Earthquakes with magnitude of about 2.0 or less are usually called microearthquakes; they are not commonly felt by people and are generally recorded only on local seismographs. Events with magnitudes of about 4.5 or greater— there are several thousand such shocks annually—are strong enough to be recorded by sensitive seismographs all over the world. Great earthquakes, such as the 1964 Good Friday earthquake in Alaska, have magnitudes of 8.0 or higher. On the average, one earthquake of such size occurs somewhere in the world each year. Although the Richter Scale has no upper limit, the largest known shocks have had magnitudes in the 8.8 to 8.9 range. Recently, another scale called the moment magnitude scale has been devised for more precise study of great earthquakes. The Richter Scale is not used to express damage. An earthquake in a densely populated area which results in many deaths and considerable damage may have the same magnitude as a shock in a remote area that does nothing more than frighten wildlife. Large-magnitude earthquakes that occur beneath the oceans may not even be felt by humans.

Given below is a table of earthquakes that were recorded during the 1990s.

Place	Year	Magnitude
San Francisco	1996	7.8
Athens	1999	5.8
Taiwan	1999	7.6
Netherlands	1992	5.9
Los Angeles	1994	6.7
Mexico	1999	6.5
Columbia	1999	4.2

Your assignment is to explain how the Richter Magnitude Scale works. Reading and doing Problem 35 in Section 4.4 (see [2]) may help you. Problem 35 defines magnitude or Richter scale rating, M, of an earthquake with seismic waves of size W as $M = \log(W/W_0)$, where W_0 is the strength of a standard earthquake. The problem then leads toward the fact that a one-point increase on the Richter scale corresponds to a tenfold increase in energy output. In your write up, assume that your reader knows what an earthquake is but not much more. Make sure to address the following points in your report:

- What does the Richter Scale measure?

- What does it mean that the scale is logarithmic?

- To find the magnitude of an earthquake you have to take the logarithm of what data?

- Which logarithm is used, i.e., which base are we talking about: 2, 10, e?

- How does an earthquake of magnitude 4 compare to an earthquake of magnitude 5, etc.?

- Compare the severity of several (at least two pairs) of the earthquakes in the table.

- Thinking about the energy that is released during an earthquake: How much more energy was released during the Los Angeles earthquake than during the earthquake in the Netherlands?

- Include a possible graph that represents the Richter Scale: Magnitude is a function of ...?

Goals: For a lot of students, logarithms are one of the reasons that they are taking precalculus and not calculus. This lab is designed to improve students' algebraic skills involving logarithms and to show them a practical application of logarithmic scales.

Advantages: The key benefit of this lab is to tie the idea of logarithms to a practical concept like the Richter scale, which most students have heard of. Even if students struggle intensely with the algebraic computations involved in this lab, most of them will remember that a one-point increase in earthquake intensity represents a tenfold increase in energy output. The instructor can then refer back to this example repeatedly when explaining the properties of logarithms in general. This lab is an excellent exercise in reading and writing about science.

Challenges: The biggest difficulty to overcome in this lab is the frustration encountered by students as a result of their lack of comfort with logarithms. Some students may not know where to begin on this lab or exactly what is expected of them. It is therefore helpful for the instructor to spell the expectations out to students ahead of time and to be prepared to clarify or rephrase the wording of some of the questions. A review of the basic properties of logarithms before the lab is assigned can also be helpful.

Revenue Lab

Furniture Barn is a chain of furniture stores in the northeastern corner of South Dakota. In the past few weeks, they have been disappointed by the revenue produced from the sale of their luxury recliners. In an effort to find a solution to their revenue problems, they have compiled the data below showing the combined daily demand for their recliners at various selling prices.

Recliner Price, $	399	459	499	569	599
Demand (recliners)	62	58	56	52	50

All levels of management seem to agree that revenue would increase by raising the price of the recliner somewhat, but there is disagreement as to how far the price should be raised. Your first job is to decide whether or not raising the price will increase revenue. Then, your job is to decide on a selling price which will maximize Furniture Barn's daily revenue from recliner sales.

Begin by plotting the demand for the recliners as a function of price. Does the collection of data points appear linear? Use linear regression to find an approximate formula for the number of recliners demanded, q, as a function of the price for the recliner, p, in dollars. Does your equation fit the data points exactly? Explain. Use your formula to predict the number of recliners which would be demanded at selling prices of $650, $700, and $750. According to your formula, is there a price at which the recliner is so expensive that there would be no demand?

Now, focus on the question of daily revenue earned from sales of the recliner at various prices. Explain how you would compute Furniture Barn's daily revenue from recliner sales at selling prices of $399 and $650. Do you think that revenue will continue to increase as the price of the recliner is raised higher and higher, or is there evidence to suggest otherwise? Justify your answer by constructing a table showing revenue, r (in dollars) as a function of recliner price, p (in dollars), for an appropriately chosen list of selling prices.

Finally, find a formula $r = f(p)$ that gives approximate revenue as a function of recliner price. Construct an accurate plot of your function on a well-chosen scale. What type of a function do you end up with? Based on your revenue function, find the selling price of the recliner that maximizes Furniture Barn's daily revenue. What is the maximum daily revenue that they can expect?

Note: Keep in mind that this is a lab *report*, not just a disjoint collection of questions to be answered. That means that it should clearly communicate the problem and solution procedure to the reader, with one idea following the next in a smooth and natural fashion. The vast majority of your report should consist of words, not calculations! It must also contain an introduction which presents the problem to the reader and describes the ultimate goal, and a conclusion which sums up the results obtained. Graphs and tables should be placed at the appropriate places in the report, or at least referred to at the appropriate places in the report. As in any other class where you write reports, neatness, organization, and presentation style will definitely weigh into your grade.

Goals: This lab provides a practical application of finding the vertex of a quadratic function. It is hoped that students will see that multiple approaches to the problem of finding the maximum value of a function can work and give them practice in explaining the problem and its solution in detail to a reader.

Advantages: One of the most positive features of this lab is that it allows students to create a simple mathematical model for an everyday quantity on their own. If they do the lab in the intended way, they will discover for themselves that revenue is modeled by a quadratic function in the context of this problem and will be forced to recognize the practical significance of the vertex. It also gives students another opportunity to gain comfort with linear regression and with using a modeling function to make predictions. If extra practice with completing the square is desired, the instructor can ask students to find the vertex of the modeling quadratic algebraically.

Challenges: While most students will have no trouble using linear regression to model demand as a function of price, many of them will automatically assume that revenue is also a linear function and try to use linear regression to find a formula for revenue. This is a great opportunity for the instructor to ask the students if a linear model makes sense for revenue, pointing out that this would mean that revenue grows without bound as the price of a recliner is increased. Also, while students seem to have no trouble at all calculating revenue when given a specific recliner price, they have great difficulty in generalizing this to a formula. It helps for the instructor to be prepared to ask the students leading questions to help them make the leap from specific to general.

Other resources for precalculus labs

We cite two excellent resources for use as labs in precalculus courses. Although they were written to accompany specific texts, they work very well with most textbooks on precalculus.

The first citation is *Projects for Precalculus* by J. Andersen, T. Swanson, and R. Keeley [1]. It is a set of 26 projects to accompany *Contemporary Precalculus* by T. Hungerford [9]. The projects promote similar themes to ones presented in this paper: viewing concepts from several points of view, emphasizing writing skills, use of technology, use of collaborative learning, and stressing the application of mathematics to other disciplines.

The second citation is *Precalculus in Context* by M.J. Davis, J.F. Moran, and M.E. Murphey [3]. This lab manual is a companion to the precalculus text by the authors. It contains 10 labs which emphasize the use of graphing technology as an integral component. Here, too, the labs are collaborative, emphasizing real-world situations, culminating in a written report.

Other reform texts

Reform texts lend themselves well for the use of labs in precalculus. They tend to emphasize understanding in multiple ways and they encourage explanations in written form. In addition to the precalculus book we use, *Functions Modeling Change: A Preparation for Calculus* by Connally, Hughes Hallet, et. al. [2] we cite two other reform texts. *Functioning in the Real World: A Precalculus Experience* by S. P. Gordon, F.S. Gordon, A.C. Tucker, and M. J. Siegel [4] is the outgrowth of an NSF-funded project. The text focuses on the following goals: mathematical thinking involving geometric, numerical, symbolic, verbal, and algebraic approaches; the mathematics in a scientific oriented society emphasizing applications and models; providing the skills and knowledge needed for subsequent courses; and the appropriate use of technology. The text recommends that students do several mathematical projects, individually or in groups, similar to the concept of a lab. The projects are intended to get students involved in formulating mathematical questions, collecting and analyzing data, drawing conclusions and presenting them in a written report.

The Contemporary Precalculus Through Applications text is written by the North Carolina School of Science and Mathematics [6]. The text provides a beginning knowledge of real-world applications in the areas of mathematical modeling, data analysis, discrete mathematics, and numerical algorithms. It adopts an investigative approach to the development of these topics. It lays the foundation to support future course work in subsequent courses such as calculus and statistics. Technology, both calculators and computers, is encouraged as a tool for investigation and discovery. The authors have also developed material for calculus.

Conclusion

In summary, we found the lab assignments to have benefits that are not usually found in a beginning level course. First and foremost, the labs provide students with the opportunity to tackle some real-life problems, to find their solutions, and then to write a report detailing their findings and conclusions. They do this in a small group setting which emphasizes teamwork. Second, they learn how to debate strategies, explain their thinking, and communicate their ideas to fellow students. These are skills that are important to their future employers. Moreover, the labs help to address a big challenge in our precalculus course, namely, how to teach a diverse student audience with varied mathematical background. Requiring students to turn in written lab reports brought these very different groups of students together and encouraged them to communicate mathematical concepts to one another. Finally, another positive aspect of the labs was to break the monotony of the traditional library of functions approach to precalculus by using these functions to model real-life situations.

While we found the labs to have a positive influence on the precalculus course, we discovered that significant preparation was required to successfully implement them. For example, the Hours of Daylight Lab and the Richter Scale Lab assignments require careful thinking before any meaningful calculations can be made. Since students at this level are inherently uncomfortable with open-ended questions, many of them have no idea how to start such a project without a carefully-led discussion. Another challenge we encountered is introducing students to the idea of writing in a math course. We found that giving students a particular audience to address helped them overcome their initial resistance. For example, the Revenue Lab asks students to assume the role of a consultant hired by the owners of a furniture store to write a business proposal for a specific purpose.

Acknowledgement: We would like to take this opportunity to acknowledge the contributions of Kate McGivney to the institutionalization of labs in precalculus. These contributions took place while she served in a teaching postdoctoral position at Arizona from 1997 to 2000.

References

1. Andersen, J., T. Swanson, R. Keeley, *Projects for Precalculus*, Saunders, 1997.

2. Connally, E., D. Hughes Hallett, et. al., *Functions Modeling Change: A Preparation for Calculus*, 2nd edition, Wiley, 2004.

3. Davis, M.J., J.F. Moran, and M.E. Murphey, *Precalculus in Context*, PWS-Kent, 1993.

4. Gordon, S.P., F.S. Gordon, A.C. Tucker, and M.J. Siegel, , *Precalculus in Context: Functioning in the Real World*, 2nd edition, Addison-Wesley, 2004.

5. Hughes Hallett, D., A. Gleason, et. al., *Calculus*, 3rd edition, Wiley, 2002.

6. North Carolina School of Science and Mathematics, *Contemporary Precalculus Through Applications*, 2nd edition, McGraw Hill/Glencoe, 2000.

7. Solow, A., ed., *Preparing for a New Calculus*, MAA Notes #36, Mathematical Association of America, Washington, DC, 1990.

8. *Severity of an Earthquake*, USGS webpage, http://pubs.usgs.gov/gip/earthq4/severitygip.html

9. Hungerford, T. W., *Contemporary Precalculus*, 3rd edition, Harcourt, 1999.

34

The Fifth Rule: Direct Experience of Mathematics

Gary Simundza
Wentworth Institute of Technology

Introduction

In recent years, the use of multiple representations has become an important part of the teaching of mathematics at all levels. This paper describes how a laboratory approach to precalculus instruction can allow students to achieve mathematical insights using direct sensory experience of quantitative phenomena. Such experience can complement and inform the graphical-numerical-analytical-verbal models students use in problem-solving.

Since the beginning of the calculus reform movement, the Rule of Three has become an essential feature of mathematics education. Facilitated by the availability of technological aids to constructing graphs and tables, the use of multiple approaches in describing mathematical functions and processes is a cornerstone of current pedagogical practice throughout the K–16 mathematics curriculum.

Shortly after the Rule of Three construct was widely adopted and became part of the mathematics educator's lexicon, it evolved into the Rule of Four [1]. If analytical, graphical, and numerical representations are all different lenses through which mathematical relationships can be viewed, they are, in many cases, language-independent substitutes for connections or processes that can also be described in words. Thus, verbal description became a co-equal partner to the original trinity. In particular, describing a mathematical model in English (or any other language) can make mathematics more accessible for those who have not typically been successful with traditional symbol-intensive mathematics. Using language (written or oral) to articulate mathematical modes of thought has now become an integral part of the way we expect our students to approach mathematics [2]. This emphasis on language has also forced mathematics teachers to be more precise in their own interactions with students. "Explain your answer" has given way to such instructions as "How is this property reflected in the graph of the function?", as both tests and textbooks have acknowledged that, in some respects, to talk intelligently about mathematics is to understand it.

Government funding initiatives beginning in the late 1980's resulted in, among other things, a number of projects that emphasized laboratory approaches to college calculus and precalculus [3, 4, 5]. Although many of these emphasized computer or calculator investigations of contextual problems, a few involved hands-on experimentation to collect data for analysis. A 1993 National Science Foundation workshop that led to the creation of the Advanced Technological Education program urged that "technician education programs should . . . make experiential, contextual, and collaborative learning an integral part of education" [6]. And one of the basic principles of the AMATYC standards is, "Mathematics must be taught as a laboratory discipline" [7]. A number of elementary and secondary materials that emphasize activities have also become available [8, 9, 10, 11, 12].

Experiencing mathematics

Emphasizing that there are multiple ways of seeing mathematical relationships is consistent with recognizing that students exhibit a diversity of preferred learning styles. Theories of multiple intelligences [13] suggest that some students are more likely to understand any subject through visual presentation, others through verbalization, and so on. As seen in the examples that follow, there is reason to suggest that the Rule of Four should be broadened to a Rule of Five: the fifth component of mathematics understanding is direct experience of mathematical processes. That is, in addition to providing opportunities for obtaining usable data through active experimentation, the act of experimenting itself can promote mathematical concept acquisition. There are numerous opportunities at the precalculus level for students to strengthen their grasp of mathematics by seeing, hearing, and feeling phenomena that reveal mathematical connections in real time. And there is some evidence that such an approach can improve student performance in subsequent courses that rely on mathematical problem-solving [14].

The Mathematics for Technology project at Wentworth Institute of Technology created a series of Mathematics Laboratory Investigations for courses preceding calculus [15]. Intended primarily for students of engineering technology, architecture, and related fields of study, most of the laboratory investigations are interdisciplinary and focus on technical problem situations. Each activity embeds mathematics within an authentic engineering or design application. One of these, entitled "Strength of Materials," uses laboratory measurements of material strength to introduce the concept of slope. Students measure the force required to stretch a rubber band by various amounts, compute stress and strain values from measured data, and graph stress versus strain. The slope of the resulting linear graph is equal to the modulus of elasticity for the material. This experiment is similar to activities that have been widely used in physics courses for many years. Before students perform the experiment, they are each given a length of extruded plastic (polyethylene rod) and asked to pull on it and describe what they observe. If pulled with a moderate amount of force, the plastic resists, but it is elastic enough to return to its original state when the force is removed. However, if the force is increased, there is a point at which the plastic suddenly yields, and without additional force it will stretch a great deal. Indeed, it permanently deforms, visibly necking down to a smaller diameter. (In fact, monofilament fishing line can be produced by such controlled stretching of extruded plastic.) *After* students have taken their rubber band data and graphed it, they are shown the graphs of stress-strain data from samples of a different type of plastic that were stretched using a Universal Testing Machine, with the accompanying instruction (see Figure 1).

"Based on your observations of what happened as you stretched the polyethylene, and assuming that a similar process happened with the plastic specimens that produced the graphs, *see if you can explain the shapes of the graphs by relating them to the behavior of the polyethylene.*"

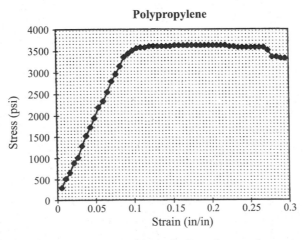

Figure 1. Stress-strain graph for polypropylene

Students who made careful observations of the polyethylene's behavior in response to being pulled are able to relate the linear portion of the graph to the elastic resistance of the polyethylene, the stress increasing with strain. They can also recognize the yield point on the graph where stress levels off, just as it suddenly became easier to stretch the plastic. Here we see the Rule of Five in action. Students experience the response of the polyethylene; they collect data in a table; they graph their own data, and find an equation for the linear portion, as well as identifying the slope as a real physical quantity. And finally they verbalize connections among the representations with reference to their experiences.

While it is certainly possible for students to do much of the important mathematical analysis from instructor- or text-provided data, with numerical-graphical-analytical connections, the experience of stretching the plastic rod makes the exercise come alive in a way that a pencil-and-paper (and graphing calculator or computer) analysis cannot.

Another example relates to properties of exponential functions. In "Getting a Charge Out of Math," a simple circuit (Figure 2a) is constructed that allows a large electrical capacitor (a charge-storing device) to be charged by a battery and then discharged.

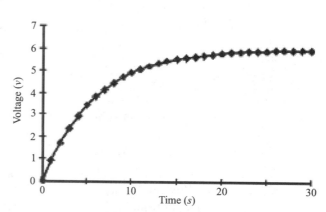

Figure 2. (a) A capacitor charging circuit   (b) Capacitor charging curve

Students use both digital and analog meters to observe the capacitor's voltage increase from 0 to 6 volts when charged, and then decrease quickly to 0 when shorted out with a wire across its terminals. If a resistor is placed between the battery and capacitor, the process is slowed down, and students can observe the change in voltage with time. As with the plastic stretching experiment, they are asked to write down their observations. What becomes obvious is that the rate of charging (or discharging) is very fast at first, becoming gradually slower and seeming to stop. Students find that if they watch the analog meter, the speed of rotation of the needle decreases in either case.

When the data are graphed (Figure 2b), the speed of the needle can be related to the absolute value of the slope (for charging or discharging).

An interesting thing happens when students watch the behavior of the digital meter, however. They know the capacitor should charge to 6 volts. But as they continue to take data until the capacitor reaches full charge, they find that, while the voltage may reach, say, 5.8 or 5.9 volts fairly rapidly, the value continues to increase, albeit slowly, for quite a while. After several minutes, the voltage may still be approaching 6 volts through values like 5.987, 5.988, 5.989, etc. The movement of the analog meter's needle has appeared to stop by this time, but the digital meter continues its ever slower increase toward 6 volts. Students who have observed this type of behavior directly develop a much deeper understanding of the nature of asymptotic behavior. And the equipment is simple enough that no specific technical knowledge is

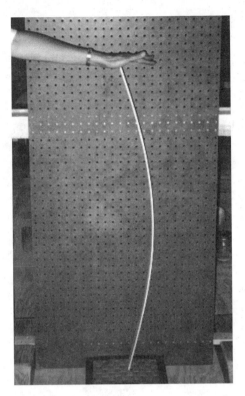

Figure 3. Buckling a thin column

needed to understand the application—architectural majors do as well with this lab as electronics majors.

Perhaps an even more personalized acquaintance with such behavior comes from buckling a thin column. If a vertical load of sufficient magnitude is placed on top of a thin column, it will buckle sideways. If this happens in a structural column, the result can be sudden and catastrophic failure. (Buckling of the thermally weakened steel columns supporting the World Trade Center towers was directly responsible for their collapse following the September 2001 terrorist attacks.) Using plastic columns and an ordinary bathroom scale, students can measure the force required to buckle columns of various lengths (Figure 3), and can experimentally investigate the relationship between column length L and Euler Load P embodied in the formula $P(L) = \pi^2 EI/L^2$.

By observing how difficult it is to buckle columns of shorter and shorter length, they gain intimate knowledge of the meaning of an asymptote (as well as, for the particularly zealous, a sore hand).

On the other hand, some functions exhibit decreasing rates of change without being asymptotic. Ask a student how much sound is produced by two alarm clocks, as compared to one, and the answer will certainly be "twice as much." But the loudness of two alarm clocks as perceived by the human ear is not double that of one alarm clock, and use of a meter measuring loudness in decibels confirms that fact. If the number of alarm clocks is increased, the loudness continues to increase as well, not linearly or asymptotically, but as a function of the logarithm of the total sound intensity. Students can experience this aurally and confirm their impressions with a sound meter.

A model for laboratory-based mathematics

In developing this approach to precalculus mathematics, our group at Wentworth settled on a paradigm for creating laboratory investigations that is shown in Figure 4.

Beginning with a goal-oriented contextual situation, problem-solving involves active experimentation,

The "Ideal" Math Lab

Engineering/Design Problem

Real World Application

Cooperative Group Work

Imitates Insustrial Situation/Peer Learning

Hands-On Activity

Data Collection/Computer Simulation

Technology Used Appropriately

Critical Thinking

Concrete Results

Reinforced Math Application

Verbal Expression

Clear Exposition of Results

Revisit the Problem

In a New Mathematical Context

Figure 4. Characteristics of an "ideal" math lab investigation

using technology where appropriate. The problem should require applying previously acquired mathematics knowledge to the learning of a new concept. While it is not always possible to incorporate all of the features of an ideal lab in a laboratory activity, this list is a useful set of guidelines for creating activities that engage and motivate students while they learn to think mathematically. An overarching goal of this project is to foster a mindset that reaches for mathematical approaches in solving problems that students will confront later in other courses and careers.

The last listed feature of the paradigm, "revisiting the problem in a new mathematical context," implies a problem with a mathematical core that transcends the specific problem situation. A good example is a "Building Site Excavation" lab in which students are given a topographical map of a proposed home-building site on a sloping piece of land (Figure 5).

The goal is to determine whether, in excavating for the foundation of a house, enough earth will be removed from the high side to build up the low side to the necessary height. If not, more fill must be trucked in. On the other hand, if there is excess dirt excavated, it must be trucked away from the site. Students can begin by creating or examining a three-dimensional model based on the topographical map (Figure 6).

They go through an estimating process that culminates in coordinatizing the foundation region and using linear interpolation to find the height of each grid point. A spreadsheet analysis of the volume contained in each element of the resulting collection of rectangular prisms (Figure 7) then allows students to solve the volume fill problem and also construct a 3-D graph of the region (Figure 8).

The connection between the 2-D site map and the 3-D model and graph can give students a concrete sense of multivariable functions. Having seen a geometrical third dimension as a function's output, they

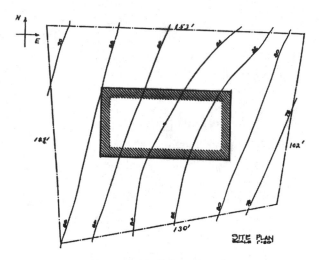

Figure 5. Building site plan

Figure 6. A physical model of the site

	X								
Y	0	10	20	30	40	50	60	70	80
0	87.2	86.2	85	83.8	83	82.2	81.2	80.9	80.4
10	87.3	86.4	84.6	84.3	83.4	82.6	81.8	81.1	80.2
20	87.5	86.7	85.9	84.9	83.9	83	82	81.4	80.5
30	87.8	87	86.2	85.3	84.4	83.4	82.5	81.5	81
40	88	87.3	86.5	85.8	85	84.2	83.3	82.4	81.5

Figure 7. Elevations on the house footprint

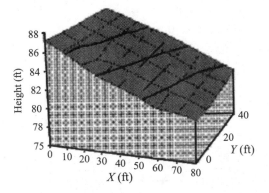

Figure 8. A computer model of the house site

are then prepared to consider level curves of non-geometrical quantities in other contexts such as temperature, pressure, electrical field strength, or material stress, and to understand how functions involving these quantities can be represented. It's not necessary to wait for a Calculus III course to gain such an appreciation.

Another feature of the building-site problem is that it is robust: it works on many mathematical levels. Even elementary school children can explore topographical maps to begin to understand their use in representing three-dimensional space [16]. Precalculus students can compare their site map and model to help them determine the direction of the land's greatest slope at any point, laying the groundwork for the later study of directional derivatives and gradients. And of course the computation of volume by adding small elements is a first step toward numerical integration.

One final example shows how a relatively abstract concept, composite functions, can be made accessible with the proper context. An inexpensive electrical device called a diode can be used as the basis for a digital temperature-measurement system in the range of $0^0 - 100^0$ Celsius. In a simple battery-powered circuit (Figure 9), the voltage across the diode is linear with temperature.

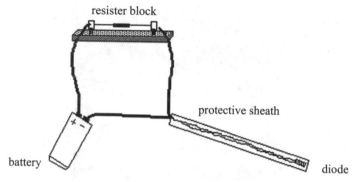

Figure 9. "Temperature Sensing Diode" circuit

Students can easily measure the diode voltage at room temperature, and then hold the diode between their fingers to provide a higher skin-temperature reading. The linear relationship can be confirmed with a third temperature-voltage pair (ice water being convenient). However, at higher temperatures the voltage is lower: an equation like $V_D = -0.0021T + 0.66$ is typical. This is inconvenient from the standpoint of using the voltage display to provide a temperature reading. The engineering design problem is as follows: Given the negatively-sloped linear function that characterizes the diode, how can an output voltage V_O be produced that is directly proportional to Celsius temperature? For example, how can the function $V_D = -0.0021T + 0.66$ be turned into the function $V_O = 0.1T$? Then a mere shift in the placement of the decimal point in the display would accomplish the direct readout of temperature.

Mathematically, what is needed is another linear function $V_O = mV_D + b$ such that $V_O(V_D) = 0.1T$. Students can show algebraically that the required function is $V_O = -48V_D + 31.7$. The engineer would build a signal-conditioning circuit (SCC) with output described by this function. Of course, the SCC completes a composite function (Figure 10).

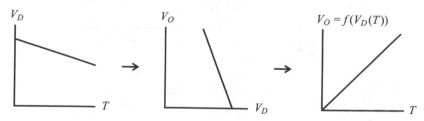

Figure 10. A composite function

Students have observed directly that as they warmed the diode with their fingers, its voltage decreased. They recognize that finding the missing link in a composite function achieves the mathematical solution of an important design goal. Instead of seeing $f(g(x))$ as merely a group of symbols, they can now connect f and g with real systems.

Conclusion

For our better students, the Rule of Four serves as an appropriate and often sufficient organizing principle for the study of mathematical relationships. For many others, particularly those who are not ready to begin their college mathematics program with calculus, inclusion of an experiential component in mathematics courses can help to make those relationships more accessible.

Acknowledgement: The Mathematics for Technology project was supported by a grant (DUE-9553704) from the Advanced Technological Education program of the National Science Foundation.

References

1. Gordon, Sheldon and Deborah Hughes Hallett, "Lessons from the Calculus Reform Effort for Precalculus Reform," in *Preparing for a New Calculus*, MAA Notes #36, Anita Solow, ed., Mathematical Association of America, Washington, DC, 1990.

2. Sterrett, Andrew, ed., *Using Writing to Teach Mathematics*, MAA Notes #16, Mathematical Association of America, Washington, DC, 1990.

3. Leinbach, L. Carl et al, eds., *The Laboratory Approach to Teaching Calculus*, MAA Notes #20, Mathematical Association of America, Washington, DC, 1990.

4. Hastings, Nancy Baxter, *Workshop Precalculus: Discovery with Graphing Calculators*, Key College Publishing, Emeryville, CA, 2002.

5. Pazdar, John S. et al, *Mathematics Explorations I and II*, Capital Community Technical College, Hartford, CT, 1998, 2000.

6. *Gaining the Competitive Edge: Critical Issues in Science and Engineering Technician Education*, National Science Foundation, Arlington, VA, 1994.

7. Cohen, Don, ed., *Crossroads in Mathematics: Standards for Introductory College Mathematics Before Calculus*, American Mathematical Association of Two-Year Colleges, Memphis, TN, 1995.

8. Consortium for Mathematics and Its Applications, *Mathematics: Modeling Our World* Courses 1–4, W.H.Freeman & Co, 1998–2000.

9. Burns, Marilyn, *About Teaching Mathematics: A K–8 Resource*, 2nd edition, Math Solutions Publications, Sausalito, CA, 2000.

10. Charlesworth, Rosalind and Deanna J. Radeloff, *Experiences in Mathematics for Young Children*, 2nd edition, Delmar, Albany, NY, 1991.

11. Education Development Center, *Mathscape: Seeing and Thinking Mathematically*, Glencoe McGraw-Hill, New York, NY, 1998.

12. *Everyday Mathematics*, 2nd edition, Everyday Learning Corporation, Chicago, IL, 2000.

13. Gardner, Howard, *Frames of Mind: The Theory of Multiple Intelligences*, Basic Books, New York, NY, 1983.

14. Simundza, Gary M., unpublished results.

15. Simundza, Gary M. et al, *Precalculus Investigations: A Laboratory Manual*, Prentice-Hall, Upper Saddle River, NJ, 1999.

16. Charbonneau, Manon P., *Learning to Think in a Math Lab*, National Association of Independent Schools, Boston, MA, 1971.

Ideas and Projects that Work: Part 2

This volume focuses on challenges that need to be met and changes that need to be made. Repeatedly authors have stressed the importance of emphasizing conceptual understanding , instead of rote manipulations; the importance of focusing on situations where mathematics is used in the real world, instead of on rinky-dink word problems; the importance of utilizing the methods of data analysis, instead of relegating it to the last chapter (if it's there at all); and the importance of fostering an active learning environment, where students ask what-if type questions, make connections, explore mathematical ideas, and work collaboratively instead of sitting passively copying notes off the board. The good news is that our colleagues are grappling with how to do this. Projects are being developed. New texts are being written. Alternative pedagogies are being implemented. The authors of the following papers are at the forefront of this movement. But, they are not alone. They are simply the ones who came to mind when we started this project over three years ago. We apologize to those of you who do not appear in this collection. You are here in spirit and are an important part of this movement of change.

The *Mathematics in Action* project aims to empower college mathematics students with a real-world mathematical literacy that will provide a solid foundation for future study in mathematics and other disciplines.... The project's goal to empower students mathematically focuses on developing desired student outcomes in five main areas: number sense, symbolic sense, a general function sense, a thorough linear function sense and a sense of nonlinear relationships. The word "sense" in each of these areas certainly conveys developing requisite skills, but far more than that, it means generating mathematical intuition and building techniques of reasoning.

Mathematics in Action: Empowering Students with Introductory and Intermediate College Mathematics,
Ernie Danforth, Brian Gray, Arlene Kleinstein, Rick Patrick, and Sylvia Svitak

Precalculus: Concepts in Context is a combined text and lab/project manual. The text portion is unusual in that it uses a mathematical modeling approach and requires students to interact with the text during reading (there are fill-ins within the text).

Precalculus: Concepts in Context
Marsha Davis

[*Functions and Change, A Modeling Approach to College Algebra* contains] applications from business, the natural sciences, social sciences and many other areas. The mode of presentation is not to look at abstract, algebraic formulas of a certain type, look at their properties and then go looking for applications. Rather, physical situations are presented, discussed, and the need for a new idea or type of function made evident.

Rethinking College Algebra
Benny Evans

[COMAP has produced] two new texts, unimaginatively tilted, *College Algebra* and *Precalculus*. It should be noted that the texts cover the mathematical content one would expect in courses with these titles, with a few notable differences. Not surprisingly, these differences all relate to the fact that our approach is applications and modeling based. This has always been at the foundation of our approach to mathematics education and it directly addresses the contextual relevance issue emphasized again and again throughout the *Standards* documents.

From The Bottom Up
Sol Garfunkel

[*Functioning in the Real World: A Precalculus Experience* emphasizes] the qualitative, geometric and computational aspects of mathematics within a framework of mathematical modeling at a level appropriate to precalculus students. All the mathematical knowledge and skills students will need both for calculus and for quantitative courses in most other disciplines are introduced, developed and reinforced in the process of applying mathematics to model and solve interesting and realistic problems.

The Functioning in the Real World Project
Florence S. Gordon and Sheldon P. Gordon

If students are to remember what they learn, the courses they take must tell a coherent story. This story provides a framework onto which they can hang their newly acquired knowledge.... The central theme we chose for our precalculus course is how functions can be used to model change. This theme provides a framework into which all the prerequisites for calculus naturally fit (functions, graphing, algebra, trigonometry, numerical approximation), while at the same time illuminating a central concept of calculus—the rate of change.

The Importance of a Story Line: Functions as Models of Change
Deborah Hughes Hallett

Workshop Precalculus: Discovery with Graphing Calculators, seeks to provide students with a bridge to the study of calculus by helping them develop the confidence, understanding and skills necessary to continue their study of mathematics.... In the workshop environment, students learn by doing and reflecting on what they have done. The guiding principle is that instructors try not to *talk* in depth about a concept until students have had an opportunity to *think* about it first.

Using a Guided-Inquiry Approach to Enhance Student Learning in Precalculus
Nancy Baxter Hastings

Since several college algebra/precalculus reform projects were already underway, we concentrated on course materials for three precursor courses: arithmetic review, elementary algebra, and intermediate algebra. Materials for these courses are commercially available as *The Maricopa Mathematics Modules* and *Beginning Algebra with Arithmetic Review*. From 15 available modules, a course instructor can select three or four modules to match their own course objectives.... The *Modules* embody four attributes: student centered, activity driven, context centered, and technology inclusive.... In the *Modules*, the guiding principle is Investigate–Generalize–Practice.

Maricopa Mathematics
Alan Jacobs

For many years, the head of Continuing Education at the University of Massachusetts, Boston had been receiving complaints from industry that their employees couldn't think quantitatively. She called together a handful of faculty from the Mathematics Department, Graduate School of Education, and Academic Support, and asked us to a design a course she could market to these companies. Having an academic bias, we took this as an opportunity to revise our traditional college algebra course—with the thought that later some of the modules could be used to serve the corporate world.... The result is the text *Explorations in College Algebra*, which we now use in an alternative college algebra course called Quantitative Reasoning.

College Algebra/Quantitative Reasoning at the University of Massachusetts, Boston
Linda Almgren Kime

Our project has attempted to move away from the procedural orientation in mathematics, which focuses on getting the correct answers and which students have learned to value above all. Instead, our curricular materials [Applying Algebraic Thinking to Data and Mathematical Investigations: Concepts and Processes for the Introductory Algebra Student] offer an alternative approach to learning algebra for students who have taken one or both of beginning and intermediate algebra—in high school or at college—and who have failed to place into a college-level mathematics course. What it means to learn mathematics, the nature of mathematics, the development of flexible thinking, and the ability to see and value connections are explicit goals of instruction.

Developmental Algebra: The First Mathematics Course for Many College Students
Mercedes McGowen

In this project—*Workshop Precalculus: Functions, Data and Models*—[we] are developing materials for a course that integrates ideas of data analysis and mathematical modeling into the study of precalculus. The primary goal of the materials is to prepare students for calculus, primarily by helping them to develop a deep understanding of the crucial concept of function. A secondary goal, especially important considering that many precalculus students never go on to study calculus, is to develop skills of data analysis and mathematical modeling, which will be valuable for courses in other disciplines.

Workshop Precalculus: Functions, Data, and Models
Allan J. Rossman

The primary goal of *Contemporary College Algebra* is to empower students to become exploratory learners, not to master a list of algebraic rules. Some of the means that are used to establish an exploratory environment for the students include: (1) queries for engaging students in questioning and exploring the material being presented; (2) exercises that explicitly ask students to: explore, ask what-if type questions, make up examples, further investigate worked examples, or iterate for the purpose of recognizing a pattern and developing a sense for the behavior of the solution; and (3) graphically fit a curve to a data set.

Contemporary College Algebra
Don Small

We wanted *Precalculus: A Study of Functions and Their Applications* to be a non-traditional precalculus text that treats functions as the object of study while focusing on important mathematical concepts. We did this by introducing each of the basic types of functions (linear, exponential, logarithmic, periodic, and power) early in the book rather than relegating each type to a separate chapter. Doing so allows us to emphasize the commonalities and differences between the various types of functions. The properties of each type of function are developed throughout the remainder of the text....The text was written in a conversational format addressed to the student and students are expected to read it.

Precalculus: A Study of Functions and Their Applications
Todd Swanson

Our materials address essentially all the topics found in traditional college algebra and precalculus books. However, we designed several innovative features to meet our objectives. To connect mathematics to the world around our students, we created a large collection of what we call Mathematical Looking Glasses, each of which is a discussion of mathematics in a physical context, accompanied by exercises....To encourage students to read and learn actively, we weave our discussion of most topics around a Mathematical Looking Glass and intersperse exercises at several points in each lesson.

Successes and Failures of a Precalculus Reform Project
David M. Wells and Lynn Tilson

Since 1991 [we] have developed unique materials for use in mathematics courses ranging from algebra through calculus. These projects have resulted in three books, *Earth Algebra* (college algebra), *Earth Angles* (precalculus), and *Earth Studies* (applied calculus), that all have applications to environmental issues that affect students' lives. They are designed to generate more interest in the use of mathematics as a tool to analyze real situations. The authors are currently working on a new curriculum development project that is an extension of the work described above. The goal is to produce versatile, technology-intensive materials for classroom use and teacher training.

The Earth Math Projects
Nancy Zumoff and Christopher Schaufele

35

Mathematics in Action: Empowering Students with Introductory and Intermediate College Mathematics

Ernie Danforth
Corning Community College

Brian Gray
Howard Community College

Arlene Kleinstein
Farmingdale State University of New York

Rick Patrick
Adirondack Community College

Sylvia Svitak
Queensborough Community College

The Mathematics in Action project aims to empower college mathematics students with a real-world mathematical literacy that will provide a solid foundation for future study in mathematics and other disciplines. The project was developed by the Consortium for Foundation Mathematics, a team of fourteen SUNY and CUNY faculty, with support from the National Science Foundation (DUE 9455638), and is based on the AMATYC *Crossroads* Standards.

The project's goal to empower students mathematically focuses on developing desired student outcomes in five main areas: number sense, symbolic sense, a general function sense, a thorough linear function sense, and a sense of nonlinear relationships. The word "sense" in each of these areas certainly conveys developing requisite skills, but far more than that, it means generating mathematical intuition and building techniques of reasoning.

To achieve desired student outcomes, project materials are written with the expectation that by completing the course, students would be able to perform tasks beyond the basic skills/knowledge level. These tasks include extracting relevant data to solve realistic problems, analyzing and interpreting graphical and tabular data, recognizing and expressing, in verbal, numerical, graphical, and symbolical format, the patterns displayed by linear data, and identifying equivalent variable relationships in numerical, algebraic, and graphical format and translating those relationships from one representation to any of the others.

The project's objectives extend to developing general education competencies through its realistic contextual approach to learning mathematics. The project aims

- to help students to use proportional reasoning appropriately in contextual applications,

- to read technical material with facility to determine information relevant to solving a problem,

- to recognize that solution processes are multifaceted,

- to express and communicate solutions to problems verbally and in writing by using appropriate vocabulary, sentence structure, and persuasion,

- to demonstrate an improved comfort level when applying critical thinking and problem-solving skills,

- to work collaboratively with others to enhance mathematical understanding through problem-solving, and

- (last if not least) to appreciate the importance of mathematics in everyday life and in the work place.

The Mathematics in Action project materials form a three-book sequence [1, 2, 3] designed to serve a very large population of college students who, for many reasons, have not succeeded in learning mathematics. The original two books (now in second editions) were developed for the elementary algebra and the intermediate/college algebra audience. The newest book, designed for the prealgebra audience, became available in fall 2003.

The project is based on the belief that students learn mathematics best by doing mathematics. The books present a series of realistic situations from which the crucial need for mathematics arises. Students are guided to take an active role, to develop a sense of independence, and to take responsibility for their own learning. This approach is illustrated below in the introduction to rates of change, which provides the crucial link between slope and linear functions.

Suppose you are a member of a health and fitness club. A special diet and exercise program has been developed for you by your personal trainer. At the beginning of the program, and once a week thereafter, you are tested on the treadmill. The test consists of how many minutes it takes you to walk, jog, or run 3 miles on the treadmill. The following data gives your time, t, as a function of weeks, w, over an 8-week period.

Weeks, w	0	1	2	3	4	5	6	7	8
Time, t, (in min)	45	42	40	39	38	38	37	39	36

Note that $w = 0$ corresponds to the first time on the treadmill when you started this exercise program, $w = 1$ is the end of the first week, $w = 2$ is the end of the second week, and so on.

This situation is followed by a series of guided questions designed to aid the student in developing a conceptual understanding of rate of change. The lesson leads to the slope concept and ultimately to linear functions. Examples of some of the questions follow.

1. Is time, t, a function of weeks, w? If so, what are the input and output variables?

2. Plot the data points using ordered pairs of the form (w, t).

3. Your time decreased during each week of the first four weeks of the program.

   a. Determine the total change in time, t, during the first four weeks of the program (i.e., from $t = 45$ to $t = 38$). Why should your answer contain a negative sign? Explain.

   b. Determine the change in weeks, w, during this period (that is, from $w = 0$ to $w = 4$).

Although the approach and order may be nontraditional, our third book covers the topics routinely found in a college algebra and trigonometry curriculum. In particular, this book [3] includes the following topics:

- Functions

- The algebra of functions (including composition)

- Exponential and logarithmic functions

- Quadratic and higher order polynomial functions

- Rational and radical functions

- An introduction to trigonometric functions

Technology is integrated throughout so students can interpret real life data numerically, symbolically, and graphically. Access to the TI-83 Plus calculator or its equivalent is assumed. Students use the technology to investigate graphs, to solve equations, to create tables for numerical investigation, and to produce scatterplots and regression equations. Students are frequently asked to solve problems either algebraically, graphically or numerically, and then to check using one of the other techniques. This would not be practical, or in some cases possible, without the technology.

The major differences between the Mathematics in Action approach and the traditional approach include the following:

- Lessons begin with a contextual problem or situation to motivate a new skill or concept. The mathematics arises naturally from the context.

- The materials are designed to work well in group situations where students are actively involved in the learning process. They learn not only from the teacher but by exploring situations on their own and sharing insights with classmates.

- There is a decreased emphasis on topics that have few practical applications, e.g., factoring and simplifying algebraic expressions. For many students, this is the last mathematics course they will take, and we believe it is vital that they understand the relevance of mathematics in the contemporary world.

- The use of alternative assessment techniques such as group exams and portfolios is encouraged.

The key to bridging the gap between abstraction and application, and the basis for transfer of learning is to encourage students to construct, reflect on, and apply their own mathematical models. We are confident that all students can succeed with our approach and materials to achieve the following goals:

- Develop mathematical intuition along with a relevant base of mathematical knowledge.
- Connect classroom learning with real-world applications.
- Be efficiently and thoroughly prepared for additional college experiences in mathematics and related disciplines.
- Work both in collaborative groups and independently.
- Increase mathematical literacy through meaningful applications and explorations.
- Build techniques of reasoning, regardless of level of preparation.
- Challenge and, at the same time, foster positive attitudes that build confidence in abilities to learn and use mathematics.
- Apply and display their learning and understanding through multi-faceted assessment.

Mathematics in Action has changed the emphasis in our own classes from "covering the syllabus" to "uncovering the mathematics" with our students and it is our goal to extend the benefits of this approach beyond our own classrooms.

Acknowledgements: We would like to note the other members of the Consortium for Foundation Mathematics:

Ralph Bertelle, Columbia Greene Community College
Judith Bloch, University of Rochester
Roy Cameron, SUNY Cobleskill
Carolyn Curley, Erie Community College–South Campus
Kathleen Milligan, Monroe Community College
Patricia Pacitti, SUNY Oswego
Renan Sezer, LaGuardia Community College
Patricia Shuart, Polk Community College
Assad Thompson, LaGuardia Community College

References

1. Consortium for Foundation Mathematics, *Mathematics in Action: Prealgebra Problem Solving*, Addison Wesley Longman, Reading, MA, 2003.

2. Consortium for Foundation Mathematics, *Mathematics in Action: An Introduction to Algebraic, Graphical, and Numerical Problem Solving*, Second Edition, Addison Wesley Longman, Reading, MA, 2004.

3. Consortium for Foundation Mathematics, *Mathematics in Action: Algebraic, Graphical, and Trigonometric Problem Solving*, Second Edition, Addison Wesley Longman, Reading, MA, 2004.

36

Precalculus: Concepts in Context

Marsha Davis
Eastern Connecticut State University

Precalculus: Concepts in Context [1] was developed in response to calculus reform and to the authors' general dissatisfaction with results in traditional precalculus courses. Believing that traditional precalculus instruction failed to prepare students for reform calculus (and we would argue, for traditional calculus as well), Judy Moran, Mary Murphy and I set out to reform precalculus with the publication of a laboratory manual, *Precalculus in Context: Functioning in the Real World* [2]. The lab manual was designed to supplement a standard precalculus course by providing opportunities for students to work collaboratively on lengthy, context-based problems. However, we soon discovered that the supplement was more powerful than the course it was meant to serve. During labs, students were actively grappling with mathematical problems from real-world contexts and they were learning to think, speak, and write mathematics. Furthermore, by listening to their discussions, we learned much about what students really understood or failed to understand. As a result, our more global fix for precalculus, *Precalculus: Concepts in Context,* was a combined text and lab/project manual. The text portion was unusual in that it used a mathematical modeling approach and required students to interact with the text during reading (there are fill-ins within the text).

In writing the second edition, we have changed neither our original goals nor our strategies for achieving those goals. Our major goal has been to make the text more student and instructor friendly. Key points are summarized in the margins for easy reference. Only one lab is essential to text development, which gives instructors greater latitude in lab selection. Chapters in the second half of the text are unlinked and, thus, can be taught in any order.

Student background

Precalculus: Concepts in Context assumes that students have an algebra II background (with or without trigonometry). However, we recognize that many students' algebraic skills are rusty. Rather than break up the flow of the text with algebra review, we include an algebra appendix. Margin notes in the text send students to specific sections of the algebra appendix as they are needed.

Project description/goals

We have rejected the more traditional approach of compartmentalized layout of topics. Instead, several themes run throughout the text: mathematical modeling of real-life phenomena, choice of a convenient and meaningful scale/variable (including exponential and logarithmic), constant and nonconstant rates of change (and how to interpret them in context), and relationships between algebraic statements and geometric representations. Our pedagogy is based on four principles:

- writing about mathematics deepens understanding

- student exploration is at least as valuable as teacher explanation

- collaboration, rather than competition, promotes genuine learning

- graphing technology widens the range of questions students can consider, supports development of connections between algebraic and geometric representations, and provides alternative strategies for problem solving.

Modeling approach

Precalculus: Concepts in Context subscribes to a modeling approach. As an example, one of our text scenarios is based on a 1992 newspaper article with the headline: "The Decade Will Add a Billion." In the accompanying article, the executive director of the United Nations Population Fund explains that the world's population is increasing by approximately a quarter of a million people every day and that she anticipates a total increase in population of one billion people during the decade of the 1990s. Students recognize that the assumption of a constant daily increase leads to a linear model. Predictions based on this linear model, however, fall short of the one billion increase indicated by the newspaper's headline.

During the analysis, students consider whether the assumption of constant annual growth regardless of population size seems reasonable. Their answer, a resounding No, motivates the adoption of a new assumption, constant annual percentage growth, and the development of an exponential model.

When students compare the graphs of their two models (see below), they realize that for the first half of the decade the two graphs are practically indistinguishable.

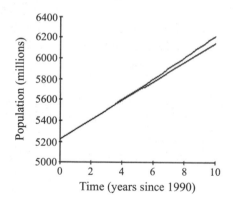

Noticeable separation occurs only toward the end of the decade. From this example, students learn about rate of change assumptions associated with linear and exponential functions and observe that an exponential function can be approximated locally by a linear function.

Lab activities

Each chapter contains one or more labs relevant to the mathematical topics in the chapter. The labs are designed to be cooperative ventures that culminate in a written report submitted by the group. Some labs can be used to introduce new topics while others can be used for assessment. For example, the *Graph Trek* lab is an exploration. This lab helps students to discover how they can affect the graph of a function by modifying its formula—that is, by performing an algebraic transformation. It covers translations (vertical and horizontal), reflections (in the x-axis and in the y-axis), and stretches and compressions (vertical and horizontal). The *Atmospheric Pressure* lab, on the other hand, serves as an assessment of text material. This lab asks students to write a linear model and an exponential model for atmospheric pressure as a function of elevation. Because elevation is measured in 1000-foot units in the linear model and 900-foot units in the exponential model, students must reconcile the variables before they can compare the two models.

Projects and explorations

In addition to labs, every chapter contains projects and explorations. These are non-routine exercises that can be used in different ways: as take home exam questions, small-group class activities, or opportunities for independent work. For example, the project *Copycats* is an exploration of how the parameters A, B, C, and D affect the graphs of sinusoidal functions, $f(x) = A\sin(B(x - C)) + D$. This is a great small-group activity to assign prior to a class presentation of period, amplitude, and phase shift. The project *Fuels Rush In* presents gas usage and temperature data (from one of the author's gas bills). Students fit sinusoidal models to the gas-usage-and-time data and to the temperature-and-time data. This makes an excellent take-home exam question.

Classroom experiences

The second edition of *Precalculus: Concepts in Context* has been field tested at Smith College (Northampton, MA) and Eastern Connecticut State University (Willimantic, CT), two very different institutions. At Smith College, precalculus is designated as a writing intensive course. Students complete 10 labs but relatively few projects. At Eastern Connecticut State University, students complete three to four labs and many more projects. (In particular, the Just Algebra projects are always assigned at ECSU and rarely assigned at Smith.)

Students are generally more comfortable with the step-by-step approach of the projects than the extensive-word-problem approach of the labs. They often complain about the labs, particularly early in the semester. Many are not sure that they should be expected to think through a math problem on their own without first seeing an example. Students worry that they cannot check their work; there are no back-of-the-book solutions for labs. Over time students learn to rely on their group members to determine whether their answers are reasonable. Finally, students complain about how much time it takes to write the lab reports, particularly the first lab report. However, after all this complaining, it is not unusual to find comments, such as the ones below, on course evaluations at the end of the semester.

- Writing lab reports ... helped me not only learn appropriate and correct mathematical language, it forced me to confront areas of understanding about which I was unsure. Writing about larger concepts and how math is applied made me clarify connections on an entirely different level.

- I liked the AIDS, SAD, and Turtles labs because they put math in the context of situations that were not mathematical.... These labs forced me to understand all the information (graphs, equations, derived functions, etc.), but then it was another, separate process to tie in each bit of info ... to explain its relevance.

Based on classroom experiences both at Smith College and at ECSU, we are very excited about the new and improved second edition. Without sacrificing the exploratory, hands-on pedagogy that was a hallmark of the first edition, we have found the second edition easier for students to use and more adaptable to individual instructors' course goals.

References

1. Moran, J., M. Davis, and M. Murphy, *Precalculus: Concepts in Context*, Brooks/Cole Publishing Company, 1996.

2. Davis, M., J. Moran, and M. Murphy, *Precalculus in Context: Functioning in the Real World*, PWS Publishing Company, 1992.

37

Rethinking College Algebra

Benny Evans
Oklahoma State University

There are many difficulties with traditional college algebra. Oklahoma State University mirrors many other campuses in that virtually every student on campus must take some mathematics course, and for the vast majority that course is college algebra. By and large the students don't like the course, and they perform miserably. Success rates are embarrassingly low; perhaps the lowest of any course on campus. Such courses quickly draw the attention of the upper administration. Even worse, for most of the students, this is the last mathematics course they will ever see, and it shapes their perception of what mathematics is. Far too many come away with the idea that mathematics is no more than the manipulation of meaningless formulas into other meaningless formulas. Little wonder that much of the population thinks mathematics is stupid and pointless. As mathematicians, we see mathematics as just the opposite; a beautiful discipline whose manifestations abound in nature, science and many other fields. I am personally on a campaign to give a much larger segment of the population a little taste of the real beauty and utility of mathematics.

In 1995, through a National Science Foundation Grant, Bruce Crauder, Alan Noel, and I started a project designed to answer some basic questions about college algebra, at least on our own campus. Our first question, "Why do students enroll in the course?" was easily answered. Their majors require it for graduation. The second question, "Why do other departments send their students to college algebra, and what do they expect them to learn?" proved a bit more difficult but a great deal more interesting. We were most interested in students who are not headed for engineering calculus. (Statistics show that at many places, that may be as high as precalculus.) Over a two-year period, we interviewed people from virtually every department on campus that had college algebra as a requirement, but not engineering calculus. We wanted to know what skills they expected their students to gain from the course. The results were interesting and sometimes surprising.

- Understand the basic relationship among formulas, verbal descriptions, graphs, and data tables.
- Understand and use linear and exponential functions. (Other functions such as logarithmic and trigonometric functions were also mentioned, but linear and exponential functions were the overwhelming winners.)
- Perform linear algebraic manipulations.
- Have a qualitative understanding of rates of change.
- Understand how to make and use elementary models.

Items not mentioned by *anyone* we talked with included simplifying complex fractions, solving quadratic equations, radicals, and so on. This may not be scientific proof that we should deemphasize algebraic manipulations, but we took it as a pretty good argument.

The next step was to scrap the traditional college algebra curriculum and design a new course from the ground up using the information we had gleaned. We settled on the following guidelines:

- Create a modeling course based on material drawn from other disciplines and from the scientific literature.

- Emphasize linear, exponential, and power functions.

- De-emphasize nonlinear symbol manipulation and supplant where possible by calculator or spreadsheet.

- Put qualitative idea of rates of change at the heart of the course.

- Emphasize mathematical reasoning and clear communication.

Course content

The project resulted in a text [1] emphasizing the following topics:

Calculator arithmetic. A brief introduction to basic calculator use and order of operations.

Functions. Functions are introduced and it is emphasized that they often serve as models of real-world situations. Functions are commonly presented via formulas, tables of values, graphs, or verbal descriptions, and often the key to understanding is to convert from one type of presentation to another.

Graphical and tabular analysis. Concentration is on functions given by formulas or verbal descriptions and using the calculator or spreadsheet to produce graphs or tables of values to analyze the function. Solutions of equations and optimization are emphasized.

Straight lines and linear functions. A linear function is introduced as one with a constant rate of change. When a linear function is used as a model of a real-world situation, the meaning of the slope is of key importance. Linear regression is introduced as a crucial topic.

Exponential and logarithmic functions. Exponential functions are presented as functions with a constant proportional or percentage rate of change. Logarithms are the inverses of exponential functions. Exponential regression is a key topic.

A survey of other common functions. Polynomial, rational, and other common types of functions are presented along with the usual methods of combining functions.

Trigonometric functions. An elementary look at trigonometric functions. Emphasis is placed on modeling periodic phenomena.

Rates of Change. Rates of change occur throughout the course, but here the idea is qualitatively formalized.

Mathematics of Population Ecology. Population dynamics, life tables, survivorship curves.

It is worth emphasizing once more that the informal notion of rates of change is at the heart of the course. Once the idea is introduced, it occurs almost daily. The flavor of the course is perhaps best illustrated by the types of problems that drive it.

Examples

Running speed versus length

The following table gives the length L (in inches) of an animal and its maximum speed R (in feet per second) when it runs [2]. (For comparison, 10 feet per second is about 6.8 miles per hour.)

Animal	Length L	Speed R
Deermouse	3.5	8.2
Chipmunk	6.3	15.7
Desert Crested lizard	9.4	24.0
Grey squirrel	9.8	24.9
Red fox	24.0	65.6
Cheetah	47.0	95.1

1. Based on this table, is it generally true that larger animals run faster?

2. Plot the data points. Does it appear that running speed is approximately a linear function of length?

3. Make a model for R as a function of L by finding the equation of the regression line, and explain in practical terms the meaning of its slope. Add the plot of the regression line to the data plot.

4. Based on the plot in Part 3, which is faster *for its size*, the red fox or the cheetah?

One feature of this problem is common to virtually every problem in the book. If the indicated reference is consulted, it will be found that it is actual data taken from the scientific literature, and what the student is asked to do is part of what was done in the paper. Biology majors can see how the mathematics they are learning really does get used. The question of "when will I ever need to know this?" just never arises. The regression line is calculated using the calculator or spreadsheet. The instruction to "explain in practical terms the meaning of its slope" is repeated ad infinitum in the course. Certainly, students need to know how to get their hands on the slope, but more importantly in physical situations the slope has an important meaning, and understanding what that meaning is can be key to understanding the physical situation. Part 4 is really the fun part. Students have to do some serious mathematical thinking to figure out how to answer it. This also is typical of the exercises in the text. It's more fun to find out the answer yourself than have the instructor tell you.

Holling's functional response curve

The total number P of prey taken by a predator depends on the availability of prey. C. S. Holling proposed a function of the form:

$$P = \frac{cn}{1 + dn}$$

to model the number of prey taken in certain situations [3]. Here n is the density of prey available, and c and d are constants depending on the organisms involved as well as other environmental features. Holling took data gathered earlier by T. Burnett on the number of sawfly cocoons found by a small wasp parasite at given host density. In one such experiment conducted, Holling found the relationship:

$$P = \frac{21.96n}{1 + 2.41n}$$

where P is the number of cocoons parasitized and n is the density of cocoons available (measured as number per square inch).

1. Make a graph of P versus n. Include values of n up to 2 cocoons per square inch.

2. What density of cocoons will insure that the wasp will find and parasitize 6 of them?

3. There is a limit to the number of cocoons that the wasp is able to parasitize no matter how readily available the prey may be. What is the upper limit?

A couple of remarks are important here. First of all, the model for the situations is the one chosen by the expert, not by us. In this case, the reasons for choosing this particular model are more appropriate for a biology course than for an elementary mathematics course, and we simply offer it as the one the expert chose. This is not an unusual occurrence in the course. Where possible, some rationale for why a particular model is chosen is offered. But often, this is not realisitc since the explanation may involve fairly deep concepts from other disciplines. Students do Part 2 by solving an equation. In this case the equation can easily be turned into a linear equation, and students might be asked to solve it by hand. Often equations that need to be solved are quite complicated, perhaps with no closed form solution at all, and students are expected to solve those using the calculator or spreadsheet. Part 3 is one that occurs often. Of course, it is a limit, but the students don't see the usual calculus treatment of limits. Rather, they are expected to understand that they can approximate the answer by looking at the tail end of the graph.

These examples perhaps give a taste of the kinds of things that the course emphasizes. There are applications from business, the natural sciences, social sciences, and many other areas. The mode of presentation is not to look at abstract, algebraic formulas of a certain type, look at their properties, and then go looking for applications. Rather, physical situations are presented, discussed, and the need for a new idea or type of function made evident.

Conclusions

After running test sections at OSU, my preference was to replace college algebra with this course. We did not do that. We continue to teach a traditional college algebra course and, in parallel, run the modeling course. (There is actually a third course taught from *For All Practical Purposes* [4] which is taken by a small minority of students.) Some departments require one or the other, some accept either, and a few require both. (The courses are sufficently different that this makes sense.) The enrollments in the two courses are about equal.

The results at OSU have been dramatic. The success rate for the modeling course is much better than was the case with college algebra and, in general, students actually like the course. They emphasize that the topics make sense, and they see most of them as part of the world they live in. This is an outcome we had hoped for. What we did not anticipate was that the success rate in traditional college algebra also improved. With hindsight, one can conjecture that when students are placed in courses which suit their preparation and interests, they do better.

There are in fact several alternative college algebra texts on the market today which I think are very good. All of them emphasize modeling in one way or another. But it is probably the case that none of them, including our own, has got it right. Probably there is no single right text, but I believe as we learn more about what works and what doesn't, better texts will be coming. What is most important today is to get the mathematics community interested and talking about college algebra, much as what happened with calculus. There have been some very good conferences over the past few years providing forums for discussion of the pertinent issues. I would like to see greater participation in the discussion. Get involved, and by all means, get thinking about what that better text should look like.

References

1. Crauder, B. Evans, and A. Noell, *Functions and Change, A Modeling Approach to College Algebra*, 2nd edition, Houghton Mifflin, 2003.

2. The table is adapted from J.T. Bonner, *Size and Cycle*, Princeton University Press, Princeton, NJ, 1965.

3. "Some characteristics of simple types of predation and parasitism," *Can. Ent.* 91, 1959, pages 385–398.

4. Garfunkel, Solomon et al., *For All Practical Purposes*, 6th edition, W. H. Freeman and Company, 2002.

38

From The Bottom Up

Sol Garfunkel
COMAP (Consortium for Mathematics and its Applications)

In 1989 the National Council of Teachers of Mathematics (NCTM) published a rather remarkable document, known to all now as the NCTM *Standards*. The *Standards* recognized and announced that we have done a rather poor job of teaching mathematics at the K–12 level. It set out a rather bold agenda for new curricula and pedagogy built around a number of principles focused upon improving the mathematics education of all students. To a surprising extent this document went relatively unnoticed at the undergraduate level. That state of affairs changed rather dramatically in 1997–8, when new curricula built to embody the *Standards'* philosophy began appearing in published form.

It is important to note and underscore the fact that the NCTM *Standards* used the words mathematics education to mean the education of our nation's students in mathematics—not the education of mathematicians. This distinction underlies much of the later criticism, which has become akin to a religious dispute, played out in a variety of political arenas. Needless to say, COMAP embraced the *Standards* mission as our own.

In fact, we received one of the National Science Foundation grants to produce a comprehensive four-year secondary school mathematics curriculum, published under the title *Mathematics:Modeling Our World* (M:MOW). This project took six years to complete and the texts began appearing in 1998.

But we were struck by the fact that much of the material in these books was a significant part of the college algebra and precalculus syllabi. We feel strongly that reform should not be level dependent. College students can and should benefit by changes in content and pedagogy as well as students in K–12. Moreover, it is no secret that college algebra is a euphemism for remediating high school failures. And repeating material louder and faster has never been a successful strategy, no matter what the motivation.

As a consequence, COMAP decided to work on two new texts, unimaginatively titled, *College Algebra* [1] and *Precalculus* [2]. It should be noted that the texts cover the mathematical content one would expect in courses with these titles, with a few notable differences. Not surprisingly, these differences all relate to the fact that our approach is applications and modeling based. This has always been at the foundation of our approach to mathematics education and it directly addresses the contextual relevance issue emphasized again and again throughout the *Standards* documents. Moreover, modeling is a life skill. And we must be honest with ourselves. There is a schizophrenia about the college algebra—precalculus sequence. On the one hand, we truly hope that students in these courses will be successful and move on to take and succeed in calculus. On the other hand, we know that for the great majority of students, one or the other of these courses is the last math course they will take. So, we have the dual responsibility of giving students the best possible calculus preparation and preparing them (hopefully) to make intelligent use of mathematics and quantitative reasoning.

As a natural consequence, both texts contain a significant amount of data analysis and experiences in using mathematics to model real problems. Many of the other unique features are pedagogical in nature. Every lesson has an activity. Activities often involve the use of appropriate technologies—graphing calculators or spreadsheets or geometric utility programs. And the applications are real. Given that, there is more reading for the student. Modeling a real problem, means learning about that problem and the application field from which it comes. Frequently that means reading. While some may find this an impediment, we feel strongly that reading mathematics and about mathematics is a necessary step to being able to write and communicate mathematics, a core feature of the *Standards*.

Sample investigation

The following investigation taken from the college algebra text gives a feel for the flavor of the course.

> You live in an information age. Whether it's CDs, digital cameras, or computers, information is stored as collections of 0s and 1s in some kind of binary code that is interpreted by the specific application. But what is information? How much binary code is needed to convey information? The basic measure of information is a bit. For the purposes of this investigation, you may think of one bit as the information contained in the answer to one yes-no question.

Part 1: Exploration

a. Consider a simple situation in which you must select one particular CD from 2 that are gift-wrapped. You know that one of them is the latest recording by your favorite artist, and you don't have it yet. What's the smallest number of questions you can ask in order to identify the CD you want? What question(s) would you ask? How many bits of information are needed?

   Let f denote the information function, where $f(n)$ represents the number of bits of information needed to identify 1 item from among n items. Thus, you just computed $f(2)$.

b. Consider an even simpler situation. Suppose there is only 1 CD from which to select, and you know it's the latest recording by your favorite artist. How many questions do you have to ask now? How many bits of information are needed here? That's $f(1)$.

c. Henry needs to identify a particular CD from among 4 CDs. He saw your solution to the 2-CD problem and reasons as follows, "Let me group the 4 CDs into groups of 2 CDs per group. That leaves me with 2 groups, and I just saw how to identify one item (group) from a pair. After that, I'll have one group (2 CDs), but that's just the 2-item problem again. So I should need exactly $f(2) + f(2)$ questions (bits)." Comment on Henry's reasoning, then find $f(4)$.

d. How many questions do you have to ask if there are 8 CDs? That is, find $f(8)$. Explain your reasoning carefully. If possible, provide more than one explanation of your computation.

e. Maria has to select from among 32 CDs. She heard Henry's explanation for $f(4)$ and reasons that she could arrange the 32 CDs into 4 groups of 8 CDs each. Then $f(4)$ questions would identify the correct group and an additional $f(8)$ questions would find the right item. Compute $f(32)$ using this approach, then check it directly.

Part 2: Modeling

a. Based on its contextual meaning, identify a reasonable domain for the information function, f, used in Part 1.

b. Should the information function be increasing, decreasing, or neither? Explain based on the contextual meaning.

c. Generalize the observations made by Henry and Maria to write $f(MN)$ as the sum of two values of f. Explain your reasoning.

d. Use the values of f that you computed in Part 1 to begin its graph. Label your axes carefully. Compute more values, either directly or by using properties you identified above, to extend the graph until you can identify it as a member of your tool kit.

e. Through your work in Parts 1 and 2 (a)–(d), the following properties have been attributed to the information function, f.

- domain: $N > 0$
- initial condition: $f(1) = 0$
- additional observed value: $f(2) = 1$
- increasing: $f(N) > f(M)$ whenever $N > M$
- addition property: $f(MN) = f(M) + f(N)$

Verify that the function you named in Part 2(d) satisfies all these properties.

Conclusion

The important point here is that the new *Standards*-based curricula represent serious change in K–12 mathematics. Yet, they are extremely slow in being accepted into the undergraduate marketplace, even when dealing with precisely the same subjects. It is our hope that this approach will become the norm and that research and assessment will show that it is an approach that works... if not for all... then for more.

References

1. Crisler, Nancy, Gary Froelich, and Jerry Lege, *College Algebra; Modeling Our World*, W. H. Freeman and Company, New York, 2001.

2. Crisler, Nancy and Gary Froelich, *Precalculus; Modeling Our World*, W. H. Freeman and Company, New York, 2001.

39

The Functioning in the Real World Project

Florence S. Gordon   and   **Sheldon P. Gordon**
New York Institute of Technology   *Farmingdale State University of New York*

The calculus reform movement of the last decade or more has led to major changes that include an emphasis on geometric and numerical ideas as a balance to symbolic manipulations, student projects, realistic applications via mathematical modeling, the use of technology, and a more active learning environment. These efforts were intended to transform calculus into a *pump, not a filter*.

But if we are to change calculus, we must also consider how we "fill the tank"; that is,

- how do we increase the numbers of students who proceed on to calculus?

- how do we improve the mathematical experience of both those students and the ones who have no intention of going on to calculus?

Each year approximately three-quarters of a million college students take some variety of precalculus course; yet only a small fraction of them ever go on to *start* calculus. Most of those who do take calculus display a singular lack of retention of the material they were taught and often cannot complete calculus. This is a dreadful indictment of the effectiveness of traditional precalculus courses. They neither motivate the students to go on in mathematics nor adequately prepare them when they do continue, especially in view of the changing curricula in calculus and also the client disciplines.

What is needed is a precalculus experience that extends the common themes in the calculus reform efforts—an experience that focuses heavily on mathematical concepts and student understanding, that provides students with an appreciation of the importance of mathematics in a quantitative oriented society, that provides the skills and knowledge that students will need for subsequent mathematics courses or courses in other disciplines, and that makes appropriate use of technology.

The *Functioning in the Real World* project [1] addressed this challenge by developing an alternative to standard precalculus courses under a series of NSF grants. Our goal was to emphasize the qualitative, geometric and computational aspects of mathematics within a framework of mathematical modeling at a level appropriate to precalculus students. We chose to capitalize on the fact that most students are more interested in the applications of mathematics than in the mathematics itself, so that the applications drive all the mathematical developments. Thus, all the mathematical knowledge and skills students will need both for calculus and for quantitative courses in most other disciplines are introduced, developed and reinforced in the process of applying mathematics to model and solve interesting and realistic problems. We believe that such an approach excites the students and encourages them to go further with mathematics by showing them some of the payoffs that mathematics provides.

Our goal was to develop a set of materials that serve a multiplicity of audiences:

- A one-semester course that lays a different, but very effective, foundation for calculus;

- A one- or two-semester course that stands as a contemporary capstone to the mathematics education of students who do not plan to continue on to calculus. As a bonus, we have found that the course encourages a surprising number of these students to change their minds and go on to calculus and other quantitative courses.

- A course in math modeling as an alternative to traditional college algebra or precalculus courses.

Course contents

The following is an annotated description of the contents of the project materials.

Functions in the Real World introduces students to the function concept from graphical, numerical, symbolic, and verbal points of view as functions arise in daily life. The emphasis is on the behavior of functions (increasing or decreasing, concave up or concave down, point of inflection, periodicity).

Families of Functions includes linear, exponential, logarithmic, and power functions, with emphasis on their applications and their qualitative behavior. The intent is to have the students learn to identify and distinguish the different families from algebraic, graphical, and tabular representations.

Fitting Functions to Data includes linear and nonlinear curve fitting. These ideas and methods reinforce the properties of the different families of functions, develop algebraic skills in working with the properties of those functions, and connect the mathematics to the real world.

Extended Families of Functions includes polynomial functions, fitting polynomials to data, the nature and relative frequency of the roots of polynomial equations, finding polynomial patterns (including sums of integers and sums of squares of integers), building new functions from old (shifting, stretching, sums, differences, products, quotients, and composition of functions), and the logistic and surge families of functions.

Modeling with Difference Equations includes the development and analysis of models for describing population growth, including logistic (inhibited) growth, eliminating drugs from the body, radioactive decay, Newton's laws of heating and cooling, geometric sequences and their sums, iteration and chaos, etc. The pre-eminent mathematical tool today in virtually every discipline is the spreadsheet and recursion is the mathematical language of spreadsheets.

Modeling Periodic Behavior stresses using trigonometric functions to model phenomena such as the number of hours of daylight as a function of day of the year, the temperature over the course of a year, and the height of tides over time. Relationships between trig functions (identities) are introduced and used to approximate the sine, cosine and tangent functions with polynomials. The trig functions are also used to examine the properties of complex numbers and chaotic phenomena.

Geometric Models includes analytic geometry, the conic sections, parametric and polar curves, and realistic applications.

Matrix Algebra and its Applications includes a variety of applications of matrices, such as Markov chains in the spirit of a finite math course, not merely the use of matrices for solving systems of linear equations.

Probability Models includes binomial probability and the binomial expansion, geometric probability, estimating areas of plane regions using Monte Carlo simulations, waiting time models, and estimating the frequency of real and complex roots of polynomials with non-integer coefficients.

Illustrative examples and problems

The following examples and problems illustrate the philosophy of the project. They also indicate the nature of problems that encourage group work and collaborative learning. The problems also indicate the necessity of incorporating some type of technology—graphing calculator and/or spreadsheet—into the course to enhance both the teaching and learning of the mathematics.

1. Identify each of the following functions (a)–(n) as linear, exponential, power or other. In each case, *explain* your reasoning.

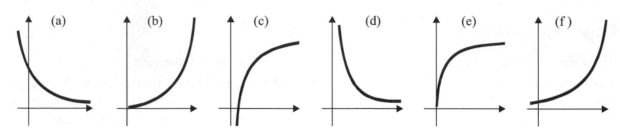

(g) $y = 1.05^x$ (h) $y = x^{1.05}$

(i) $y = (0.7)^x$ (j) $y = x^{0.7}$

(k) $y = x^{(-1/2)}$ (l) $3x - 5y = 14$

(m)   (n)

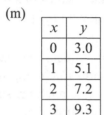
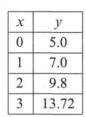

x	y
0	3.0
1	5.1
2	7.2
3	9.3

x	y
0	5.0
1	7.0
2	9.8
3	13.72

2. The world-wide wind power generating capacity, in megawatts, in different years is shown.

Year	1980	1985	1988	1990	1992	1995	1997	1999
Wind Power	10	1020	1580	1930	2510	4820	7640	13840

   a. Which variable is the independent variable and which is the dependent variable?

   b. Explain why an exponential function is the best model to use for this data.

   c. Find the exponential function that best fits this data.

   d. What are some reasonable values that you can use for the domain and range of this function?

   e. What is the practical significance of the base in the exponential function you created in (c)?

   f. What is the doubling time for this exponential function? Explain what it means.

   g. According to your model, what do you predict for the total wind power generating capacity in 2010?

3. Biologists have long observed the fact that the larger the area of a region, the more species that inhabit it. The table on the next page gives some data on the area A (in square miles) of various Caribbean islands in the Greater and Lesser Antilles and estimates of the number N of amphibian and reptile species living on each island.

   a. Which is the independent variable and which is the dependent variable?

   b. The overall pattern in the data suggests either a power function with a positive power $p < 1$ or a logarithmic function, both of which are increasing and concave down. Explain why a power function is a better model to use for this data.

Island	Area	N
Redonda	1	3
Saba	4	5
Montseratt	40	9
Puerto Rico	3459	40
Jamaica	4411	39
Hispaniola	29418	84
Cuba	44218	76

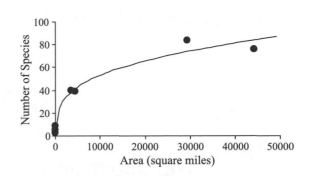

   c. Find the power function that models the relationship between the number of species, N, living on one of these islands and the area, A, of the island and find the correlation coefficient.

   d. What are some reasonable values that you can use for the domain and range of this function?

   e. The area of Barbados is 166 square miles. Estimate the number of species of amphibians and reptiles living there.

4. The average daytime high temperature in New York City as a function of the day of the year varies between 32°F and 94°F. Assume the coldest day occurs on the 30th day of the year and the hottest day on the 214th day of the year.

   a. Sketch the graph of the temperature as a function of time over a three year time span.

   b. Write a formula for a sinusoidal function that models the temperature over a year.

   c. What are the domain and range for this function?

   d. What are the amplitude, vertical shift, period, frequency, and phase shift of this function?

   e. What is the most likely high temperature on March 15?

   f. What are *all* the dates on which the high temperature is most likely 80°?

Conclusion

In conclusion, examples and problems such as these change the entire dynamic of the course and the classroom environment. There is classroom discussion of mathematics and its implications, and the classroom comes alive. The mathematics becomes something valuable to the students, because it is obviously applicable to realistic situations all around them. The simultaneous emphasis on conceptual understanding also changes students' attitudes—mathematics becomes a subject that one should understand, not simply a set of rules that one performs by rote to solve seemingly meaningless problems that are minor variations of worked examples. After all, the students see that virtually any routine operation can be programmed; it is the emphasis on the intellectual aspects that gives them a valuable and useful commodity to take away from the course.

References

1. Gordon, Sheldon P., Florence S. Gordon, Alan C. Tucker, and Martha J. Siegel, *Functioning in the Real World: A Precalculus Experience*, 2nd Edition, Addison-Wesley, 2004.

40

The Importance of a Story Line: Functions as Models of Change

Deborah Hughes Hallett
University of Arizona

If students are to remember what they learn, the courses they take must tell a coherent story. This story provides a framework onto which they can hang their newly acquired knowledge. Without such a framework, teachers find themselves having to repeat material. Precalculus courses often run the risk of not being memorable because they are defined as the skills needed in calculus rather than telling a coherent story. Thus, the first decision in designing a new precalculus course is to choose the story it will tell.

The central theme we chose for our precalculus course is how functions can be used to model change. This theme provides a framework into which all the prerequisite material for calculus naturally fit (functions, graphing, algebra, trigonometry, numerical approximation), while at the same time illuminating a central concept of calculus—the rate of change [1]. Choosing a family of functions to represent a real situation requires students to think about the qualitative behavior of different types of functions. A good way to decide, for example, whether an exponential function fits a particular set of data is to look at a plot. The shape of the plot suggests the family; the values of the parameters are then determined from the data.

We have found that introducing the rate of change as the slope of a line is an excellent springboard for comparing the behavior of linear and exponential functions (absolute versus relative rate of change), and for introducing the concepts of increasing, decreasing, and concavity. Thus, even without the advantage of the derivative, precalculus students can experience some of the central ideas of calculus.

No matter what flavor of calculus course they take, students benefit from a precalculus course that emphasizes interpretation as well as calculation. Particularly for students who are repeating material that they did not fully master before, a focus on meaning is an essential part of making the ideas fit together and finally stick. For example, the following problem asks students to think about the meaning of function notation:

1. The number of gallons of paint, n, needed to cover a house is a function of the surface area, A, measured in ft^2, of the house. That is, $n = f(A)$. Match each story below to one expression.

   a. I figured out how many gallons I needed and then bought two extra gallons just in case.

   b. I bought enough paint to cover my house twice.

   c. I bought enough paint to cover my house and my welcome sign, which measures 2 square feet.

         (i) $2f(A)$ (ii) $f(A + 2)$ (iii) $f(A) + 2$

A precalculus course also needs to provide a context for reinforcing skills—though which skills are chosen may vary widely from instructor to instructor. Some will want to focus on algebraic and graphical fluency, others on the ability to model a real situation. Thus, we have put together a stock of varied problems. Problems that involve the use of parameters are particularly useful in our view. Working with functions expressed in terms of parameters is seldom familiar to students, and provides valuable experience both with algebraic manipulation and with understanding the behavior of an entire family of functions. As an example, consider the following problem:

2. Consider the exponential functions graphed in Figure 1 and the six constants a, b, c, d, p, q.

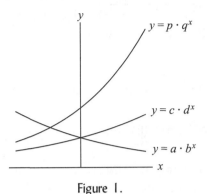

Figure 1.

a. Which of these constants are definitely positive?

b. Which of these constants are definitely between 0 and 1?

c. Which of these constants could be between 0 and 1?

d. Which two of these constants are definitely equal?

e. Which one of the following pairs of constants could be equal?

   a and p b and d b and q d and q

As in calculus, we believe students should be encouraged to make a connection between their calculations and reality. Since many students benefit from the experience of solving longer problems, we have written both problems and projects. An example of each follows.

3. Hong Kong shifted from British to Chinese rule in 1997. Figure 2 shows the number of people who emigrated from Hong Kong during each of the years from 1980 to 1992.

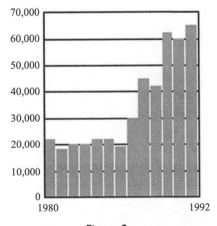

Figure 2.

a. Find an exponential function that approximates the data.

b. What does the model predict about the number of emigrants in 1996?

c. Write a short paragraph explaining why this model is or is not useful to predict emigration in the year 2000.

4. The table shows the population of Ireland at various times between 1780 and 1910.

Population of Ireland, 1780–1910, where 0 corresponds to 1780.

Years since 1780	0	20	40	60	70	90	110	130
Population (millions)	4.0	5.2	6.7	8.3	6.9	5.4	4.7	4.4

a. When was the population increasing? Decreasing?

b. For each successive time interval, construct a table showing the average rate of change of the population.

c. From the table you constructed in part (b), when is the graph of the population concave up? Concave down?

d. When was the average rate of change of the population the greatest? The least? How is this related to part (c)? What does this mean in human terms?

e. Graph the data in the table and join the points by a curve to show the trend in the data. From this graph, identify where the curve is increasing, decreasing, concave up, and concave down. Compare your answers to those you got in parts (a) and (c). Identify the region you found in part (d).

f. Something catastrophic happened in Ireland between 1780 and 1910. When? What happened in Ireland at that time to cause this catastrophe?

We have successfully used these materials for several years both at the college and at the high school level. Some of the high school students using the materials were sufficiently enthusiastic that they made a presentation to their school board about the mathematics they were learning. Since then, many have gone on to do excellent work in AP and college calculus.

References

1. Connally, Eric et al, *Functions Modeling Change: A Preparation for Calculus*, New York: John Wiley, 2000.

41

Using a Guided-Inquiry Approach to Enhance Student Learning in Precalculus

Nancy Baxter Hastings
Dickinson College

The Workshop Mathematics Program

The Workshop Mathematics program [2, 3, 5] broadens student access to university-level mathematics by providing multiple entry points into the discipline. Courses in the program—*Workshop Statistics, Workshop Precalculus,* and *Workshop Calculus*—seek to enable students, who might otherwise "fall through the cracks," to develop the skills and understanding necessary to use mathematics in other disciplines and to continue their study of mathematics. Like other reform courses, workshop courses seek to encourage students to read, write and talk about mathematical ideas and develop confidence in their abilities to think about and do mathematics. They seek to promote student learning through guided hands-on investigations. But most importantly, workshop courses seek to provide a supportive environment in which students enjoy learning and studying mathematics and feel comfortable asking questions and taking risks. [1]

The workshop approach

In the workshop environment, students learn by doing and reflecting on what they have done. The guiding principle is that instructors try not to *talk* in depth about a concept until students have had an opportunity to *think* about it first. The workshop approach makes no formal distinction between classroom and laboratory work. Rather than using lectures, instructors follow an interactive teaching format that includes the following components:

- **Summary discussion:** Typically, workshop instructors devote the beginning of each class to summarizing what happened in the last class, reviewing important ideas, and presenting other related material. Although this segment of a class may take only ten minutes or so, many students claim that it is one of the most important parts of the course, as it helps them make connections and focus on the overall picture. Students understand, and consequently, value the discussion because it relates directly to work they have done.

- **Introductory remarks:** The summary discussion leads into a brief introduction about the material that students will cover next. The purpose of this initial presentation is to help guide students' thoughts in appropriate directions without giving away too much information. Instructors introduce new ideas and concepts in an intuitive way, without providing any formal definitions, proofs of theorems, or detailed examples.

355

- **Collaborative activities:** The major portion of the class consists of students working collaboratively in groups of two to four on the activities in their Workshop Mathematics text. The activities are designed to help students think like mathematicians—to make observations and connections, ask questions, explore, guess, learn from their errors, share ideas, and read, write and talk mathematics—as they work with their peers. As students work together, the workshop instructor moves from group to group, guiding discussions, posing questions, and responding to queries.

Workshop Precalculus

There are two Workshop Precalculus books—an applied version and a standard version. Both versions incorporate the effective workshop-based educational practices utilized in the other workshop texts [2, 3]. Both versions seek to meet the challenge of preparing students for calculus, while providing a positive learning experience for students who choose not to take another mathematics course. Both versions emphasize traditional precalculus concepts, including a study of linear, polynomial, exponential, logarithmic, and trigonometric functions. However, in the applied version, *Workshop Precalculus: Functions, Data and Models*, new concepts are motivated by real-world applications and basic data analysis concepts are integrated throughout the materials. This text, which I am co-authoring with Allan Rossman, is now in the writing phase and is described in Allan's paper in this volume [4]. The standard version, *Workshop Precalculus: Discovery with Graphing Calculators*, seeks to provide students with a bridge to the study of calculus by helping them develop the confidence, understanding and skills necessary to continue their study of mathematics [5]. The remainder of this paper describes the standard version.

Workshop Precalculus: Discovery with Graphing Calculators—or *Workshop Precalculus*—consists of a sequence of in-class activities and a related set of follow-up activities. The in-class activities are designed to help students explore new concepts and discover ways to solve problems. The steps in the activities provide students with a substantial amount of guidance—in other words, they use a guided-inquiry approach. Students make predictions, do calculations, and enter observations directly in their book. The conclusion of each activity includes a summary of the main ideas, and then students receive a brief overview of what they will be doing in the next activity. The follow-up activities provide students with an opportunity to review important ideas from algebra, utilize new techniques that were introduced in the section, think more deeply about new concepts, and tackle applications.

Students finish each unit by reflecting on what they have learned and recording those thoughts in a journal. They are asked to describe in their own words the concepts they have studied, how those concepts fit together, and which ones were easy and which were hard. Students are also asked to reflect on the learning environment for the course. We view journal writing as one of the most important activities in each unit. Not only do journal entries provide us with feedback and enable us to catch any misconceptions, but more importantly, they provide the students with an opportunity to think about what they have learned and write about their observations.

As students begin to use the *Workshop Precalculus* text, they are encouraged to tear out the pages for the current section and place them in a three-ring binder, so that they can intersperse the pages with lecture/discussion notes, responses to follow-up activities, supplemental activities, and projects. During the course, students put together their own book.

Sample activities

Using technology to create mental images associated with fundamental concepts

Technology plays an important role in *Workshop Precalculus*. Students use technology not only to do numerical and graphical manipulations, but also to form mental images associated with abstract mathematical ideas. For example, students use a motion detector connected to a Computer- or Calculator-Based Laboratory interface (CBL or MBL) to create graphs of distance versus time functions and to analyze their behavior. As students walk back and forth in front of a motion detector, they make a mental connection between how they move and the shape of the graph representing their movement.

In the first activity in the book, students are asked to increase their distance from the detector while walking at a slow, steady pace. Then they are asked to fit a line to the graph; they are asked to do this by hand and to turn to the members of their group for help recalling how to read a graph, to calculate the slope of a line, and to find the equation of a line. Next, students are asked to create a graph by walking at a slightly faster constant pace, fit a line to the graph, and compare the new graph to the original one. In addition to helping one another remember some basic algebraic tools, students observe that when they walk at a constant pace, the graph representing their movement is linear; when they increase their distance from the detector, the graph rises from left to right and the value of its slope is positive; and when they walk faster, the value of the slope increases. Finally, with much less guidance, students walk toward the detector at a steady pace and make relevant observations.

To us as mathematicians, these seem like very simple ideas—ones that we could easily tell students about and then quickly move on. However, by doing this activity, students develop mental images associated with slope and with the concepts of increasing and decreasing. Students take ownership of the ideas, and the ideas mean something to them. The activity also serves to set the tone for the course. The classroom is a hubbub of activity. Students start talking with each other about mathematical ideas (using their own jargon). They begin to understand the importance of reading individual questions carefully. And they begin to develop confidence in their ability to figure out and explain things on their own. More importantly, they have more fun than listening to a lecture, and we do, too.

Building a bridge to calculus

The activities in *Workshop Precalculus* also seek to help students develop a conceptual understanding of fundamental calculus terminology and concepts. After exploring the concept of slope, students develop squiggly curves by walking back and forth in front of a motion detector. They observe that when they increase their distance and then decrease their distance from the detector, the curve representing their motion opens down, and similarly, when they decrease their distance and then increase their distance from the detector, the curve representing their motion opens up. In the process, students are introduced to the terms local maximum and local minimum, and they develop, on their own, a statement of the first derivative test (without, of course, using the word "derivative").

Students do similar activities to help them develop mental images associated with the concept of concavity. For example, students explain how they would walk to create functions with the following shapes:

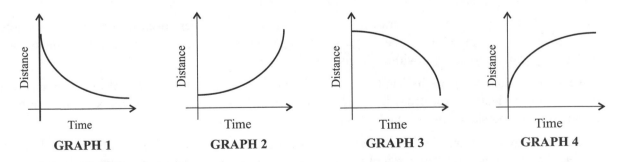

In subsequent activities, students develop an intuitive understanding of the concept of tangent line in terms of local linearity by zooming in on a point on a smooth curve. They explain why a tangent line cannot exist at a cusp or a sharp peak. They examine the behavior of the tangent line as it travels along a curve and note the relationship between the sign of the slope of the tangent line and the shape of the curve, in terms of whether it is increasing or decreasing. They also investigate the relationship between the location of the tangent line to the curve at a point (above, below, passes through) and the concavity of the curve at that point.

After observing how useful the tangent line is, we note that in the case of a squiggly curve it is hard to find the actual value of the slope of the tangent line at a particular point P, since the slope formula requires knowing the coordinates of two points on the line, but they only know the coordinates of one, namely P. We mention that they will learn how to find the exact value in calculus, but that in the meantime they can approximate the value by considering a nearby point on the curve and calculating the slope of the associated secant line.

In a follow-up activity, students examine the limiting behavior of the slopes of some secant lines to the graph of $y = x^2 + 1$ at $P(2, 5)$. They are asked to make a large scale graph of the function, sketch the secant line determined by a point near P, and then find its slope. They repeat the process for several points closer to P and list their results in a table. Finally, they make a best-guess sketch of the tangent line at P and use the slopes of the secant lines to approximate the value of its slope.

These types of activities serve as pointers to calculus. Moreover, since they appear in the first unit in the text, students are able to utilize these ideas as they study other classes of functions in subsequent units.

Implementing the workshop approach

Assessment data show that the workshop approach is an effective pedagogical tool. However, utilizing workshop-based materials, especially for the first time, can be challenging. Based on our experiences and those of others, Allan Rossman created the following list of suggestions for workshop instructors.

- Take control of the course. It is a mistake to think of a workshop course as a self-paced class in which the instructor plays only a minor role.

- Keep the class roughly together. Students in a workshop class spend most of their time working through activities in small groups. The instructor needs to set the pace, not letting some groups get too far ahead or lag behind.

- Allow students to discover. Workshop instructors need to resist the temptation to tell students too much. They need to let students discover ideas for themselves.

- Promote collaborative learning among students. Instructors should ask students to work on activities in pairs or groups of three.

- Encourage students to guess and to develop their intuition. Instructors should encourage students to think and make predictions about issues before analyzing them in detail.

- Lecture when appropriate. By no means do we propose that you never speak to the class as a whole. As a general guideline, however, instructors should try not to lecture on an idea until students have had an opportunity to grapple with it themselves.

- Have students do some work by hand. While technology is a powerful tool for exploring mathematical phenomena, students benefit from becoming competent at performing computations, doing symbolic computations, and sketching graphs by hand.

- Use technology as a tool. The counterbalance to the previous suggestion is that students should come to regard technology as a valuable tool for modeling situations and tackling complex problems.

- Provide plenty of feedback. Some students are fearful of the workshop approach because they worry about reaching wrong conclusions. Instructors need to provide students with regular, consistent feedback by interacting with the student groups in class and collecting a sampling of activities.

- Stress good writing. Many activities call for students to write interpretations and explain their findings. In addition to writing in complete, coherent sentences, students should relate their findings to the context at hand.

- Motivate students to read well. Instructors need to help students realize that they do a great service to themselves by taking their time and reading the questions and the brief discussions carefully.

- Have fun! We enjoy teaching with the workshop approach, principally because we get to know the students better and we love seeing them actively engaged with the material. We genuinely enjoy talking with the students on a regular basis. We hope that other workshop instructors will have as much fun.

Acknowledgement: The work reported in this paper was supported by the Division of Undergraduate Education of the National Science Foundation under grants DUE-995442483. Views and conclusions expressed are those of the author and not necessarily those of the Foundation.

References

1. Baxter Hastings, Nancy, "Workshop Mathematics: Abandoning Lectures," *Student-Active Science: Models of Innovation in College Science Teaching*, McNeal, A. P. and C. D'Avanzo, eds, Saunders Publishers, 1997.

2. Rossman, Allan, *Workshop Statistics: Discovery with Data*, Springer-Verlag, 1996.

3. Baxter Hastings, Nancy, *Workshop Calculus with Graphing Calculators: Guided Exploration with Review*, Volumes 1 and 2, Springer-Verlag, 1999.

4. Rossman, Allan, "Workshop Precalculus: Functions, Data and Models," in this volume.

5. Baxter Hastings, Nancy, *Workshop Precalculus: Discovery with Graphing Calculators*, Key College Publishing, 2002.

42

Maricopa Mathematics

Alan Jacobs
Scottsdale Community College

Faculty in the Maricopa Community Colleges began the project, The Maricopa Mathematics Consortium (NSF grants: DUE9352897 and DUE9602386), in 1993, at the convergence of two significant reform movements: Calculus reform and the implementation of the 1989 NCTM *Standards*. We believed that the mathematics curriculum before calculus would need to change, not only to prepare students for a reformed calculus course, but also because our entering students will have had a different preparation in their school mathematics. We decided to reconstruct the entire curriculum below calculus and to write appropriate course materials. Since several college algebra/precalculus reform projects were already underway, we concentrated on course materials for three precursor courses: arithmetic review, elementary algebra, and intermediate algebra. Materials for these courses are commercially available as *The Maricopa Mathematics Modules* [1] and *Beginning Algebra with Arithmetic Review* [2]. From 15 available modules, course instructors can select three or four modules to match their own course objectives. *The Modules* won the AMATYC 2000 Input Award, highlighting exemplary mathematics programs revitalized in accordance with the *Crossroads in Mathematics: Standards for Introductory College Mathematics Before Calculus* [3].

Features and content

The Modules embody four attributes: student centered, activity driven, context centered, and technology inclusive. These attributes flow from our belief that students need to gain a strong conceptual foundation of mathematics and that learning mathematics means building connections among various mathematical topics. We recognize that students do not build conceptual knowledge quickly, nor do they build conceptual knowledge by merely becoming proficient at template exercises. Rather, students need to think and reflect; they need to explore a topic numerically, symbolically, graphically and verbally.

Most traditional mathematics textbooks are organized around the guiding principle: Rule-Example-Practice. In *The Modules*, the guiding principle is Investigate-Generalize-Practice. A typical lesson begins with a student investigation, usually in small groups. Students are guided to make conjectures, draw upon knowledge from life experience, and discuss with each other to reach a conclusion or generalization. We see several valuable consequences of this approach. Students view mathematics less as an externally imposed system because they have had a role in its development. Since the investigations are set in context, students make meaning in what they learn as they connect the new knowledge with their own prior experience outside the classroom and with related mathematics. Because the investigations take place in groups, students gain much practice in expressing mathematics to each other verbally and in writing.

We illustrate the Investigate-Generalize-Practice model in a lesson on logarithms in the Exponential Growth and Decay module, in intermediate algebra. This lesson introduces students to logarithms through

Source	Acoustic power watts/square meter	Scientific notation	dB
whisper	0.000000000032	3.2×10^{-11}	15
rustling leaves	0.000000000064	6.4×10^{-11}	18
bedroom at night	0.00000000064	6.4×10^{-10}	28
library	0.00000000128	1.28×10^{-9}	31
refrigerator	0.00000005	5.0×10^{-8}	47
dishwasher (in next room)	0.0000001	1.0×10^{-7}	50

Figure 1.

their use in decibels, based on a news magazine article that explores the health risks of loud noises. Students are asked to complete a data table of sound intensities and their corresponding decibel reading. Part of that table is shown in Figure 1.

Students are asked to study the table and report the patterns they notice. They observe that as the sound intensity doubles, the decibels go up by 3; as the sound intensity is multiplied by 10, the decibels go up by 10. After confirming several instances of this in the table, they practice this pattern (if one fan has 75 dB, what will the decibel reading be for 2 fans? for 8 fans? for 10 fans?) to solidify their understanding. Now that students have been introduced to the concept of logarithms by a table of data, set in a real-life context, they have a foundation for exploring other related ideas. For example, the decibel table has the characteristic that a multiplicative change in the input (number of fans) resulted in an additive change in output (decibel level). Students have a basis for exploring whether this might be the inverse of the exponential function, for which an additive change in input means a multiplicative change in output.

Many of the modules set the mathematics in a specific context. For example, in *Data and Graphs*, students learn to make and interpret graphs while investigating fast-food franchises. In *Systems*, students solve systems of linear equations in the context of commercial fishing and ecology, leading to a mathematical view of sustainable harvest.

The goal of *The Modules* is to draw students into the habit of using mathematics to learn about the world, while preparing them for their future coursework. As the following results show, when students are in the habit of using mathematics to investigate problems, they are well-prepared for their next courses.

Results

We have collected evidence that suggests that community college students benefit from using *The Modules*. At three community colleges in the southwest, some students used published versions of *The Modules* for a course; most used the conventional text for comparison purposes. Our basic measure is an extension of "pass-rate"

$$\frac{\# \ of \ students \ who \ passed \ a \ course}{\# \ of \ students \ who \ started \ the \ course}$$

which applies to a single course. We extend the concept to the pass-rate through two successive courses, calling it "throughput-rate." In this study, we compared the throughput-rate of concurrent cohorts of students at each college, where throughput-rate is defined to be

$$\frac{\# \ of \ students \ who \ passed \ both \ course \ 1 \ and \ course \ 2 \ in \ successive \ semesters}{\# \ of \ students \ who \ started \ course \ 1}$$

We believe that this measure is an appropriate indicator of the quality of course 1 because it captures the core of what the mathematics program intends: students learn enough in the first course to prepare them to successfully learn the material in the second course. As an indicator of the quality of course 1,

Table 1. Aggregate throughput-rate results for three course transitions

	Start Course 1	Earn grade of ABC	Register for Course 2	Pass Course 2 with ABC	Two-semester throughput-rate
Module	975	627 (0.64)	453 (0.72)	312 (0.69)	0.32
Conventional	8262	4595 (0.56)	3323 (0.72)	1870 (0.56)	0.23

Table 2. Intermediate algebra to college algebra throughput-rate results

	Start Intermediate Algebra	Earn grade of ABC	Register for College Algebra	Pass College Algebra with ABC	Two-semester throughput-rate
Module	457	313 (0.68)	213 (0.68)	168 (0.79)	0.37
Conventional	3113	1807 (0.58)	1166 (0.65)	683 (0.59)	0.22

throughput-rate is most valid when students scatter to a variety of instructors in the second course. That way, the varieties of influences of the second course are averaged. Most of the students in this analysis, both module and conventional, scattered to different instructors for the second course. All courses were taught by both residential and adjunct faculty.

In Table 1, the aggregate throughput-rate results of all the students at the three colleges are shown for three first course to second course transitions: arithmetic review to elementary algebra, elementary algebra to intermediate algebra and intermediate algebra to college algebra.

The students using the module had a two-course throughput-rate of 32% compared to conventional student throughput-rate of 23%. In other words, from a hypothetical initial class of 31 students, 10 module students passed the successor course, whereas 7 conventional students did.

In Table 2, the throughput-rates for the transition from intermediate algebra to college algebra are shown. The throughput-rate for module students was 37% compared to the throughput–rate for conventional students of 22%; that is, from a hypothetical initial class of 31 students in intermediate algebra, 11 passed college algebra the following semester, compared to 7 conventional students. Furthermore, the pass-rate in college algebra was 79% for students with a module background, compared with 59% for the conventional students.

These throughput-rate results confirm that *The Modules* are successful in preparing students to succeed in their next mathematics courses. The results also serve as an indicator that the approaches used in *The Modules* (student centered, activity driven, context centered, and technology inclusive) benefit students. Anecdotal evidence corroborates the empirical results. One intermediate algebra student expressed it this way, which we think typifies what we hope to accomplish in *The Modules*.

I have never taken a math course that actually requires you to think about the answers! Most classes focus on memorizing equations and steps; it always seemed there was very little logic in what you were doing. It made math highly anxiety provoking. This class was a learning experience for me. I feel my anxiety level has gone down. I always thought people were good at math or they were not good at it. This class has given me the confidence to take a moment to look at a problem before I am convinced I can't do it.

Acknowledgements: Maricopa Author Team
    Scott Adamson, Chandler Gilbert Community College
    Paula Cheslik, Glendale Community College
    Anne Dudley, Glendale Community College
    David Dudley, Phoenix College

Teri Glaess, Scottsdale Community College
Karen Hay, formerly of Mesa Community College
Alan Jacobs, Scottsdale Community College
Keith Worth, Scottsdale Community College

References

1. Adamson, Scott, et al., *Maricopa Mathematics Modules*, Houghton Mifflin Co., Boston, 2000.

2. Jacobs, Alan, et al., *Beginning Algebra with Arithmetic Review*, Houghton Mifflin Co., Boston, 2004.

3. Cohen, Don, ed., *Crossroads in Mathematics: Standards for Introductory College Mathematics Before Calculus*, American Mathematical Association of Two-Year Colleges, Memphis, TN, 1995.

43

College Algebra/Quantitative Reasoning at the University of Massachusetts, Boston

Linda Almgren Kime
University of Massachusetts at Boston

Motivation

For many years, the head of Continuing Education at the University of Massachusetts, Boston had been receiving complaints from industry that their employees couldn't think quantitatively. She called together a handful of faculty from the Mathematics Department, Graduate School of Education, and Academic Support, and asked us to a design a course she could market to these companies. Having an academic bias, we took this as an opportunity to revise our traditional college algebra course—with the thought that later some of the modules could be used to serve the corporate world.

The discontent (at least for some) with the college algebra course was for the usual reasons:

1. It was a fast forward version of a skill and drill high school course in which many of our students had fared poorly the first time around.

2. The students viewed the topics as irrelevant and boring.

3. College algebra had the largest enrollment in the Math Department, but only because it was the penalty foisted upon most students as a result of distribution requirements.

4. The failure/withdrawal rate was unacceptably high.

5. In an informal survey, less than 2% of the students in calculus came from college algebra.

Judy Clark and I received an NSF grant to rethink college algebra. The result is the text *Explorations in College Algebra* [1], which we now use in an alternative college algebra course called Quantitative Reasoning.

Math/quantitative reasoning requirement

New general education requirements mandated, for the first time, a math/quantitative reasoning requirement. The Mathematics Department agreed that our traditional college algebra should be narrowly targeted towards math/science majors and was not the appropriate course for most students. The university then needed to decide what course the bulk of the students—liberal arts, social sciences, management, nursing, education majors—should take to meet the new requirement. A special general education subcommittee was set up to study the QR issue. After some debate, the subcommittee rejected certain models.

- They did not want the a la carte approach, where the QR requirement is met by taking one or more selections from an unrelated list of courses (in multiple departments) each of which contains some element of quantitative reasoning.

- They did not want a terminal course, in which a selection of random topics is taught, often at the discretion of the instructor, with no integration into future (or past) courses.

Since we would offer many sections of QR, we decided, for efficiency's sake, to create one basic multi-sectioned course that the bulk of our 9,000 undergraduates could take (or place out of). After much discussion, the course ended up in the Mathematics Department, but is taught by faculty from several different disciplines. We debated setting up a separate program, but felt that it would not have the clout to ensure that there were enough faculty to teach the course. Other departments are encouraged to create QR courses and a handful have been created, but nowhere near enough to meet the university's needs.

None of the QR courses is terminal. So the challenge was to define the basic skills (algebraic and technological) that would feed smoothly into other quantitatively based courses (statistics, economics, future math courses, and so on) and be helpful in understanding, constructing and critiquing quantitative arguments throughout life.

QR course design

Philosophical/pedagogical premises

The QR committee decided that all QR courses should share the following premises:

1. Close links with other departments

   This ensures that appropriate topics are being covered and enables us to "borrow" ideas to incorporate into the QR course.

2. Class sizes restricted to 25 students

3. Student–centered focus

4. Data-driven approach

   We use real data from the Internet, Statistical Abstracts, medical journals, newspapers, etc.

5. Systematic integration of technology

   Currently all the sections are taught in a computer lab and utilize Excel and some specially created software. In the past, some sections have used graphing calculators. All students utilize the Internet, email, and class websites.

6. Group work

7. Emphasis on communication of ideas through:

   - Active class discussions
   - Writing

     Students regularly create 60-second summaries, translating algebraic results into common English.

   - Formal group presentations

     Students work in small groups of 2 to 3 to prepare at least one formal class presentation. They prepare overheads and have a peer evaluation. However, students write up individual reports that are graded.

- Reading and analyzing articles

  The text includes related articles and students are encouraged to bring in current event examples.

8. Basic skills not neglected

   A mathematical spine runs throughout. Students regularly do algebra aerobics and are given regular short skill quizzes.

9. Periodic evaluations of the courses

Course topics

In addition, the QR committee decided that all QR courses should cover the following topics:

1. Basic descriptive statistics

   We include measures of central tendency, histograms, and pie charts, so that students are able to create or interpret the basic kinds of graphs that occur in the press.

2. Linear and exponential models

   Students should leave with a thorough understanding of the two most common mathematical models. For example, students should be able to articulate clearly the differences between the statement that the average house cost has been "increasing by $20,000 each year" versus it has been "increasing by 10% each year." They should be able to construct appropriate functions (by hand or using technology), along with their corresponding graphs, and make future predictions.

3. Optional topics

   Instructors can choose from among power functions, proportionality, laws of scale, quadratics, polynomials, logarithms (including semi-log and log-log plots), probability, and so on.

Sample homework problems

1. The following population pyramids show the age distribution in the United States and Tanzania in the year 2000. The U.S. is considered an industrialized country and Tanzania a developing country. Construct a 60-second summary comparing the differences in population distribution between the two countries. (You may want to go to the web site and compare these charts with those of other industrialized or developing countries.)

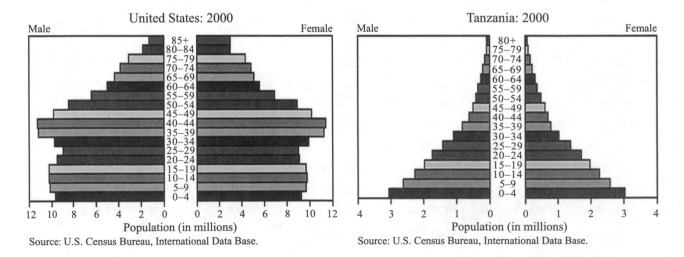

Source: U.S. Census Bureau, International Data Base.   Source: U.S. Census Bureau, International Data Base.

2. One part of the Exploration "Having it your way," includes the following assignment:

   With your team, use the data on the distribution of high school rank and SAT scores in the current Student Statistical Report for UMass/Boston to make the following opposing arguments:

   a. You are the president of the student body, arguing before the trustees that UMass/Boston is becoming a more elite institution, and hence is turning its back on its urban mission.

   b. You are the head of the Honors Society at UMass/Boston, writing a letter to mass media proclaiming that the university is lowering its academic standards and is in danger of compromising its academic credibility.

3. A patient was admitted to Brigham and Women's Hospital in Boston for a bone marrow transplant. The transplant was needed to cure myelodysplastic syndrome, a disease in which the patient's own marrow fails to produce enough white blood cells to fight infection. The patient's own bone marrow was intentionally destroyed using chemotherapy and radiation, and on Oct. 7th the donated marrow was injected. Each day the hospital carefully monitored the patient's white blood cell count to detect when the transplant took hold and the new marrow became active. The patient's white cell counts are listed in the table (omitted here), available in electronic form, and plotted in the accompanying figure. Normal counts for a healthy individual are between 4,000 and 10,000 cells per milliliter.

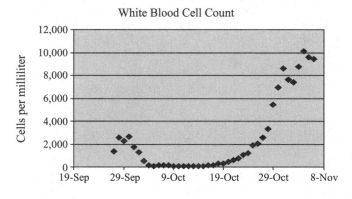

White Blood Cell Count

   a. Identify the time frame in which the white blood cells appear to be growing exponentially. This will be the domain for your model.

   b. Using technology, generate a best-fit exponential function to the section of data you identified in part (a). You may want to reinitialize your starting day as day 0.

   c. Interpret the equation in this context. In particular how fast were the number of white blood cells increasing? At that rate, how long would it take for the white blood cells to double?

Conclusions

The QR course has been in place for over eight years now. We currently run 10-12 sections per semester. The course is assessed regularly using, among other factors, a combination of student portfolios, student questionnaires, and common final exam questions. We have learned that about 35% of the QR students have taken a developmental mathematics course at UMass/Boston. Over time, students' facility with technology has increased significantly. In spring 2001, 90% of entering QR students were familiar with word processing, 60% with spreadsheets, 92% with the Internet, 93% with email, and 52% with graphing calculators. As a result, we now have to spend substantially less time on technology training. Few of the students (less than 25%) entered the course because they enjoyed math, or wanted to learn more about quantitative reasoning. However, by the end of the course, most would recommend it to friends. Almost

80% thought the course was at the right level. The majority felt that the features that enhanced the course were integration of technology, use of real world data, working in groups, and giving formal presentations.

The anecdotal comments from faculty—especially in other disciplines—have been favorable. QR students seem to do well in follow-up courses. The College of Management is considering dropping its calculus requirement and instituting a sequel to the current QR course. The chancellor has highlighted the QR course as one of our cutting edge programs.

The math/QR requirement is not perfect yet. We still need to work out a better relationship between the revised college algebra/QR course, the traditional college algebra course, and precalculus. We need to communicate with other disciplines to ensure that the QR skills are actually used throughout the curriculum. Our long-term goal is not one QR course, but the creation of quantitatively literate citizens.

References

1. Kime, Linda, Judy Clark, and Beverly Michaels, *Explorations in College Algebra*, 3rd Edition, John Wiley and Sons, 2005.

44

Developmental Algebra:
The First Mathematics Course for Many College Students

Mercedes A. McGowen
William Rainey Harper College

Since 1980, increasing numbers of students are repeating their high school mathematics courses as undergraduates and enrollment in the developmental courses has continued to grow. In Fall 2000, more than three million students were enrolled in undergraduate mathematics courses taught in departments of mathematics. Thirty-one percent of these students (981,000) were enrolled in remedial mathematics courses (arithmetic, algebra I, algebra II). Of these students, 763,000 were at two-year colleges (57% of the total two-year math enrollment) [1].

The large number of students who enroll in remedial courses suggests that the traditional emphasis on showing students how to use a rule to get the answer has failed many students. They learned the rules and passed the course(s) but, despite having learned the rules, they don't know when to apply them. They have not made sense of symbolic notation, nor have they learned to think for themselves. For these students, algebra is nothing but rules applied, often incorrectly, to manipulate symbols, which are meaningless marks on paper [2, 3, 4].

Based on their prior mathematical experience, most students expect to be told which formulas to use, and how to get the correct answers. This narrow approach is where many students stop in their understanding of mathematics—this strict utilitarian perspective too often limits their mathematical vision. Students' descriptions of their prior mathematical experiences and their views of mathematics are remarkably similar:

> I was used to having a formula and all I cared about was getting the right answer.

> Ever since I could remember the teacher would sit in front of the class and my friends and I would listen to try to figure out where to plug a number in or what formula was used. I would raise my hand and say, "I don't get it," and he would give me the explanation of how to use the formula. This caused problems for many like me because when it came to something that didn't fit the exact rule, I became confused.

> We were taught the rules and how to use them. There are many times I remember seeing the book show how they came about an equation and most teachers would ignore that part of it.

Changing what students value in mathematics and how they view learning mathematics are frequently much harder challenges than teaching them mathematical procedures and application of formulas. Essential components for success in learning mathematics include the ability to convert mathematical material into formulas, to generalize, to think flexibly, to operate with numerals and symbols, and to develop a proficiency in sequential, properly-segmented logical reasoning. Flexible thinking has been defined as an ability to switch rapidly from one mental operation to another, from one method of approach to another, from one method of solution to another. A particular case of flexibility of thinking is the ability to transfer from a direct to a reverse train of thought [5, 6, 7]. Setting expectations for growth in flexible thinking at the beginning of a course plays a major role in determining how students grow mathematically. Students also need opportunities to make choices about which techniques make the most sense in a particular situation and to check their answers for reasonableness.

Our project has attempted to move away from the procedural orientation in mathematics, which focuses on getting the correct answers and which students have learned to value above all. Instead, our curricular materials offer an alternative approach to learning algebra for students who have taken one or both of beginning and intermediate algebra—in high school or at college—and who have failed to place into a college-level mathematics course. What it means to learn mathematics, the nature of mathematics, the development of flexible thinking, and the ability to see and value connections are explicit goals of instruction.

With the concept of function as an organizing lens, this developmental algebra curriculum incorporates the results of research on how mathematics is learned [8, 9, 10]. Development of the materials was initially funded by NSF DUE Grant #9354471. Following extensive national field testing, preliminary editions were published in 1996, with the materials currently in second editions [11, 12]. The curriculum builds on fundamental ideas linked to each other and prepares students for continued study of increasingly sophisticated mathematical ideas. Important elements such as terminology, definitions, and notation are investigated and explicitly discussed. Conceptual understanding and skills emerge in the process.

Examples

In the intermediate algebra course, the unifying concepts of functions and difference equations are used to develop mathematical models from data. Students investigate rates of change with arithmetic and geometric sequences, with constant finite differences and ratios used to introduce various types of functions. These explorations are followed by examining differences between parameters and variables. Initially, students analyze functional processes.

Example 1. Consider the function machine for the function shown below.

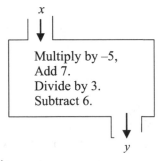

a. Create a table of five input/output pairs.

b. Graph the distribution.

c. Write the algebraic representation (equation).

d. Find the output if the input is 2.6.

e. Write an equation that displays the case for the output of -13.

f. Find the input if the output is -13.

g. Write an inequality that represents the case when the outputs are less than or equal to 2. Solve the inequality.

Building on their initial investigations using function machines, students focus on making sense of mathematical notation and arithmetic operations.

Example 2. Create a function machine for $y = x^2 - 3x + 5$.

a. Find the outputs for this function if the inputs are 4, -5, and $2/3$.

b. When you use -5 as input, do you write it as $y = -5^2 - 3(-5) + 5$ or as $y = (-5)^2 - 3(-5) + 5$? Does it make a difference? Why or why not?

We emphasize investigations that focus on key characteristics of various types of functions, using constant finite differences and ratios to find parameter values. Students investigate the role of parameters and make interesting discoveries and connections.

Example 3. Consider the function $y(x) = x^2 + 2x - 3$.

a. Identify the values of the parameters a, b, and c for this specific quadratic. (Note: the general quadratic: $y(x) = ax^2 + bx + c$.)

b. Predict the effect on the graph if only the sign of c is changed.

c. Predict the effect on the graph if only the sign of b is changed.

d. Predict the effect on the graph if only the sign of a is changed.

e. Graph the given function along with the function in which the sign of c is changed. Describe the effect if only the sign of c is changed.

f. Graph the given function along with the function in which the sign of b is changed. Describe the effect if only the sign of b is changed.

g. Graph the given function along with the function in which the sign of a is changed. Describe the effect if only the sign of a is changed.

Students develop conjectures about the change that occurs when only the sign of the quadratic coefficient is changed. They create additional quadratic functions, testing and modifying their conjectures when necessary.

Course structure

After investigating key characteristics of various basic types of functions, students select and create appropriate models that mathematically describe real-world situations based on their analysis of the data. Figure 1 provides an overview of the course structure.

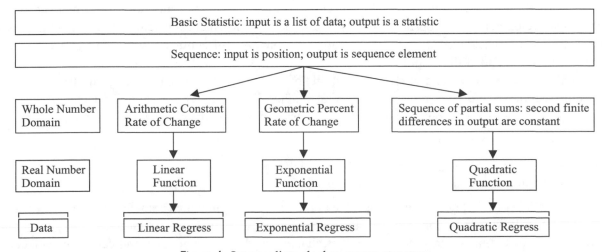

Figure 1. Intermediate algebra course structure

Conclusions

The function machine representation is a generic image that embodies the salient features of the idea of function—including process (input-output) and object—with various representations seen as methods of controlling input-output. It provides both an object-like status and the process aspect from input to output. The usual representations of function (table, graph, formula, procedure, verbal formulation, etc.) may be seen as ways of representing or calculating the input-output relationship.

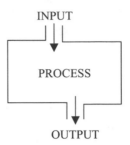

We use the function machine as a central unit of core knowledge from the first introduction of the function concept up through the most sophisticated ideas presented. Recent studies indicate that the introduction of the function machine as an input-output box enables students to have a mental image of a box that can be used to describe and name various processes [13, 14]. A student's ability to utilize algebraic representations of functions is correlated with the ability to form mental images of functions as machines. Combined with the use of finite differences and finite ratios along with use of a graphing calculator as a concrete manifestation of a function machine, students develop a more meaningful understanding of where equations come from and make sense of parameters from given data.

For students who have taken algebra previously, starting with a process notion of function and a function machine representation seems to help them make sense of notation. Students believe they need to be able to manipulate symbols successfully. As a consequence of this belief, they focus on being able to *do* algebra—that is, manipulate symbolic expressions. A function machine representation provides them with a means to organize their thinking so they can be more successful in dealing with notation than in their past experiences with algebra.

Students develop confidence in their ability to make sense of mathematics with contextual problem situations that involve various ways of thinking about algebra and multiple representations. They develop the ability to reason mathematically by looking for patterns, generalizing those patterns, and justifying results. After students have had the opportunity to think about and to investigate a problem, they talk about it—first in small groups, then in whole-class discussions in an interactive, supportive environment. A key issue is the development of appropriate skills for future courses—whether for calculus or non-calculus based courses. Skills are introduced in both mathematical and real-world contexts, with connections between mathematical ideas and skills emphasized.

Growth in understanding and improved flexibility of thought on the part of students enrolled in courses using our materials have been reported in several published studies. These studies document students' explanations and work based on an input-output process and give evidence of their improved ability:

(i) to interpret and use ambiguous function notation;

(ii) to translate between and among various function representations; and

(iii) to view a function as an object in its own right [15].

David Clarke et al [16] defined meaningful learning as "learning in which students are actively involved in integrating, or linking, new concepts and skills into an already existing conceptual framework, not simply accumulating isolated facts and procedures." It is characterized by evidence that indicates the students:

• claim to have learned something new;

- can articulate what it is they think they have learned with some degree of clarity and accuracy;
- are able to demonstrate formation of links with an existing framework that they already possess.

This definition combined with students' comments and work offers a useful framework by which we assess student learning. An examination of student work together with students' claims of understanding made in their mid-term and final course self-evaluations provides evidence of meaningful learning [17]. A journal problem given to students near the end of the semester was used by a student in his final self-evaluation as one piece of evidence, along with other work to support his claim of understanding quadratic functions:

My understanding of the quadratic function is quite clear. I now know when to use the quadratic formula and when not to. Actually, when it's more efficient to do so. I have also learned to solve a quadratic equation (if it crosses the x-axis) using a table and a graph. And when it doesn't cross the x-axis, to use the quadratic formula because there are no x-intercepts. Knowing that a quadratic function is to the 2nd degree also tells me that there are supposed to be 2 solutions.

The journal problem:

Given $f(x) = x^2 - 6x + 25$, determine the values of x so that the output is 4. Show all work and justify your response.

The student initially investigates the problem algebraically, substituting the value of 4 for $f(x)$:

$$x^2 - 6x + 25$$
$$4 = x^2 - 6x + 25$$
$$\text{Eq} = 0 \qquad x^2 - 6x + 21 = 0$$

He concludes: "This equation cannot be factored out so I put it into $y =$ and graphed it." He sketches the graph and records the view window settings (Figure 2), writing:

At the vertex the output is 12 which is the smallest number the output can produce. Given this problem, the value of x cannot produce an output of 4.

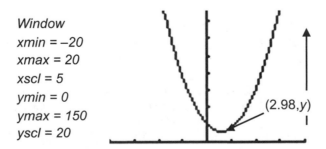

Window
xmin = −20
xmax = 20
xscl = 5
ymin = 0
ymax = 150
yscl = 20

(2.98,y)

Figure 2. Student work: graph and view window settings

The student then uses the discriminant to algebraically confirm his conclusion that there are no real number values for x that produce an output of 4 (see Figure 3).

What does the student know? How well does the student know it? What constitutes evidence of a student's understanding? Efforts to answer these questions focus attention on the need to clarify our beliefs, our values, our instructional goals and our expectations of students. In order to provide some guidance, our answers are based on analysis of student's work combined with descriptions of what they

What am I looking for?   i) graph   or   ii) $b^2 - 4ac$
   2 sol $a = 1; b = -6; c = 25$
   ~~1 sol (perfect square)~~ $(-6)^2 - 4(1)(25) = 36 - 100$
   No real sol   No real solutions
     $x^2 - 6x + 21 = 0$ 1/2 of b and sq it
     $x^2 - 6x + (-3)^2 = -21 + (-3)^2$ $\sqrt{(x-3)^2} = -12$
   Square root of both sides: $x - 3 = \pm\sqrt{-12}$
   Simplify: $x = 3 + 2\sqrt{3}i; x = 3 - 2\sqrt{3}i$

Figure 3. Student's algebraic confirmation of his graphical analysis and initial response

claim to have learned in their journals and self evaluations. Individual interviews with students provide additional information and evidence of understanding.

With encouragement and support, students work past their initial discomfort and discover they can make sense of mathematics. This is accomplished by directing students' attention to relevant information, explicitly discussing ambiguous notation, selecting and sequencing tasks that provide students with opportunities to build connections, and incorporating reflective writing and revisions of submissions into these courses. Students experience great personal satisfaction in discovering that they are indeed able to "do mathematics"—sharing strategies and conjectures, discussing the results of investigations with group members and the class, and convincing others that the approach and results are valid. Building on the foundational ideas of functional relationships and patterns, students find that algebraic symbols and language become more meaningful and purposeful means of expressing relationships between objects and processes. In turn, this provides students with mathematical experiences that deepen their understanding of mathematical concepts as well as improve their skills.

References

1. Lutzer, D., Maxwell, J. and Rodi, S., *Statistical Abstract of Undergraduate Programs in the Mathematical Sciences in the United States: Fall 2000 CBMS Survey*, American Mathematical Society, 2002.

2. Thompson, P., "Students, Functions, and the Undergraduate Curriculum," In Dubinsky, E., Schoenfeld A. and Kaput, J.J., (Eds.), *Research in Collegiate Mathematics Education, I. CBMS Issues in Mathematics Education*, 1994, Vol. 4, pp. 21–44.

3. Tall, D., Gray, E., Ali, M., Crowley, L., DeMarois, P., McGowen, M., Pitta, D., Pinto, M., Thomas, M., and Yusof, Y. "Symbols and the Bifurcation between Procedural and Conceptual Thinking," In *Canadian Journal of Mathematics, Science and Technology Education*, 2000, Vol 1, No. 1, pp. 81–104.

4. Gray, E. M. and Tall, D. O., "Success and Failure in Mathematics: Procept and Procedure: Secondary Mathematics," University of Warwick Mathematics Education Research Centre, Coventry, UK, 1991.

5. Krutetskii, V.A., *The Psychology of Mathematical Abilities in School Children. Kilpatrick*, J. & Wirszup, I. (Eds.), (Teller, J. Jr.), University of Chicago Press, Chicago, IL, 1996.

6. Skemp, R., *The Psychology of Learning Mathematics, Expanded American Edition*, Lawrence Erlbaum & Associates, Publishers, Hillsdale, NJ, 1997.

7. ——, *Intelligence, Learning, and Action,* John Wiley & Sons, New York, 1979.

8. Davis, R. B., "Conceptual and Procedural Knowledge in Mathematics: A Summary Analysis," In Hiebert, J. (Ed.) *Conceptual and Procedural Knowledge: The Case of Mathematics*, Erlbaum, Hillsdale, NJ, 1986, pp. 265–300.

9. Tall, D. O. and Vinner, S., "Cognitive Conflict and the Learning of Mathematics," In *Proceedings of the First Conference of the International Group for the Psychology of Mathematics Education*, Utrecht, Netherlands, 1981.

10. ——, "Concept Image and Concept Definition in Mathematics, with Special Reference to Limits and Continuity," In *Educational Studies in Mathematics*, 12, 1981, pp. 151–169.

11. DeMarois, P., McGowen, M. and Whitkanack, D., *Applying Algebraic Thinking to Data*, 2nd Edition. Kendall Hunt Publishers, Dubuque, IA, 2000.

12. ——, *Mathematical Investigations: Concepts and Processes for the Introductory Algebra Student*, 2nd Edition. Kendall Hunt Publishers, Dubuque, IA, 2000.

13. Tall, D., McGowen, M., and DeMarois, P., "The Function Machine as a Cognitive Root for the Function Concept." In *Proceedings of 22nd Annual Meeting of the North American Chapter of the International Group for the Psychology of Mathematics Education.* Tucson, AZ, 2000, pp. 255–261.

14. McGowen, M., DeMarois, P., and Tall, D., "Using the Function Machines as a Cognitive Root," In *Proceedings of the 22nd Annual Meeting of the North American Chapter of the International Group for the Psychology of Mathematics Education,* Tucson, AZ, 2000, pp. 247–254.

15. Davis, Gary E. and McGowen, M., "Function Machines & Flexible Algebraic Thought." In *Proceedings of the 26th International Group for the Psychology of Mathematics Education.* University of East Anglia, Norwich, UK, 2002.

16. Clarke, D., Helme, S., and Kessel, C., "Studying Mathematics Learning in Classroom Settings: Moments in the Process of Coming to Know," Paper presented at the National Council of Teachers of Mathematics Research Presession, San Diego, CA, 1996.

17. McGowen, M. and Tall, D., "Concept maps and schematic diagrams as devices for documenting the growth of mathematical knowledge." In O. Zaslavsky (Ed.), *Proceedings of the 23rd International Group for the Psychology of Mathematics Education*, Haifa, Israel, Vol. 3, 1999, pp. 281–288.

45

Workshop Precalculus: Functions, Data, and Models

Allan J. Rossman
California Polytechnic State University

The Workshop Mathematics program [1], developed at Dickinson College, leads students to discover, explore, and apply fundamental concepts of introductory mathematics and statistics courses. Active learning is the distinguishing feature of the "workshop" pedagogical approach, which replaces lectures with activities through which students interact with each other, with technology, and with the instructor. This program has been extended from its beginnings in Calculus [2] and Statistics [3] to the precalculus level, supported by a grant from the National Science Foundation (#9952483). In this project, Nancy Baxter Hastings and I are developing materials for a course that integrates ideas of data analysis and mathematical modeling into the study of precalculus. The primary goal of the materials is to prepare students for calculus, primarily by helping them to develop a deep understanding of the crucial concept of *function*. A secondary goal, especially important considering that many precalculus students never go on to study calculus, is to develop skills of data analysis and mathematical modeling, which will be valuable for courses in other discliplines.

Some of the principles guiding the development of these materials are:

1. to focus on the concept of function as essential preparation for calculus;

2. to emphasize students' facility with multiple representations of functions (graphical, tabular, symbolic, verbal) throughout the course;

3. to integrate the study of data analysis and mathematical modeling with more standard precalculus topics;

4. to use genuine data and applications for motivating topics, activities, and exercises;

5. to include applications from areas of general appeal, from a wide range of subject fields, from probability and statistics, and from other areas of mathematics to make connections and strengthen students' mathematical understandings;

6. to emphasize general problem-solving skills;

7. to use technology both as a learning tool and as a problem-solving tool;

8. to promote development of students' number sense and of fundamental skills of quantitative reasoning.

Course content

An outline of the content of the course follows:

1. Introduction to Functions, Data, and Models

2. Modeling Data with Linear Functions

3. Power, Polynomial, and Rational Functions

4. New Functions from Old

5. Exponential, Logarithmic, and Related Functions

6. Periodic Functions

7. Functions as Recursive Sequences

The first unit prepares students for the entire course by establishing the interactive nature of the learning environment and expectations for student involvement, along with the emphasis on conceptual understanding, problem-solving skills, and effective communication. The first unit also aims to convince students that the mathematics they are about to learn is applicable, accessible, and fun. It introduces them to fundamental ideas related to the concept of function that will recur throughout the course. These include the definition of a function, understanding a function both as a process and as an object, multiple representations (graphical, tabular, symbolic, verbal) of functions, inverses of functions, and behavior of functions. One activity through which students investigate these ideas involves walking in front of a motion detector where the student's distance from the detector is graphed as a function of time. Students create graphs of functions with various characteristics, such as increasing at a constant rate or increasing at an increasing rate.

In this first unit, students also begin to encounter the interplay among functions, data, and models that permeates the book. Students learn that a key to analyzing data is recognizing the importance of variability. They discover how to construct and to interpret graphical displays of the variability in data; they also learn to recognize the usefulness and limitations of numerical summaries of data. Students gather and analyze data about themselves as well as available data from various fields of application. For example, in one activity, students record the number of letters and the number of "Scrabble points" in their names. They recognize that the number of letters and number of points are both functions of the name, but the number of points is not a function of the number of letters. They also create a scatterplot of number of points versus number of letters and discover the concept of statistical association: names with more letters typically tend to have more points than names with fewer letters, although there are certainly exceptions to that tendency.

In the second unit, students begin to learn principles of approximating data by mathematical models. They study criteria by which functions of "best fit" can be chosen to summarize the relationship between variables, and they learn to analyze residual plots and to consider the influence of individual observations. For example, students analyze data on the distance to a destination and the cheapest available airfare to that destination. They usually find a fairly strong, positive association between these variables, and the relationship is often well summarized by a linear model. Students use technology, particularly the Fathom dynamical software package from Key Curriculum Press, to choose a line that seems to fit the data well, and then they use the least squares criterion to compare the performance of different lines. Students examine residual plots and what they reveal about the adequacy of the model, and they investigate the influence that individual data values (in this case, destinations) have on the line.

The remaining units introduce other types of functions that can be used to model real-world behavior, as well as discussions of combining functions to form new ones. The idea of recursion is also introduced.

Sample activity

In the following sample activity, students analyze genuine data on how the Consumer Price Index has changed over the years. In the process they reinforce their understanding of the concept of function and also apply what they have learned about analyzing the behavior of functions, particularly step functions. Moreover, students learn about how to compare monetary amounts in different time periods, adjusting for inflation. The questions asked of students in this activity, along with some passages of interspersed expository text (in italics), appear below.

The annual salary for the President of the United States has been increased only five times. It started at $25,000 in 1789 and stayed there for 84 years before climbing to $50,000 in 1873. Then it rose to $75,000 in 1909, $100,000 in 1949, $200,000 in 1969, and $400,000 in 2001.

(a) Sketch a graph of the President's salary as a function of year, from 1789 to 2003. [As always, be sure to label both axes.]

   This type of function is called a step function.

(b) How many steps occur in this function? In which years do the steps occur?

(c) What does the height of the step reveal (in the context of these data)?

(d) Why would it not be reasonable to pay today's President the same annual salary of $25,000 that George Washington received? In other words, explain why it is not fair to compare monetary amounts across so many years.

The Consumer Price Index (CPI), determined by the U.S. Bureau of Labor Statistics, measures changes in the prices of goods and services over time. The CPI therefore enables us to draw comparisons between the real buying power of monetary amounts at different times.

The following graph displays the CPI as a function of year, from 1913 through 2002:

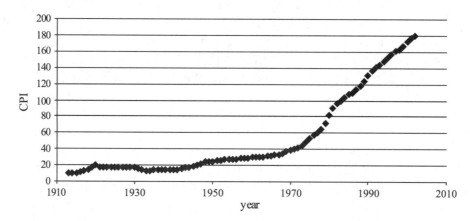

(e) Describe the behavior of this function:

   – Has it generally increased or decreased over the years, or does it run in cycles?

   – During what years did the CPI decrease? Why does this make sense?

   – During what years did the CPI increase most dramatically?

You can translate a monetary amount at one time period into its worth at another time using the expression: dollars at time B = dollars at time A × (CPI at time B) / (CPI at time A). Since the CPI's at both time periods are constants, this expression treats the worth at time B as a function of the monetary amount at time A.

The following table reports the value of the CPI for selected years between 1915 and 2002:

Year	CPI	Year	CPI	Year	CPI	Year	CPI
1915	10.1	1950	24.1	1970	38.8	2001	177.1
1920	20.0	1969	36.7	1984	103.9	2002	179.9

(f) When the President's salary was raised in 2001, suppose that it had been decided to raise it to what the $200,000 salary in 1969 was worth in 2001 dollars. What would this amount have been?

(g) Now suppose that the salary had been raised to what the $100,000 in 1950 was worth in 2001 dollars. What would this amount have been?

(h) Finally, suppose that the salary had been raised to what the $75,000 in 1915 was worth in 2001 dollars. What would this amount have been?

(i) In 1915 and in 1920 the President's salary was $75,000. Convert both of these amounts into constant 2002 dollars.

(j) What do these calculations reveal about how well the President's salary has kept pace with inflation over the years?

The following graph displays the President's salary in "constant 2002 dollars" as a function of year:

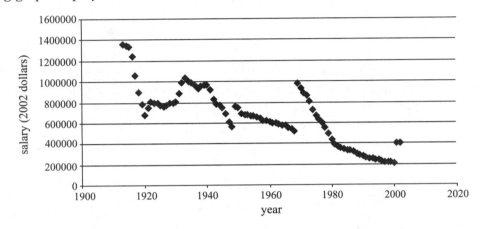

(k) Write a paragraph describing the behavior of this function. Be sure to relate your comments to the context, and be sure to address the following questions:

  – Is there an overall upward or downward tendency?

  – Are there upward or downward tendencies within certain periods?

(l) In 1915 the cost of a U.S. postage stamp was 2 cents. Express this amount in constant 2002 dollars.

(m) Is the cost of a stamp much more expensive now than in 1915, after adjusting for inflation?

(n) In 1970 the cost of a stamp was 6 cents. Express this amount in constant 2002 dollars. Is the cost of a stamp much more expensive now than in 1970, after adjusting for inflation?

(o) Convert the current cost of a stamp into constant 1984 dollars.

(p) In 1984 the cost of a stamp was 20 cents. Is the cost of a stamp much more expensive now than in 1984, after adjusting for inflation?

Acknowledgements: The Workshop Precalculus materials will be published by Key College Publishing. Sample materials and further information may be found at http://calc.dickinson.edu.

References

1. Hastings, N.B., "The Workshop Mathematics Program: Abandoning Lectures," in *Student-Active Science: Models of Innovation in College Science Teaching*, C. D'Avanzo and A. McNeal (eds.), Saunders Publishing Company, 1997.

2. Hastings, N.B., *Workshop Calculus: Guided Exploration with Review*, Volumes I and II, Key College Publishing, 1997.

3. Rossman, A. and Chance, B., *Workshop Statistics: Discovery with Data*, 2nd ed., Key College Publishing, 2001.

46

Contemporary College Algebra

Don Small
U.S. Military Academy

Contemporary College Algebra is designed to educate students for the future rather than to train them for the past. The course, developed in collaboration with faculty in several disciplines as well as with people in the workplace, provides a strong base for quantitative literacy programs.

The primary goal of the course is to empower students to become exploratory learners, not to master a list of algebraic rules. Some of the means that are used to establish an exploratory environment for the students include:

- Queries for engaging students in questioning and exploring the material being presented
- Exercises that explicitly ask students to explore, ask what-if type questions, make up examples, further investigate worked examples, or iterate for the purpose of recognizing a pattern and developing a sense for the behavior of the solution
- Graphically fitting a curve to a data set
- Small group work with both in-class activities and out-of-class group projects. In-class activities culminate in student presentations to the class and out-of-class projects culminate in written reports.

Other goals of the course include:

- Improve communications skills: Students read about, write about, listen to, and present mathematical ideas.
- Use technology: Every student has daily access to a graphing calculator and/or computer.
- Model situations: Students are empowered to use mathematics to quantify real-life situations.
- Build confidence: Students develop personal confidence as problem solvers. Specifically they develop confidence in the iterative process: try something, note the errors, modify previous attempt to lessen the errors, and try again.
- Provide a positive learning experience: Students enjoy applying mathematics to meaningful situations.

Concepts and techniques are introduced and motivated by real-life situations. Computational techniques are introduced in response to the need to solve real-life problems. For example, the quadratic formula is introduced in order to solve motion problems involving quadratic equations. The course emphasizes the importance of understanding elementary data analysis, the ability to extract a function relation from a data set, and the ability to mathematically model real-life situations in different disciplines.

Graphical analysis and problem solving in the modeling sense pervade the text. It is fundamentally important that students understand and experience the three stage problem-solving/modeling process applied to real-life problems. Because these problems usually involve dependency on many factors, the first stage of the process, creating a model, involves making assumptions in order to make the problem tractable. The second stage is the analysis-computation stage that leads to a mathematical solution. The results of this stage are often accomplished using technology. The third stage is the interpretation of the mathematical solution in the setting of the original problem. If the solution is not applicable or meaningful, the process is repeated after modifying the assumptions. This process is illustrated by the following diagram.

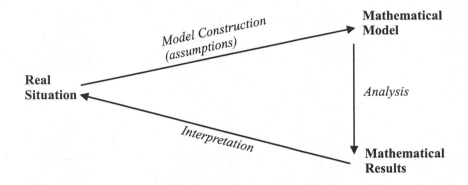

A nice example of this process is developing a vehicle stopping distance model for a road network serving residential and business areas as well as highway travel. The first stage is accomplished by accepting the assumptions and data presented in the 1995 *Texas Drivers Handbook* for dry, paved roads.

Speed (mph)	20	30	40	50	60	70
Stopping Distance (feet)	45	78	125	188	272	381

Students are expected to form a scatter plot of the data and then fit a curve to the plot (second stage). An acceptable curve is given by the regression equation $y = 0.9343x^2 - 1.8429x + 46$ with $R^2 = 0.9997$. The third stage, interpretation, raises serious questions about the applicability of the model particularly for slow speeds. According to the solution, the stopping distance for zero speed is 46 feet! Because speed limits of 5, 10 and 15 miles per hour are common near schools, hospitals and in congested areas, the model must be applicable to these speeds. Therefore changes in the assumptions and/or model need to be made and the process repeated. Two possible modifications of the assumptions would be to include $(0, 0)$ as a data point or to require that the regression curve pass through the origin. Requiring the latter yields the regression equation $y = 0.0715x^2 + 0.334x$. The R^2 value is 0.9977, only slightly less than for the previous model.

Vehicle Stopping Distance

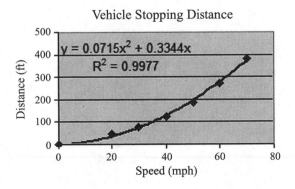

Features of the course

Overview. The analysis of data is the starting point for most of the topics. Instructors are encouraged to begin the course by having students fill out a chart of individual characteristics, such as height, weight, shoe size, eye color, male/female, hair color. This data chart is then reproduced and distributed to the students as their "Class Data." Questions involving the Class Data are asked throughout the semester. For example, compile a class profile based on the Class Data.

Data and variables. Study how to read and display data—table, pie chart, scatter and line plots, and bar charts. Students learn the meanings, use, and methods to compute the three principle summary measures of a data set—average (mean), median, mode. The understanding that data is information about a variable, gives meaning to the notion of variable and its use as a mathematical pronoun. The exploration of relations between variables leads to the study of straight lines. Applications of linear equations lead naturally to systems of linear equations, linear inequalities, and their applications in linear programming.

Functions. The concept of function is one of the most important ideas in mathematics. The concept is introduced informally through discussions of academic grades, modeling water level in a well, and warming a can of soda. Definitions of function and related terms are then clearly presented and illustrated. Graphically extracting function relations from data sets introduces the shapes of the basic functions—power, radical, exponential, logarithmic, and periodic (sine, cosine). The skill to graphically fit a curve to a data plot is enhanced by studying the function transformations of shifting and scaling. The development of linear regression for lines passing through the origin is used to illustrate the regression concept. However, students are expected to use regression programs in their calculators when computing a regression function. The algebra of functions is developed graphically, symbolically, and numerically. The ability to display data and to graphically approximate numerical solutions of equations and zeros of functions is an important thread throughout the course.

Modeling. The concluding chapter in the text for the course contains several sections, each of which is devoted to modeling real-life situations in a particular discipline (e.g., business, life sciences, economics, the arts, etc.). The principle techniques used are graphical approximations and recursive sequences (e.g., *New Situation = Old Situation + Change*). An example of how this chapter is used as a capstone for the course is given by Carrington Stewart of Texas Southern University. He divides his class into seven groups and assigns a particular section to each group. The group's responsibility is then to learn the material in their section well enough to present it to the class.

Fun projects. These are small group, out-of-class projects. Two or three Fun Projects are assigned each semester. Each project involves an inquiry component as well as a written report.

Sample projects

The following three Fun Projects illustrate the scope and the flavor of these projects.

Recommended daily amount of sodium

On a plane ride to Montana, a flight attendant gave Don a can of Welch's Orange Juice. The nutrition label on the can listed 15 mg of sodium representing 1% of the recommended daily value (DV) based on a 2,000 calorie/day diet. The person in the next seat had a can of Canada Dry Ginger Ale, which listed 90 mg of sodium representing 4% of the DV. Can both of these labels be correct? Can you tell the DV for a 2,000 calorie/day diet from the information on a nutrition label? How accurate is the information on nutrition labels? In particular, would it make a difference if the weight or the percentage figure were rounded off to a full integer?

Your tasks are:

1. Record the weight (mg) and the percentage amounts listed for sodium on at least ten other varieties of soda.

2. For each variety, determine the allowable range of DV assuming that both the weight and the percentage figures were rounded to full integers.

3. Are there any contradictions in the data that you collected? That is, could the sodium information on all of the labels be correct.

4. What information concerning the DV for sodium in a 2,000 calorie/day diet can you extract from your data?

5. What is the official Food and Drug Administration DV for sodium? (Hint: Research the FDA website.)

6. Include in your written project report a one page essay on nutrition labels. Base your essay on the FDA article "Scouting for Sodium" that originally appeared in the September 1994 FDA Consumer and was later revised and reprinted in September 1995. The article can be found at http://www.fda.gov/fdac/foodlabel/sodium.html.

The optimal dimensions of a soda can

The article "The Aluminum Beverage Can" in the September 1994 issue of *Scientific American* estimated that 100 billion drink cans are produced every year. With this number of cans, a sizeable increase in profits can be realized by making a small reduction in the amount of material that is used in producing a drink can. Today's 12-ounce beverage can weighs approximately 0.48 ounces compared to the 0.66 ounces when the cans were first introduced in the 1960s. The savings from a further reduction of one percent are approximately $20 million dollars per year. In this Fun Project, you are asked to optimize the dimensions of a 12-ounce aluminum beverage can. Assume that the can is a closed cylinder. That is, ignore the neck, domed base, and pull-tab of an actual can. Also assume that the idealized can holds 12 ounces (1 ounce = 1.8047 cubic inches).

Your tasks are:

1. Write a letter to a soft drink company asking for the approximate number of 12-ounce aluminum beverage cans the industry produced during the previous year. Include a copy of your letter and the response received in your project report.

2. Carefully measure the diameter and height of a 12-ounce aluminum soda can.

3. Assume that the material of the can has uniform thickness (bottom, top, sides). Let the thickness be one unit. Express the volume of material of a 12-ounce aluminum soda can as a function of the radius of the can and then plot the function.

4. Using the plot from Task 3, graphically determine the dimensions (height and radius) that minimize the volume of material of the can.

5. Compare your results against the dimensions of an actual soda can. If the results are considerably different, explain why the soft drink company would not use your dimensions in order to save money.

6. Change the assumption in Task 3 to be that the thickness of the top is 3 times the thickness of the sides and bottom (to allow for the pull-tab). Now repeat Task 3. Compare your results against those of an actual can. Comment on the comparison.

Teacher notes:

i. If the students have access to a physics lab, encourage them to measure the thickness of the bottom, sides, and top of a soda can.

ii. This project offers an interdisciplinary cooperation opportunity with the English Department. For example, invite an English instructor to speak to your class about writing a business letter.

Oxygen levels in the Naraguagus River

Environmentalists have grown concerned over the algae build-up in the Naraguagus River. A scientist assigned to look into the issue asked that the oxygen levels at a designated location in the river be monitored for four successive days. Due to a mix-up in communications, the field worker assigned to do the monitoring understood that he was to take oxygen readings in the river at four different times on the same day. He reported the following data. (The oxygen level is reported as the number of milligrams of oxygen for every 1,000 grams of water.)

Time	Oxygen Level
5:00 a.m.	8
11:00 a.m.	10
5:00 p.m.	17
midnight	10

The scientist, disappointed by having data for only one day rather than four days, has turned to your group for help. She comments that the day-to-day changes in the weather during the four-day period were minimal and asks your group to develop several models of the oxygen level in the water as a function of the time of day for the four-day period. She then wants you to determine which model is best and to describe your reasoning in clear statements.

Your tasks are:

1. Research oxygen levels in a river.

   a. List five factors that affect oxygen levels in a river.

   b. Determine the most important factor (explain your reasoning).

   c. Describe how oxygen gets into the water.

   d. Explain why oxygen levels may differ at different times of the day.

2. Plot the data.

3. Graphically fit a curve to your data plot. List the major characteristics you would like the curve to have.

4. Using the regression capability of your graphing calculator, determine the following regression models: (a) Linear (b) Quadratic (c) Cubic (d) Quartic (e) Sine

5. Determine which of the five regression models gives the best fit for the four-day period. Describe your reasoning for each of the five regression models.

6. Superimpose the plot of your best model on the plot of the data.

7. Use your best model to predict the oxygen level at 2:00 p.m. on the third day.

Summary

Anecdotal evidence strongly suggests that student-centered pedagogy combined with emphasis on modeling real life situations has a positive impact on student attitudes. Several instructors have commented on "how alive the class is" and on the depth of the questions students ask. The use of technology has enabled students to model and analyze situations that they had previously been unable to do because of the complexity of the algebraic manipulations involved. The emphasis on graphical analysis, rather than just symbolic, has bolstered student confidence. Group work, particularly projects, has also contributed to student confidence and self-satisfaction. Student pride is clearly evident in the project reports.

The text for the course, *Contemporary College Algebra: Data, Functions, Modeling*, 5th edition, by Don Small (McGraw-Hill College Custom Series) was developed through the HBCU Consortium for College Algebra Reform. A monthly newsletter *Vision-Potential* helps network the instructors teaching the course.

47

Precalculus: A Study of Functions and Their Applications

Todd Swanson
Hope College

In response to the calculus reform movement in the mid-1990s, we wanted to find a way to prepare students for a calculus course that was more conceptual, contained more real-life applications, and required students to view functions in multiple representations. With the aid of an NSF grant (DUE-9354741) we set out to write projects that would have students solve interesting, real-life problems that involved multiple representations as well as multiple topics from a precalculus course. These 26 projects were eventually published under the title, *Projects for Precalculus* [1] and could be used as a supplement to any precalculus course.

We, along the with the publisher, thought it would be a good idea to write a textbook based on these projects. Using some of the same philiosophy from our projects, we set out to write *Precalculus: A Study of Functions and Their Applications* [2]. We wanted this to be a non-traditional precalculus text that treats functions as the object of study while focusing on important mathematical concepts. We did this by introducing each of the basic types of functions (linear, exponential, logarithmic, periodic, and power) early in the book rather than relegating each type to a separate chapter. Doing so allows us to emphasize the commonalities and differences between the various types of functions. The properties of each type of function are developed throughout the remainder of the text. This leads to a natural review as students build on their understanding of functions. This organization addresses a common student misconception that mathematics consists of unrelated bits of information that can be forgotten once a particular topic is finished.

Just as in the projects, applications became an integral part of the text. We took great care to research every application to insure that the data given is accurate and documented. This way students can easily see that these applications come from the real world and not just from a mathematics text.

The text is written in a conversational format addressed to the student and students are expected to read it. To aid them, reading questions were incorporated throughout each section in which students are asked to write about key concepts and do simple problems.

Our original projects were written to allow students to be more involved in a problem than is possible with standard exercises. To put this idea into every section of our text, we decided to write what we call investigations. These problems are more extended than typical exercises, but shorter than projects. We included investigations along with each exercise set. These investigations can be used as part of a regular homework assignment, a group homework, a small group in-class activity, or as a class discussion.

Some projects that were written earlier were included at the end of each chapter. Many of the other projects were broken up and used as examples, exercises, and investigations.

Annotated table of contents

An Introduction to Functions. Various representations of functions as well as the language and notation associated with functions are introduced. This unit also illustrates how graphing calculators can be used and misused in the study of functions.

Families of Functions. Linear, exponential, logarithmic, periodic, and power functions are introduced. Students are shown how to recognize these functions in their various representations. Students are also shown how to obtain a formula when given a linear, exponential, or power function either numerically or graphically. This lays the groundwork for these functions and for their use throughout the remainder of the book.

New Functions from Old. The basic functions previously studied are transformed to form new functions in a variety of ways. In particular, the relationship between a transformed function in its symbolic form is compared with its graphical form. Transformations include addition and multiplication as well as composition. The relationship between a function and its inverse is also explored.

Polynomial and Rational Functions. Polynomials, introduced as transformations of particular power functions, are important enough to study as independent objects. We look at their properties as well as how they can be combined through division to form rational functions.

Trigonometric Functions. The periodic functions of sine and cosine, introduced previously, are reviewed and other trigonometric functions are introduced in this unit. These functions are introduced by using the unit circle definitions. The geometry of a circle, including arc length and area, is also explored. Transformations, are applied to the trigonometric functions and trigonometric identities are introduced.

Applications of Trigonometric Functions. Using trigonometric functions to model situations in the world is the focus of this unit. It begins by looking at problems involving triangles. Combinations of the periodic functions with other periodic functions as well as with non-periodic functions are then explored. Thus we expand the areas in which we can use trigonometric functions to model applications.

Solving Equations and Fitting Functions to Data. Different methods for solving equations and inequalities are introduced. This structure gives students a review of the functions first previously studied. The techniques of linear, exponential, and power regression are introduced as methods of fitting functions to data.

Getting Ready for Calculus. This unit serves as an introduction to calculus by exploring the concepts of limits, the derivative, and the integral. These topics are meant to help students prepare for the study of calculus and serve only as introductions.

Additional Topics. We conclude with a look at parametric equations, vectors, and multivariable functions. A property of a conic section is the focus of the project in this unit.

Examples

Reading questions

Reading questions ask students to write about key concepts and do simple problems. The following are examples of reading questions about exponential functions:

1. How is a linear function similar to an exponential function? How is a linear function different from an exponential function?
2. Let f be the exponential function $f(x) = 3 \times 4^x$. What is the y-intercept? What is the growth factor?
3. Let $f(x) = a^x$ such that $f(1) = 6$. What is the value of a? How do you know?

4. In the chicken bacteria example from this section, we noted that the bacteria doubles every six hours. Explain why this is equivalent to a sixteen-fold increase each day.

Exercises

In addition to providing sufficient practice to understanding the concepts given in the text, the exercises also provide many interesting applications. The following exercises about fitting exponential and power functions to data appear in section 7.5.

1. Two hundred thumbtacks were tossed onto a table. The ones that landed point up were removed. The remaining tacks were again tossed onto the table and again the ones landing point up were removed. This process continued until all of the thumbtacks were removed from the table. The results are shown in the following table.

Tosses	0	1	2	3	4	5	6	7	8
Number of tacks remaining	200	76	28	12	10	1	1	1	0

(a) Why does it make sense that an exponential function would fit a situation like this?

(b) Why will the last point, $(8, 0)$, make it impossible to use exponential regression on the data?

(c) Eliminate the last point and find an exponential equation where the input is the toss number and the output is the number of thumbtacks remaining.

(d) How do you think your exponential function would change if instead of removing the thumbtacks that landed point up, the ones not landing point up were removed?

2. In a book about the biology of birds, the equation $\log M = \log 89 + 0.64 \log W$ is given. In this equation, W is the weight of a bird in kilograms and M is the bird's metabolic rate in kilocalories per day. The given equation can be considered a linear equation where the input is $\log W$ and the output is $\log M$.

(a) Transform this equation into one where the input is W and the output is M.

(b) It is stated in the book that a twofold increase in the body weight of a bird is accompanied by less than a doubling of the metabolic rate. Using your transformed equation from part (a), explain why this is true.

Investigations

Investigations are extended problems in which students explore either a mathematical concept or an application. The following investigation involves rational functions.

Discover, a credit card company, has a Cashback Bonus Award Program. In this program, each year a cardholder receives a check representing a portion of the money he or she charged to Discover during the year. The percentage that each cardholder receives depends on the amount of money that was charged, according to the following plan.

- 0.25% of the first $1000 in purchases
- 0.50% of the second $1000 in purchases
- 0.75% of the third $1000 in purchases
- 1.00% of the amount of purchases in excess of $3000

1. Write a piecewise function where the amount charged, a, is the input and the cashback bonus award, c, is the output.

2. If you charge \$500 in a year, it is clear that you earn 0.25% as a cashback bonus rate. In general, if a is the amount charged, then your cashback bonus rate is r, where $r \times a$ = cashback bonus award. For any yearly charges less than \$1000, notice that your cashback bonus rate is always 0.25%. If you charge above \$1000 in a year, though, then your cashback bonus rate increases with each additional dollar charged.

(a) What is your cashback bonus rate if you charged \$1050 in a year?

(b) What is your cashback bonus rate if you charged \$1950 in a year?

(c) What is the maximum possible cashback bonus rate if you charge up to \$2000 in a year?

3. Assume someone charges over \$3000 in a given year and, because of the restrictions on the credit card, is not allowed to charge over \$2000 per month or \$24,000 per year.

(a) Determine the function where the amount charged, a, is the input and the cashback bonus rate, r, is the output.

(b) Determine the x-intercepts, vertical asymptotes, and long-range behavior of the function from question 3, part (a). Give the physical interpretation of each.

(c) In view of the physical situation, what are the restrictions on the domain and range of the function from question 3, part (a)?

(d) Suppose you heard on a commercial, "You could earn up to 1% of the amount you charged as a cashback bonus award using your Discover Card." Would you agree or disagree with this statement?

References

1. Andersen, J., T. Swanson, and R. Keeley, *Projects for Precalculus*, Brooks/Cole, 1997.

2. Swanson, T., J. Andersen, and R. Keeley, *Precalculus: A Study of Functions and Their Applications*, Brooks/Cole, 2001.

48

Success and Failures of a Precalculus Reform Project

David M. Wells   and   **Lynn Tilson**
Penn State University   *Penn State University (retired)*

Where we started

During the years from 1988 through 1996, the authors developed a set of materials ([1], [2], and [3]) for college algebra and precalculus. Our initial discussions about teaching and writing occurred at Penn State-New Kensington, a regional campus of Penn State University. The campus has an enrollment of about 1000 students, most of whom are freshmen and sophomores. The precalculus course is populated primarily by students who plan to complete degrees in engineering, technology, or science, either at New Kensington or at Penn State's University Park campus. College algebra is often taken as a terminal course or as a prerequisite for precalculus or business calculus. Failure rates in both courses are high, and neither course has lasting value for those students who pass, but choose not to take calculus. The size of the institution and the need for transferability of credits between campuses make it difficult for an individual instructor to depart radically from prescribed course content. This limitation led us to conclude that we could improve the courses most effectively by providing innovative approaches to traditional topics.

What we planned

We wanted our materials to be flexible enough to prepare students for both traditional and reform calculus courses, and to be used by instructors with different teaching styles. Our primary goals were that each course should be a valuable educational experience in its own right, and that students should recognize the value as the course was in progress. In particular, we wanted students to appreciate that they are developing skills in learning, problem solving, and communication that are independent of mathematical content and that will be useful to them in both their everyday lives and their careers. To support those goals we formulated several objectives, some of which focus on developing mathematical skills. Specifically, we wanted students:

- to learn to create and interpret mathematical models,
- to learn to communicate mathematical ideas,
- to master traditional skills with equations and graphs, both manually and with the aid of technological tools, and
- to make informed decisions about whether to use a technological tool to perform an operation, and about the reliability of the output.

Other objectives focus on developing learning skills that transcend mathematics. We wanted students:

- to acquire the habit of active learning, and
- to utilize broad problem-solving strategies that are applicable to both mathematical and non-mathematical problems.

Still other objectives focus on connecting mathematics with the world at large. We wanted students:

- to analyze realistic applications of precalculus mathematics,
- to understand that mathematics is created within larger contexts of history and culture, and
- to learn how mathematicians view the world, much as students in survey courses in other disciplines learn how practitioners of those disciplines view the world.

What we produced

Our materials addressed essentially all the topics found in traditional college algebra and precalculus books. However, we designed several innovative features to meet our objectives. To connect mathematics to the world around our students, we created a large collection of what we call Mathematical Looking Glasses, each of which is a discussion of mathematics in a practical context, accompanied by exercises. Some guide students through the solution of an actual problem that a scientist or engineer might encounter. Others trace developments in the history of mathematics. Still others illustrate mathematical views of everyday activities and the insights that can be gained by asking mathematical questions. Most involve the creation or interpretation of mathematical models, and many require students to write explanations in full sentences. The following exercises are taken from a typical Mathematical Looking Glass.

The Coast Guard maintains helicopters at several stations in the Caribbean. Each can fly 200 miles without refueling. To study a typical helicopter rescue mission, let's place a portion of the Caribbean into a coordinate system. Units are measured in miles, Miami is at $(-80, 0)$ and Nassau is at $(100, 0)$. A ship is lost in a storm after leaving Orange Cay, at $(0, -40)$. Although it may have been blown in any direction, it is probably still within 20 miles of Orange Cay.

1. Write an equation for the boundary of the region where the ship is located.
2. Write an equation for the boundary of the region that could be searched by a helicopter based in Miami. Repeat for a helicopter based in Nassau.
3. Can the helicopters in Exercise 2 reach every point where the ship might be located? How can you tell?
4. Suppose that a helicopter takes off from Nassau, conducts a search, and lands in Miami, covering a total of 200 miles. Explain why the region it can cover lies inside an ellipse with foci at Miami and Nassau, and find the equation of the ellipse.
5. You are in charge of helicopter operations out of Nassau. You have received a call from Miami, saying that a helicopter is already searching for the lost ship and will return to Miami. You can assist by sending a second helicopter, which will either return to Nassau or fly on to Miami after conducting its search. Which option do you choose? Why?

To encourage students to read and learn actively, we weave our discussion of most topics around a Mathematical Looking Glass and intersperse exercises at several points in each lesson. To introduce students to broad problem-solving strategies, we describe the process developed by George Polya in [4], apply the Polya process to both mathematical and non-mathematical problems, and frequently refer to specific Polya strategies.

We assume that students have access to graphing calculators, and our approach to many topics requires their use. We seldom give explicit instructions on when to use calculators, preferring to let students or their instructors make that decision. To guide students in the use of technology, we often illustrate both manual and calculator-assisted approaches to the same problem and discuss the advantages and limitations of each.

What we learned

From both our experiences and insights gleaned from others who have used the materials, we learned several lessons. One is that most students have a narrow idea of the purpose of their education. Students who major in engineering or technology are especially inclined to see their degree programs as career training, as opposed to a broad preparation for life. Such students do not value a mathematical idea unless it can be shown to have direct application to the job they plan to hold after graduation. For that reason, we have attempted to convince our students of the importance of all our objectives, not just those that relate to mathematical skills. For example, one exercise in our section on problem-solving strategies asks students, "Think of three problems, mathematical or otherwise, you have encountered in your 'real life' (that is, outside of mathematics classrooms). Identify at least three Polya strategies you used in solving these problems, and describe how they were used."

Most students also have narrow comfort zones. (A student evaluation of one of our courses complained, "Dr. Wells expects me to think, and I don't have time for that!") Even those students who demand real-world applications do not readily embrace the study of mathematics in a physical context. Furthermore, the presence of technology-based explorations and exercises requiring essay answers often compounds math anxiety with calculator anxiety and writing anxiety.

One reason for resistance to these new skill requirements is that most students are not very good at the skills involved. Not only do most have trouble writing about mathematics, but many have trouble writing a correct English sentence. The first few writing assignments in a course should require no more than a sentence or two. We find it helpful to provide examples of both good and poor responses to questions, and to involve instructors of English in helping students.

The strategy of starting slowly and providing plenty of guidance also applies to the development of modeling and problem-solving skills. We learned to construct early problem-solving assignments so that students would have to supply only the more obvious steps and explain why the less obvious instructor-supplied steps were valid.

We were eventually able to avoid the need for extensive guidance in calculator usage. Our experience in this area illustrates the potential advantages of cooperation with high schools. In 1990 and 1992 one of us, with another colleague, obtained grants from the Pennsylvania Department of Education to supply a large number of local high schools with graphing calculators, instruct teachers in their use, and develop application-based materials with calculator-assisted solutions. Since entering students at New Kensington come primarily from those high schools, most college algebra and precalculus students already have some familiarity with graphing calculators. Recent conversations with mathematicians from other Penn State campuses indicate that New Kensington is still unique in that respect.

References

1. Wells, David M. and Lynn Tilson, *College Algebra: A View of the World Around Us*, Prentice Hall, 1997.

2. ———, *Algebra and Trigonometry: A View of the World Around Us*, Prentice Hall, 1998.

3. ———, *Precalculus: A View of the World Around Us*, Prentice Hall, 1998.

4. Polya, George, *How to Solve It*, Princeton University Press, 1945.

49

The Earth Math Projects

Nancy Zumoff   and   **Christopher Schaufele**
Kennesaw State University   *Cortez, Colorado*

Since 1991, with support from the National Science Foundation (NSF) and the U.S. Department of Education's Fund for Improvement of Secondary Education (FIPSE), the authors have developed unique materials for use in mathematics courses ranging from algebra through calculus. These projects have resulted in three books, *Earth Algebra* (college algebra) [1], *Earth Angles* (precalculus) [2], and *Earth Studies* (applied calculus) [3], that all have applications to environmental issues that affect students' lives. They are designed to generate more interest in the use of mathematics as a tool to analyze real situations. Using mathematics to study real problems that are interesting to students has been shown to improve interest in, and understanding and appreciation of the role of mathematics in science and society. Formal evaluations have been conducted by Dr. Pamela Drummond [4] and are on file with FIPSE and NSF.

The authors are currently funded by NSF (grant #9952568) and FIPSE (grant #P116B001780) for a new curriculum development project that is an extension of the work described above. The goal is to produce versatile, technology-intensive materials for classroom use and teacher training. Reform-based applications are incorporated into platform-independent software to make them accessible to anyone with a computer. Environmental applications from the previous projects are being redesigned to be used in courses from algebra through calculus independent of the course textbook. The project features an inquiry-based format, web-based interactive materials, seamless interface with state-of-the-art technology, use of real data, and interesting applications of mathematical concepts. The materials can be used as student projects to enhance the learning of mathematical concepts, thus providing flexible classroom use. The materials are designed for use in both high school and college, and for training in-service and pre-service teachers. Appropriate software has been developed to perform the mathematics needed and the materials are published on CD-ROM and are available over the Internet, at earthmath.kennesaw.edu.

Each module, or study, can be used as an in-class or lab project that illustrates real applications of mathematics. Each study is written with three linked components: the text, the journal, and the tool chest. The *text*, written in HTML, presents the problem to be analyzed, asks questions, and provides links to relevant information and mathematical concepts. When applicable, Java applets are inserted in the text to allow the students to experiment with the concept being studied. The *journal* is a place for students to record their work and take notes—this is to be done using the word processor on the student's computer. Students can transfer the results of mathematical manipulations done with the applets to the journal. Most of the actual computations are done using the *tool chest*, a collection of applets that can perform the mathematics necessary to analyze the problems presented in the text. This is also where students can experiment with solutions.

Modules are written in a guided-inquiry format that has four stages:

1. presentation of a problem (comprehension);
2. acquisition of relevant information and skills (acquisition);
3. application of information and skills to the problem (application), and
4. investigation of the solution (reflection).

One two-part module, *Atmospheric Carbon Dioxide Concentration*, focuses on the increase in CO_2 in the Earth's atmosphere and its impact on global temperature and ocean level. Each part follows the described format. In the comprehension section, students are asked relevant questions based on the text material to initiate their own thoughts on the situation. In the acquisition section, relevant data is presented with links to websites. In the application section, students first use the data and other information provided to derive a model describing atmospheric CO_2 concentration over time. Next, equations are derived based on this model describing average global temperature increase and ocean level increase. In this section, students access applets from a side menu to perform the mathematics needed to construct and use their models. In the reflection section, students are asked questions regarding the validity of their models and the significance of the mathematics used. This module is written for use at three levels, high school algebra, college algebra, and precalculus.

Below is the text adapted from the web precalculus version of the first part of the CO_2 modules. The interested reader is invited to access the Earth Math website to see the interactive structure of the text and applets and the review topics for all the modules.

Carbon dioxide concentration

Comprehension

1. What is the greenhouse effect?
2. Which gases are greenhouse gases?
3. How has the concentration of atmospheric carbon dioxide changed in the last century?
4. What are some of the effects of increased concentration of carbon dioxide?

Acquisition

The table below provides information on carbon dioxide concentration for selected years.

Year	Concentration (ppm)
1960	316.75
1965	319.87
1970	325.52
1975	330.99
1980	338.52
1985	345.73
1990	354.04
1995	360.91
2000	369.40

1. What trends do you observe from the data?
2. Do you think these trends will continue in the near future?
3. What kinds of functions (linear, quadratic, exponential, etc.) would accurately describe the change in atmospheric CO_2 concentration over the last 40 years?

Application

1. For convenience, let t denote years since 2000 and let C denote the carbon dioxide concentration. Make points out of the data and plot them; use a (t, C) coordinate system.

2. Use exponential regression to fit the data with an exponential function. Call this function $C(t)$ and sketch its graph.

3. For each of the years for which you have data, compare the actual carbon dioxide concentration with the concentration predicted by your function. Why do you think the figures are different?

4. Use the function $C(t)$ to answer these questions.

   (a) Estimate the carbon dioxide concentration in 2001.

   (b) Predict when the carbon dioxide concentration will reach 400 ppm.

   (c) How much did carbon dioxide concentration increase between January 1, 1990 and January 1, 1991? Between January 1, 2000 and January 1, 2001? Between January 1, 1990 and January 1, 2000?

5. Compute the following:

   (a) The annual change in concentration for 2001.

   (b) The change in concentration between times $t = 0$ and $t = 5$.

   (c) The average annual change in concentration between $t = 0$ and $t = 5$.

6. Until recently, the carbon dioxide concentration stayed at about 280 ppm.

   (a) According to your model, when was the concentration at this level?

   (b) When will the concentration reach double this level?

Reflection

1. Reflect on the computation.

   (a) What did you do to predict carbon dioxide concentration at a particular time?

   (b) What did you do to predict when the concentration reached a particular level?

2. Reflect on the graph.

   (a) Which do you think is greater: the time it takes for the concentration to increase from 280 ppm to 330 ppm, or the time it takes to increase from 330 ppm to 380 ppm? Why do you think that is?

   (b) How is the graph different from the graph of a linear function? How is it similar?

3. Reflect on the interpretations.

   (a) What factors might cause a change in the trend you observed?

   (b) For how far into the future do you think your model is accurate?

   Other studies at various levels can be found on the project website covering such topics as population, streamflow and water availability, fuel economy of automobiles, emissions from automobiles, natural resource availability and usage, and food supply. Demonstration modules are posted to instruct students on format and use of the materials.

   These materials have been tested at Kennesaw State University, Diné College, Portland State University, Phoenix College, San Juan College, and at high schools in Georgia, New Mexico, and on the Navajo Reservation. [4]

References

1. Schaufele, Zumoff, M. Sims, and S. Sims, *Earth Algebra: College Algebra with Applications to Environmental Issues*, 2nd edition, Addison Wesley, Reading, MA, 1999.

2. Zumoff and Schaufele, *Earth Angles: Precalculus Mathematics with Applications to Environmental Issues*, preview edition, Addison Wesley, Reading, MA, 1997.

3. Zumoff, Schaufele, and Latiolais, *Earth Studies: Applied Calculus–A Modeling Approach*, Kendall Hunt, Dubuque, Iowa, 2000.

4. P. Drummond, "Earth Algebra, Earth Math, and Earth Studies Evaluation Reports, Summaries," <earthmath.kennesaw.edu>, full reports available from the authors upon request.